RENEWABLE ENERGY: RESEARCH, DEVELOPMENT AND POLICIES

PEROVSKITE SOLAR CELLS

PROPERTIES, APPLICATION AND EFFICIENCY

RENEWABLE ENERGY: RESEARCH, DEVELOPMENT AND POLICIES

Additional books and e-books in this series can be found on Nova's website under the Series tab.

RENEWABLE ENERGY: RESEARCH, DEVELOPMENT AND POLICIES

PEROVSKITE SOLAR CELLS

PROPERTIES, APPLICATION AND EFFICIENCY

MURALI BANAVOTH PHD
EDITOR

Copyright © 2019 by Nova Science Publishers, Inc.

All rights reserved. No part of this book may be reproduced, stored in a retrieval system or transmitted in any form or by any means: electronic, electrostatic, magnetic, tape, mechanical photocopying, recording or otherwise without the written permission of the Publisher.

We have partnered with Copyright Clearance Center to make it easy for you to obtain permissions to reuse content from this publication. Simply navigate to this publication's page on Nova's website and locate the "Get Permission" button below the title description. This button is linked directly to the title's permission page on copyright.com. Alternatively, you can visit copyright.com and search by title, ISBN, or ISSN.

For further questions about using the service on copyright.com, please contact:
Copyright Clearance Center
Phone: +1-(978) 750-8400   Fax: +1-(978) 750-4470   E-mail: info@copyright.com.

NOTICE TO THE READER

The Publisher has taken reasonable care in the preparation of this book, but makes no expressed or implied warranty of any kind and assumes no responsibility for any errors or omissions. No liability is assumed for incidental or consequential damages in connection with or arising out of information contained in this book. The Publisher shall not be liable for any special, consequential, or exemplary damages resulting, in whole or in part, from the readers' use of, or reliance upon, this material. Any parts of this book based on government reports are so indicated and copyright is claimed for those parts to the extent applicable to compilations of such works.

Independent verification should be sought for any data, advice or recommendations contained in this book. In addition, no responsibility is assumed by the Publisher for any injury and/or damage to persons or property arising from any methods, products, instructions, ideas or otherwise contained in this publication.

This publication is designed to provide accurate and authoritative information with regard to the subject matter covered herein. It is sold with the clear understanding that the Publisher is not engaged in rendering legal or any other professional services. If legal or any other expert assistance is required, the services of a competent person should be sought. FROM A DECLARATION OF PARTICIPANTS JOINTLY ADOPTED BY A COMMITTEE OF THE AMERICAN BAR ASSOCIATION AND A COMMITTEE OF PUBLISHERS.

Additional color graphics may be available in the e-book version of this book.

Library of Congress Cataloging-in-Publication Data

ISBN: 978-1-53615-858-8
Library of Congress Control Number: 2019945036

Published by Nova Science Publishers, Inc. † New York

CONTENTS

Introduction		vii
Acronyms		xi
Chapter 1	Hybrid Organo-Inorganic Perovskite Solar Cells: Architecture Evolution, Materials of Functional Layers, Photoelectric Characteristics, Properties, and Efficiency *P. P. Gladyshev, M. Banavoth, T. Swetha, N. Bingwa, Ya. B. Martynov, T. Yu. Zelenyak, V. A. Kinev and R. G. Nazmitdinov*	1
Chapter 2	Mechanisms of Radiation-Induced Degradation of Hybryd Perovskites Based Solar Cells and Ways to Increase Their Radiation Tolerance *Boris L. Oksengendler, Nigora N. Turaeva, Marlen I. Akhmedov and Olga V. Karpova*	77
Chapter 3	Perovskite: Material and Device Optimization for Solar Cell Applications *Antonio Frontera, Yaroslav Martynov, Rashid Nazmitdinov and Andreu Moìa-Pol*	117

Chapter 4	Molecular Engineering of the Perovskites: Dynamics/Kinetics of the Photovoltaic Behaviors *Foroogh Arkan and Mohammad Izadyar*	175
Chapter 5	Transition from Small-Area Devices to Large-Area Modules for Perovskite Photovoltaics *Soonil Hong, Hongkyu Kang, Jinho Lee, Hyungcheol Back, Sooncheol Kwon, Heejoo Kim and Kwanghee Lee*	207
Chapter 6	Flexible Perovskite Solar Cells (FPSCs) *Banavoth Murali, T. Swetha, Sachin G. Ghugal and Ranadeep Raj Sumukam*	241
Chapter 7	Bismuth and Antimony Based Perovskite, Perovskite-Like, and Non-Perovskite Materials for Lead-Free Perovskite Solar Cells *Ashish Kulkarni and Trilok Singh*	269
Chapter 8	Effects of NH_4Cl or PbI_2 Additions to $CH_3NH_3PbI_3$ Perovskite Solar Cells *Takeo Oku, Yuya Ohishi and Naoki Ueoka*	299
About the Editor		343
Index		345
Related Nova Publication		351

INTRODUCTION

Photovoltaic technologies continue to fuel global efforts in the advancement of new concepts and materials to promote photovoltaic (PV) cell technologies, both regarding solar-to-electrical conversion efficiency and commercial competitiveness. Despite significant improvement in silicon-based PV devices, that has reduced the energy production costs, the solar cell technology has not been able to cut a place for itself as a thoroughly viable solution to address a central open challenge – sustainable and affordable energy production by renewable means, primarily due to (still) higher production costs of conventional cells and PV modules. Therefore, cost-effective, highly efficient alternatives are need of the hour. Thanks to perovskite solar cells for being considered and establishing as viable alternative promising candidates, due to their phenomenal performances, in the future technological applications.

Recently the unrivalled potential for high power conversion efficiency (PCE) and cost-effective processes of fabrication have made organometal halide perovskite absorbers as more alluring candidates and thus have captivated tremendous attention from both academia and industry. The quest for opting the suitable architectures and employing the environmental benign, scalable, efficient procedures are of immediate need of the hour. Perovskite due to the remarkable optoelectronic properties such as high diffusion lengths, low trap densities, high absorption coefficients and tunable band gaps along the wide bandgap organic absorbers are being used as donor in the hybrids. For realising and understanding of such hybrid concepts, the integrated devices in various configurations are yet to be analysed.

Hybrid organic and inorganic perovskites (HOIP) have shown remarkable progress since the first realisation of efficient PSCs with a PCE of 3.9% in 2009; the record PCE reached 23.3% in 2018. The next step will undoubtedly be developing scale-up techniques for transitioning small-area devices to large-area modules. Most of the books outline only the necessary theoretical background, fabrication methods and applications. To bridge the gap between academia and industry, a profound understanding of the recent advancements in the HOIPs field is necessary. Experts' insights in this book present an in-depth overview of information regarding the materials synthesis methodologies, effects of dopants, optimized optoelectronic properties, suitable deposition methods, engineering and improving the stability of various device architectures using printing methods for flexible large-area PSC modules, including the module concept, discuss various challenges and issues that can open the door for the researchers towards commercialization of durable perovskite solar cells.

Moreover, this book also covers the developments on the zero-, two-, and three-dimensional non-toxic perovskite/non-perovskite materials, the radiation degradation of solar cells, the synergetics of cooperative phenomena in tandem systems and provide some recommendations to overcome the challenges for improving the photoconversion efficiency. This quality publication book will be an excellent resource for the scholars at Masters and PhD level from physics, chemistry and materials sciences, R&D researchers, engineers and scientists to broaden their knowledge on perovskite solar cells.

I am thankful and owe a tremendous debt to all the experts who have contributed to each chapter by spending long hours writing the best for this book. I dedicate this book to all the pioneers in this field, who have brought the current state of the art by successfully harvesting the solar energy and to the future generations who plan to present us a carbon-free clean society.

Finally, I express my gratitude to my family Divyanshu Raj, Murali Krishna, Adithi Sree and Rojleena Jadhav, for stealing their many hours, while working on this book.

Dr. Murali Banavoth

ACRONYMS

3TPYMB	Tris (2,4,6-trimethyl-3- (pyridin-3-yl) phenyl) borane
ALD	Atomic layer deposition
AN	Acetonitrile
APCE	Absorbed photon-to-current conversion efficiency
C_{60}	Fullerene
CVC	Current-voltage characteristic
D	Diffusion coefficient
DEH	4- (diethylamino) -benzaldehyddiphenylhydrozone
DES	Diethyl sulfide
DFT	Density Functional Theory
DMF	Dimethylformamide
DMFA	N, N-dimethylformamide
DMSO	Dimethyl sulfoxide
DPS	Di-n-propylsulfide
DR3TBDTT	(5Z,5′Z)-5,5′-((5″,5″″″-(4,8-bis(5-(2-ethylhexyl)thiophen-2-yl)benzo[1,2-b:4,5-b′]di-thiophene-2,6-diyl)bis(3,3″-dioctyl-[2,2′:5′,2″-terthiophene]-5″,5-diyl))bis(methanylylidene))bis(3-ethyl-2-thioxothiazolidin-4-one)
DSSC	Dye-sensitized solar cell
DSSCs	Dye-sensitized Solar Cells
E "V (HOMO)	Valence band (highest occupied molecular orbital)

E'C (LUMO)	Conduction band (lowest unoccupied molecular orbital)
Eg	Band gap energy
EIS	Electrochemical impedance spectroscopy
EQE	External quantum efficiency
ETL	Electron transport layer
ETM	Electron transporting materials (ESC - electron selective contact, ETL - electron transport layer)
FF	Fill factor
FTO	Fluorine doped tin oxide
GFF	Geometrical fill factor
H101	2,5-bis(4,4'-bis(methoxyphenyl)aminophen-4''-yl)-3,4-ethylenedioxythiophene
HOIP	Hybrid organic-inorganic perovskites
HOMO	Highest occupied molecular orbital
HTL	Hole transport layer
HTM	Hole transporting materials (HSC)
HTM	Hole Transporting Materials
HTM/ETM	Hole/Electron Transporting Materials
I	Dark current
ICBA	Indene-C60 bisadduct
ICT	Intramolecular charge transfer
I_L	photogenerated current
IPCE	Incident photon conversion to current efficiency
I_{SC}	short-circuit current
ITO	Indium doped tin oxide
IVC	Current-voltage characteristic
J_{SC}	Short-circuit current density
k	Boltzmann's constant
L	Length of the diffusion of charges
LHE	Light harvesting efficiency
LiTFSI-	Lithium bistrifluoromethanesulfonimidate
LUMO	Low unoccupied molecular orbital
MSSC	Meso-superstructured solar cells
$mTiO_2$	Mesoporous TiO_2

Acronyms

η	Overall efficiency
N	Diode ideality factor (1 for an ideal diode)
NBO	Natural bond orbital
NT	Nanotubes
NW	Nanowires
OCVD	Open-circuit voltage decay
OSC	Organic solar cell
P3HT	Poly(3-hexylthiophene-2,5-diyl)
PANI	Polyaniline
$PC_{61}BM$	Phenyl-C_{61}-butanoic acid methyl ester
$PC_{71}BM$	Phenyl-C_{71}-butanoic acid methyl ester
PCBM	[6,6]-Phenyl C_{61} butyric acid methyl ester
PCDTBT	Poly[N-9'-heptadecanyl-2,7-carbazole-alt-5,5-(4',7'-di-2-thienyl-2',1',3'-benzothiadiazole)]
PCE	Power conversion efficiency of solar cell
PCEs	Power conversion efficiencies
PCPDTBT	Poly[2,6-(4,4-bis-(2-ethylhexyl)-4H-cyclopenta[2,1-b;3,4-b']dithiophene)-alt-4,7(2,1,3-benzothiadiazole)]
PEDOT	poly(2,3-dihydrothieno-1,4-dioxin)
Pin	Power of the incident light
PL	photoluminescence
P_{mp}	Maximum power density
PSC	Perovskite solar cell
PSCs	Perovskite solar cells
PSS	Poly(4-styrenesulfonic acid) sodium salt
PTAA	Poly-triarylamine
Q	Elementary charge
QCRs	Quasichemical reactions
RDF	Radiation defect formation
R_S	Series resistance
RSD	Radiation-stimulated diffusion
RSDis	Radiation stimulated disordering
RSDM	Radiation-stimulated movement of dislocations
R_{SH}	Shunt resistance

RSMB	Radiation-stimulated movement of boundaries
RSQCR	Radiation-stimulated quasi-chemical reactions
SC	Solar cell
SCs	Solar cells
SILAR	Successive Ionic Layer Adsorption and Reaction - method of molecular layering
spiro-MeOTAD	2,2',7,7'-Tetrakis[N,N-di(4-methoxyphenyl)amino]-9,9'-spirobifluorene
ssDSC	Solid State Dye Solar Cell
τ	Lifetime of charges
T	Absolute temperature
T103	2,6,14-Tri(N,N-bis(4-methoxyphenyl)amino)-triptycene
V	Voltage
V_{oc}	Open-circuit voltage

In: Perovskite Solar Cells
Editor: Murali Banavoth

ISBN: 978-1-53615-858-8
© 2019 Nova Science Publishers, Inc.

Chapter 1

HYBRID ORGANO-INORGANIC PEROVSKITE SOLAR CELLS: ARCHITECTURE EVOLUTION, MATERIALS OF FUNCTIONAL LAYERS, PHOTOELECTRIC CHARACTERISTICS, PROPERTIES, AND EFFICIENCY

P. P. Gladyshev[1,*], M. Banavoth[2], T. Swetha[2],
N. Bingwa[3], Ya. B. Martynov[4], T. Yu. Zelenyak[1],
V. A. Kinev[1] and R. G. Nazmitdinov[1,5]
[1]Department Chemistry, New Technologies and Materials,
Dubna State University, Dubna, Russia
[2]School of Chemistry, University of Hyderabad, Hyderabad, India
[3]Department of Chemistry, University of Johannesburg,
Johannesburg South Africa
[4]Theoretical Department, State Scientific-Production Enterprise "Istok,"
Fryazino, Russia

---

* Corresponding Author's Email: pglad@yandex.ru.

[5]Bogoliubov Laboratory of Theoretical Physics,
Joint Institute for Nuclear Research, Dubna,

ABSTRACT

Hybrid organic and inorganic perovskites (HOIP) have shown the remarkable progress since the late 19th century and receiving more attention from the photovoltaic community because of its unique physical properties and their remarkable power conversion efficiencies above 20%. This chapter will provide valuable information regarding architecture of perovskite solar cells, preparation of the main functional layers, making of the perovskite thin films by vapor deposition methods and from solutions as well electronic processes, photoelectric characteristics and stability of the HOIPs, which open the door for the researchers towards improving the photovoltaic properties of the perovskite solar cells.

Keywords: organic-inorganic perovskites, photovoltaics, solar cell architecture, synthesis, stability, efficiency

1. INTRODUCTION

Currently, photovoltaics of the third generation have been successfully developed, the driving force of which is the desire to abandon expensive and toxic materials, as well as the transition to simpler technologies for the production of solar cells (SCs). It is possible to distinguish between the different types of third-generation SCs (sensitized dyes, organic, and polymeric inorganic SCs) based on the nature of the materials used during their fabrication, the structure of the SCs, and the principle of operation. To the third generation undoubtedly belong also hybrid organo-inorganic perovskite (HOIP) SCs (herein, HOIPs will be referred to as PSCs). PSCs promise to break the existing paradigm, combining in the long run, low cost and high efficiency. Although HOIPs have been known for quite some time, their application in photovoltaics attracted interest and inspired intensive investigations in this field. For the first time, PSCs were demonstrated in

2009 by Kojima et al. [1], and since then they are the subject of intensive study. Owing to the fact that HOIPs have a forbidden energy zone with a direct transition, they absorb light more efficiently than silicon, the main material of modern photovoltaics, and a thin layer is required to obtain PSCs, which can be obtained by precipitation from solution, which significantly reduces the cost of production of PSCs.

The development of SCs of the third generation began with the work of Gratzel (1988-1991) on dye-sensitized solar cells (DSSC) [2, 3]. In DSSC, mesoporous oxide semiconductors with a wide forbidden zone are used as an anode, which is coated with a dye as a photosensitive material. When the dye absorbs light, one of the electrons transitions from the ground state to the excited state. The excited electron moves from the dye to the conduction band of TiO_2. Then the electron diffuses through the TiO_2-film, reaches the glass electrode and then flows through the external circuit to the second electrode. The dye molecule with the loss of the electron is oxidized and its reduction to the original state occurs by obtaining an electron from the iodide ion, converting it into an iodine molecule, which in turn diffuses to the opposite electrode, receives an electron from it, and again turns into an iodide ion. With all the advantages of DSSCs, the use of the liquid phase was a significant limitation in practical terms, thus, DSSc with liquid electrolyte replaced by solid-state hole conducting materials such as spiro-MeOTAD were developed [4]. The use of such a composition increased the stability and performance of SCs. However, the efficiency of solid-state DSSCs is also limited by the rapid electron-hole recombination [5]. Work to address this problem continues today, but these studies faded into the background with the development of SCs based on HOIPs. At the initial stage, the SCs architecture was based on the DSSC concept, in which HOIPs were used instead of the organic dye. In the beginning, as in the case of dye-sensitized solar cells, the perovskite material is deposited on a charge-conducting mesoporous framework (most often TiO_2) as a light absorber [6]. These perovskite-sensitized SCs used liquid electrolytes based on iodide and bromide and had an energy conversion efficiency of 3.8% and 3.1%, respectively [1]. Between 2012 and 2013 a promising breakthrough of organo-inorganic SCs on the basis of crystal structures with a perovskite lattice was unexpectedly made. From that moment, a new, rapidly developing field of solar cells based on hybrid organo-inorganic semiconductor materials with a perovskite structure, such as $CH_3NH_3PbX_3$ (X = Cl, Br, I), unexpectedly appeared, with its tempting prospects and unresolved problems. This class of semiconductors attracted much attention due to its excellent light-absorbing characteristics. Moreover, HOIPs consist of inexpensive and easily accessible materials and allow the use of low-temperature wet technologies, including printing methods, in the formation of solar cells. In 2012, the liquid electrolyte in SC was replaced by a solid electrolyte, namely the organic material spiro-MeOTAD through which the injected perovskite absorber holes were diverted to the external contact. This immediately increased the efficiency of the deviceto ~ 9.7% [4]. It was also found that replacing the liquid electrolyte with a solid hole conductor, such as spiro-MeOTAD, largely eliminates the degradation process, and optimization of the procedure for the formation of perovskite mesostructured SC can play a decisive role in improving the efficiency of the device. The two-step application technology of perovskite precursor solutions during the formation of HOIP within the mesoporous metal oxide film described in [7] significantly increased the reproducibility of the SCs characteristics and made it possible to achieve 15% efficiency [8]. In 2014, the efficiency of such systems was brought to 20.1% [9]. These SCs also used the architecture inherent in the Gratzel cells [2].

In 2012, Lee et al. [10] showed that, contrary to the original view, the PSCs with an insulating framework (Al_2O_3) could also work efficiently in the PSCs structure of a mesoporous oxide framework with a wide forbidden zone of PSCs (TiO_2). This indicates an ambipolar charges transfer in a perovskite film with a balanced diffusion length of electrons and holes up to 1 micron [11, 12]. Moreover, devices using insulating Al_2O_3 scaffolds showed a longer lifetime of charge carriers compared to analogues with a TiO_2 scaffold because the TiO_2 participating in the collection of charges has many orders of magnitude lower electron mobility than $CH_3NH_3PbX_3$, as well as forbidden subzones, which leads to interfacial recombination [10]. It became clear that perovskites, given the good conductivity of the charges, do not require a meso-skeleton and function well in a thin-film planar structure [13]. It was shown that the perovskite structure functions in terms of charge transfer, ultimately better in planar architecture [14]. This opened the way from an architecture with a volumetric hetero-input to an architecture with a planar p-i-n type heterojunction and led to a simplification of the PSCs architecture. Currently, the efficiency of PSCs with a planar structure has reached 21% [15, 16]. Gradually, other efficient and more simpler PSCs architectures were proposed. Now the creation of SCs based on HOIPs is seen as a significant breakthrough in the development of third generation technologies. It is important to note that significant progress in the field of PSCs was achieved unusually fast compared to other types of SCs (Figure 1). This success was due to the participation in these developments of a large number of researchers and the extraordinary growth of publications (Figure 1). Power conversion efficiency (hereinafter efficiency) of PSCs increased from 3.8% in 2009 to 22.1% in 2016 [1, 12, 17-19], up to certified 22.7% [20]. In 2018, for tandem photovoltaic systems consisting only of PSCs (All-perovskite tandems), an efficiency of 23% was achieved [21].

In recent years, a large number of reviews have been published reflecting the evolution of PSCs architecture and advances in this field of knowledge and technology [14, 22-37] and also plenty of books on this topic [35, 38-72]. However, in this rapidly developing multifaceted field of knowledge, there remains a series of incompletely developed questions that constantly require a new examination and generalization. In this chapter, another attempt is made to systematically review the main achievements and prospects for the development of perovskite photovoltaics. At the same time, the authors give a full report that it was impossible to comprehensively summarize all the experience and knowledge accumulated in this field. The main focus of this chapter is the evolution of architecture, materials of functional layers, photoelectric characteristics, properties and achieved PSCs efficiency. The perovskite materials themselves are treated in a separate chapter there.

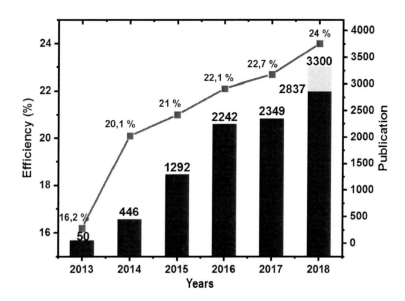

Figure 1. Increase in the number of publications on PSCs according to the publications of the Web of Science (as of October 2018, 2837 publications and a forecast for the end of 2018, 3300 publications) and the effectiveness of PSCs by year.

2. THE ARCHITECTURE OF HYBRID ORGANO-INORGANIC PEROVSKITE SOLAR CELLS

The high mobility of carriers of electric charges, the long lifetime of electrons and holes, the presence of a direct transition in HOIPs at a large perovskite absorption coefficient make them attractive for various optoelectronic devices. Unlike other semiconductors, the band gap of the HOIPs, and, consequently, their absorption and luminescence spectra can be easily controlled by manipulating their chemical composition and structure. They combine the advantages of both organic semiconductors (the possibility of obtaining from solution) and inorganic semiconductors (high mobility of charge carriers). This makes them good candidates for creating broadband absorbers of SCs.

The emergence of PSCs was preceded by the successful development of organic SCs, in particular sensitized dye solar cells (DSSC) with bulk architecture [13]. The implementation of organic SCs has overcome a number of fundamental problems [6]. Since in organic SCs carrier mobility is usually provided by excitons, it has been proposed to use the 3D bulk SCs architecture. This is achieved by mixing two organic semiconductor materials, which makes it possible to form an interpenetrating network of numerous contacts to facilitate the dissociation of excitons and the removal of charges through these contacts. In such a bulk heterojunction structure, unlike the classical planar heterojunction, the donor-acceptor boundary is distributed over the entire volume of the SC working layer, which allows an exciton excited at any point in the volume to reach the interface of two selective buffer semiconductor materials. However, in practice it is difficult to achieve complete separation of charges, since a certain fraction of them can be locked in isolated parts of the donor and acceptor phases, which is one of the drawbacks of volumetric heterojunctions.

Kojima et al. [1] in the framework of the DSSC concept, instead of the dye, a thin layer of perovskite was applied to mesoporous TiO_2, which was used as an electron collector (Figure 2a). Irradiation of the thus obtained open cell in the air caused its degradation within a few minutes. However, in 2011 the research resumed and Park and colleagues optimized the structure of the mesoporous frame of solar cells and reached an efficiency of 6.5% [73]. However, the degradation limitation of soluble perovskite nanocrystals in liquid electrolyte remained unresolved.

As was shown in Chapter 1 in the case of HOIPs, excitons have low formation energy and dissociate into electrons and holes that have a large mean free path. This ensures a high efficiency of planar architectures. Since then, both planar and bulk mesostructured architectures have continuously developed in a constant rivalry. A simpler planar p-i-n structure was developed, where the absorber absorbed by the vapor deposition of perovskite allowed to achieve an efficiency of 15.4% [75]. In 2013, an efficiency of 15.9% was reported for meso-superstructured solar cells (MSSC) formed at low temperature [76].

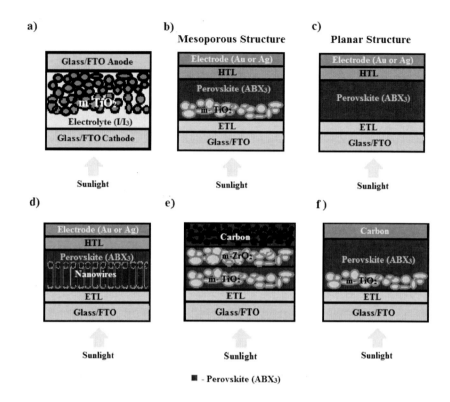

Figure 2. Architecture of perovskite solar cells [7, 8, 59, 74, 95]: a) Perovskite-sensitized solar cell with volumetric heterojunction and mesoporous TiO$_2$ anode material of n-type; b) PSC with bulk heterojunction and mesoporous TiO$_2$, ZrO$_2$ anode material of n-type; c) Classical planar PSC with p-i-n type heterojunctions; d) PSC with bulk heterojunction with ordered n-type nanostructured TiO$_2$ or ZnO anode material; e) Fully printed mesoscopic PSC. Mesoporous layers of TiO$_2$ and ZrO$_2$ with a thickness of ~ 1 and 2 microns are respectively deposited on FTO-coated glass. All layers were impregnated with a solution of HOIP precursors followed by perovskite formation, and f) PSC with one selective conductive buffer layer.

Step by step, the architecture of the perovskite solar cells was simplified from the initial architecture typical of the dye-sensitized solar cells (Figure 2b), to the architecture with a planar p-i-n type heterojunction (Figure 2c). The most typical proposed PSC architectures are shown in Figure 2. An alternative to the mesoporous metal oxide framework (Figure 2b) can be the vertical nanostructures - nanowire (NW) or nanotubular (NT) structures (Figure 2d). The NW and NT structures increase the density of the coating and reduce the recombination processes, and thus, these are more advantageous than the framework mesostructure [77]. It was shown that the composition of ZnO NW improves the electronic transport considering the short paths of electrons in the ZnO nanorods. However, this architecture did not develop, apparently because of the relatively complex technology.

An unexpected discovery was that the perovskite itself can transport holes and electrons for a considerable length [78]. It was shown that a mesoporous TiO_2 layer is not required for perovskite solar cells to transport of electrons [10]. It turned out that when replacing the mesoporous TiO_2 ETM with a porous Al_2O_3 the efficiency of the battery not only did not fall, but reached a higher value. It was proven [78] that a perovskite can transport within its volume electrons and holes without their appreciable recombination and for large distances on the order of a micron. Moreover, it was gradually realized that large areas of heterojunctions, characteristic of mesostructured bulk architecture, are not required, and it is possible to realize effective perovskite solar cells with a typical thin-film planar architecture. This is explained by the fact that the perovskite absorber itself ensures highly efficient ambipolar transfer of photoinduced charge carriers from the depth of its volume to the surface of planar buffer layers. This means that thin-film solar cells with a typical planar architecture can be realized for perovskite elements. Therefore, the logical development of perovskite SCs was the transition from an architecture with a volumetric hetero-input (Figure 2b) to an architecture with a planar p-i-n type heterojunction (Figure 2c). Thus, p-i-n PSCs were developed with an efficiency of more than 15%, which experimentally confirmed that there is no need for nanostructuring to achieve high efficiency [79]. The p-i-n PSCs with a planar architecture using TiO_2 and ZnO photoanodes with an efficiency of 12.1% and 15.7%, respectively, were demonstrated [80, 81].

Another type of architecture of perovskite SCs has been realized, in which a transparent electrode acts as a cathode, collecting photoinduced p-type charge carriers [82, 83]. This "invert" architecture, in which the HTM is located at the front illuminated side, and the collector of electrons - from the rear, is typical for organic solar cells.

In 2014, Young and co-workers demonstrated an efficiency of 19.3% and later 20.1% for planar thin-film architecture [82, 85]. In November 2014 at the 6th World Conference on Photoelectric Energy Conversion in Kyoto (Japan), without bringing detailed information was proclaimed to achieve an energy conversion efficiency of 24% for the single-junction PSC [85].

Thus, from the time when the first PSCs were developed, a significant evolution of their architectures has occurred. To date, PSCs have a very wide range of architectures, more than any other kind of solar cells. The properties of the various PSC architectures are determined by the morphology of the materials used (mesoporous, mesostructured or planar thin films), the presence or absence of mobile charges in them and, accordingly, their electrical and optical properties (n- or p-type semiconductors or insulators), the relative arrangement of layers of different materials with respect to light flux. This allows us to distinguish the n-i-p or p-i-n, p-n or n-p architectures of PSCs with mesoporous and planar structures. In view of the diversity, it is difficult to build a harmonious classification of all the proposed PSC architectures. Salim [86] identified two main architectures of PSCs: with a volumetric heterojunction (mesostructured) and thin-film (planar). In turn, mesostructured and planar SCs are subdivided depending on the location of the active layers relative to the light flux on the n-i-p and p-i-n structures (Figure 3). This classification does not consider HOIPs all the developed architectures, in particular a completely mesoscopic architecture without the HTM layer [87].

Fakharuddin et al. proposed the most detailed classification [88]. PSCs are classified according to two main characteristics: mesoporous PSCs (architecture with bulk heterojunction) using a scaffold and planar PSCs. The mesoporous scaffold can be electrically conductive, such as TiO_2 and ZnO, or an insulator such as Al_2O_3 and ZrO_2. The planar PSCs are divided into two subclasses: (1) PSCs having a perovskite layer between the two selective buffer materials by the electron-conducting material (ETM) and the hole-conducting material (HTM) and (2) the PSC having only one of these two selective buffer materials. However, this classification also does not cover all known versions of the PSC architecture. Below is a series of the most interesting PSCs architectures.

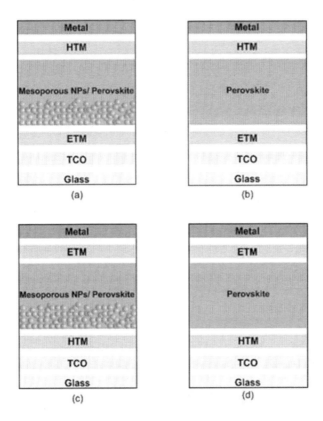

Figure 3. SCs architectures based on HOIPs a) - with a bulk heterojunction, n-i-p; b)- planar, n-i-p; c) - with bulk heterojunction; p-i-n; and d) -planar, p-i-n) [86].

2.1. Architecture with Bulk Heterojunction

In the mesoscopic SCs architecture with a bulk transition, the perovskite can either only cover the TiO$_2$ surface of the scaffold, or there may be another perovskite layer applied over mTiO$_2$. The photoactive layer is in contact with the n-type material to extract electrons and the p-type material to extract the holes. An alternative to the mesoporous metal oxide scaffold can serve as ordered vertical nanostructures - nanowire (NW) or nanotubular (NT) structures. NW and NT structures increase the density of the coating and reduce the recombination processes, and for this reason become more advantageous than the framework mesostructure [77].

In Figure 2b shows a PSC scheme in which the active layer consists of a layer of mesoporous TiO_2 coated with a perovskite absorber. The HOIP layer is in contact with the n-type material for electron extraction and the p-type material for the extraction of holes. After the absorption of light in the perovskite absorber, the photogenerated electrons are emitted into the mesoporous TiO_2, through which they are extracted. Spiro-MeOTAD with additives (LiTFSI and 4-tert-butylpyridine) [7], as well as copper thiocyanate (CuSCN) [89-92] or derivatives of poly-triarylamine (PTAA) [93-95] are the most commonly used HTM. Often PSC is terminated by a thermally sprayed thin layer of gold or silver. Researchers focused on the optimization of each layer and ensured a rapid increase in efficiency to 22.7% over a three year period [20].

2.2. Planar Architecture

In planar architecture, a perovskite photoabsorber in the form of a thin film is placed between different materials of selective buffers, which form two heterojunctions with an absorber (Figure 2c). Electron transport is carried out through the interface with the ETM which is blocked the movement of holes. The second interface with HTM performs the opposite function. In turn, selective buffer layers are conjugated to electrodes. Various ETMs, such as a traditional blocking layer of TiO_2, SnO_2, ZnO or phenyl-C_{61}-butyric acid methyl ester (PCBM) can be used as selective electron transport layers. In particular, the use of SnO_2 as ETM has attracted attention due to their increased stability and efficiency [15, 16]. As the selective hole-transporting layers, various HTMs are used, for example, spiro-OMeTAD, PEDOD [10, 96 - 98].

2.3. The Architecture with One Selective Conducting Buffer Layer

To simplify the design and reduce costs, PSCs without ETM or HTM were developed. In these PSCs, one of the selective ETM or HTM buffer layers was replaced by contacts with a modified band structure to extract the carriers. The architecture of PSCs without HTM is shown in Figure 2f. Li et al. developed a planar PSC that does not contain a layer of HTM (Hole-conductor-free planar PSC), with an efficiency of 16.0% [87]. A planar PSC with an inverted superstrate construction was fabricated. This device was highly stable and had more higher efficiency than a control device using NiO_x as a hole conductor. Simplified architecture and good efficiency make hole-conductor-free planar PSCs competitive with other SCs.

Hu et al. [99] developed a highly efficient, highly stable hole-conductor-free PSC with a potentially low cost. In this PSC, TiO_2 was used as the buffer layer of the ETM. Cheap carbon was used as the counter electrode layer. It was found that the efficiency of PSC depends on the thickness of the compact layer of TiO_2 which was received by the atomic layer deposition (ALD). The best PSCs was obtained by depositing the TiO_2 film as a result of 2000 cycles of ALD had an efficiency of up to 7.82%. This method opens the possibility of creating a stable PSCs for practical applications.

Duan et al. [63] used ultrathin graphite as a hole extractor and demonstrated efficiency of 14.07%. Although these PSCs have not achieved record efficiency, they can be quite important in the development of industrial PSCs. In addition, they can be useful for understanding the physics of the functioning of perovskite photovoltaic systems. It should be noted and other work on the creation of hole-conductor-free PSC [99, 100]. There is much less work on creating electron-conductor-free PSC.

2.4. Fully Mesoscopic Architecture without a Hole Carrier

Completely mesoporous solar cells represent the opposite of planar PSCs. They are manufactured by sequential printing of a mesoporous layer of TiO_2, a mesoporous inorganic spacer (ZrO_2 or Al_2O_3) and a mesoporous reverse contact (usually based on graphite). The three layers are sintered together and impregnated with a solution of the HOIP precursors, which fills the pores of all three layers [100-102]. Thus, a perovskite solar cell was made in which two layers of mesoporous TiO_2 and ZrO_2 as a scaffold impregnated with perovskite without a layer of a hole conductor (Figure 2e). Perovskite was formed in a pore system from a solution of PbI$_2$, methylammonium (MA) iodide and 5-ammoniumvaleric acid (5-AVA) iodide (5-AVA). Mixed perovskites (5-AVA)$_x$(MA)$_{1-x}$PbI$_3$ had a lower defect concentration and better pore filling of the TiO$_2$ framework, which provided a higher quantum yield for photoinduced charge separation compared to MAPbI$_3$. In this case there is no selective extraction of holes, which must be transported through the perovskite layer to the reverse carbon contact. Therefore, the functioning of the device depends on the rate of electron charge transfer to titanium dioxide. CS had an energy conversion efficiency of 12.8% and was stable for more than 1000 hours in air under full sunlight.

Similar fully printed three-layer MPSCs were also manufactured, except that TiO$_2$ layers were obtained from mesoporous mTiO$_2$ or P25 (commercial TiO$_2$) [95]. Mesoporous annealed crystalline (single-crystal) anatase provides a large surface area and has fewer traps for carriers than polycrystals. This method is more complicated, which prevents its use as an (ETM) for mesoscopic perovskite solar cells (MPSC). In [95], the fine-pore, mesoporous single crystal anatase with a large surface area was made in a simpler manner and achieved a mean PCE of 12.96% with good reproducibility and maximum efficiency of 13.47%. This work demonstrates that single-crystal mesoporous anatase with small particle size is a promising candidate as electronic transport materials ETM.

2.5. Tandem Organic-Inorganic Perovskite Photovoltaic Systems

The efficiency of single junction SCs is limited by the fundamental efficiency limit of Shockley-Kisser Shockley-Queisser (SQ) [103, 104]. The term single-junction SCs here means a photovoltaic system consisting of a single solar cell (a system with one semiconductor absorption material). In single-junction SCs based on a single semiconductor absorption material, the no-load voltage cannot exceed the band gap of this semiconductor, and in the case of a heterojunction it cannot exceed the width of the band gap of the narrow-gap semiconductor. The simplest way to overcome the fundamental limit of SQ efficiency is the formation of tandem (multi-junction, cascade, multi-element) SCs, in which sub-cells with different semiconductor absorption materials with different forbidden zones are arranged one after another. Tandem systems allow you to expand the spectral characteristics of the system, covering the spectral regions of solar radiation, which cannot be used for photoelectric energy conversion SCs with one absorber material. In connection with this, tandem photovoltaic systems are created in which the perovskite cell is combined with a cell with an absorber made of another semiconductor material. There are 4-terminal and 2-terminal architectures of tandemly systems. In the 4-terminal architecture, two independent mechanically connected sub-elements are electrically isolated, and in the 2-terminal architecture, the photovoltaic system represents a single monolithic layered structure. The 4-terminal tandem architecture may have slightly higher efficiency compared to the 2-terminal architecture, since the 4-terminal tandem architecture does not require equality of current in two sub-elements as in a 2-terminal tandem architecture. Nevertheless, in the opinion of the authors, monolithic 2-terminal tandems are more technically promising, since they are more compact and do not require additional commutation and equipment to coordinate the elements.

Silicon or CIGS (copper indium gallium selenide) cells are most often used as the second non-perovskite cell [105]. Such photovoltaic systems have higher efficiency [106-108]. PSCs [109-112] that were transparent in the near-infrared range were developed, which made it possible to increase the efficiency of 4-terminal tandems to 26.7% when using a silicon bottom cell [112, 113] and up to 23.9% using CIGS bottom cell [114]. At the same time, for monolithic perovskite/CIGS tandem SC, an efficiency of 22.43% was achieved [115]. This non-encapsulated device retained 88% of the initial efficiency after 500 hours with continuous illumination with one sun.

More promising and high-tech are monolithic 2-terminal tandems, the development of which began with the work of Mailoa et al. [116]. When using the single junction c-Si cell as a bottom cell, an efficiency of 13.7% was achieved. Later on, intensive research was conducted with the aim of achieving better tandem efficiency. Tandem efficiency of 18.1% [117], 21.2% [118] and 23.6% were reported [119]. At the same time, it was possible to improve the environmental safety and thermal stability of SCs [119]. In 2018, Sahli et al. demonstrated a monolithic tandem cell with an efficiency of 25.2% [121], Oxford Photovoltaics presented a cell with an efficiency of 27.3% [122].

In this chapter, the emphasis is placed on tandem photovoltaic systems consisting only of PSCs (All-perovskite tandems). The creation of All-perovskite tandems was preceded by the development of various HOIPs materials, which together cover a wide area of solar radiation. In particular, in 2016, efficient perovskite materials with a low band gap (1.2–1.3 eV) were developed [123]. This allowed developing the concept of All-perovskite tandems. In the construction of such photovoltaic systems, two layers of perovskites with different band gap located one behind the other were used. Two and four terminal architectures were developed, for which tandem perovskite systems achieved, respectively, an efficiency of 17% and 20.3%. Thus, perovskite tandems not only open the way to increase the efficiency of PSCs, but can provide competitiveness to widely used silicon and gallium arsenide SCs.

In 2017, Dewei Zhao et al. [124] made mixed Sn-Pb perovskite solar cells (PVSC) with a narrow bandgap (~ 1.25 eV) 620 nm thick, which allows to increase the graininess and to obtain a higher crystallinity to extend carrier lifetime to more than 250 ns, reaching maximum energy conversion efficiency (PCE) - 17.6%. In addition, this narrow-gap PVSC has reached an external quantum efficiency (EQE) of more than 70% in the 700-900 nm wavelength range, an important infrared region of the spectrum in which sunlight is transmitted to the lower cell. They also combined the lower cell with the perovskite upper cell with a band gap of ~ 1.58 eV and created a fully-perovskite tandem solar cell with four terminals with a PCE of 21.0%.

In 2018, Zhao et al. [21] made Four-Terminal All-Perovskite Solar Cells Achieving Power Conversion 23% by mechanically connecting semitransparent 1.75 eV wide-bandgap $FA_{0.8}Cs_{0.2}Pb(I_{0.7}Br_{0.3})_3$ perovskite top cells with 1.25 eV low-bandgap $(FASnI_3)_{0.6}(MAPbI_3)_{0.4}$ bottom cells. The top cell had transparent MoO_x/ITO electrodes and provided transmittance up to 70% above 700 nm. In this paper, the effectiveness of All-Perovskite Tandem SCs for the first time exceeded the world record of single junction PSCs. These results demonstrate the promise of all-perovskite tandem solar cells for achieving high-efficiency PSCs.

3. MATERIALS AND FORMATION METHODS OF THE MAIN FUNCTIONAL LAYERS OF HYBRID ORGANO-INORGANIC PEROVSKITE SOLAR CELLS

A high efficiency of PSCs is achieved by a successful combination of properties of HOIPs, buffer layers and ohmic contacts. The development of reliable technology for the production of thin films is determined by the following factors [125]: the effect of grain boundaries on recombination and charge transfer; contact phenomena for systems of several materials; point defects (pinholes) in thin-film SCs.

Compared to other SCs production technologies, the advantage of organic and organo-inorganic SCs is that the synthesis of precursors is inexpensive and these SCs can be created by "wet" methods at low temperatures in the process of high-speed roll-to-roll roll technology. It is important that the perovskite technology is compatible with the technologies of the first and second generation, which can contribute to a sufficiently rapid solution of the problem of the stability of solar batteries and the achievement of an industrial scale in their production [126].

Requirements for the materials of PSCs functional layers are described below in Section 3.5. "Selection of materials of buffer layers and ohmic contacts." Before this, let us consider the properties and methods of obtaining thin films of these materials.

3.1. Hybrid Organo-Inorganic Perovskite Absorbers

HOIPs are not new materials and have been extensively studied in connection with their practically useful physical properties. At the same time, from the whole variety of compounds with a perovskite-like structure in photovoltaics, hybrid organo-inorganic perovskite materials have found the greatest application, in the crystal lattice of which the alkali metal cation is replaced by a cation of an organic ammonium base. The most widely used HOIPs has the formula $CH_3NH_3PbX_3$, where X is Cl^-, Br^- or I^-. The structure of such synthetic perovskites consists of PbX_6 octahedra connected vertically to a three-dimensional perovskite framework and ammonium cations that are located in cavities between octahedra and are coordinated with 12 ligands [127]. This structure is commonly referred to as the 3D structure of the HOIP. There are also layered 2D structures of HOIPs, on which we will pause in one of the subsections.

The material of the absorber in this form of SCs are various HOIPs. As emphasized above, the prospects for using HOIPs in SCs are due, firstly, to the strong absorption of photons over a wide range of the solar spectrum, and secondly to the high mobility of photoinduced electrons and holes. A purposeful change in the chemical composition of the perovskite photoabsorber, in combination with the optimal choice of buffer materials associated with it, and the control of the quality of interfaces in the SC heterostructure, opens up additional wide possibilities for controlling the optical and transport characteristics of PSCs. It should be noted that the resistance of HOIPs to the effects of solar radiation and atmospheric humidity remains a fundamental scientific problem. The development of two-dimensional perovskites can be one of the ways to solve this fundamental problem.

Properties, methods of HOIP synthesis and films based on them are discussed in more detail in other chapters of this book. Below we do not consider solution technology. Here, only some additional information about HOIP is given, especially about the preparation of HOIP thin films by vapor deposition, which is important for analyzing the operation of thin-film PSCs.

To get the highest efficiency of the HOIPs, the morphology of film, thickness, crystallinity, and crystal size of the deposited perovskite layer have played a vital role in the photovoltaic performance [128, 129]. Therefore, researchers have conducted various techniques to produce uniform and large grain size perovskite thin films by modifying the deposition parameters and procedures. Solution-processed is an excellent technique to make perovskite films, because of its advantages of simplicity and low-cost. However, it has some disadvantages like control over the low-temperature crystallisation process and sequential film deposition from solution cannot produce perovskite-perovskite heterostructures.

Vapour-based deposition technique is the alternative technique to substitute the solution processed methods, which is a possible route to overcome the earlier problems [130-133]. This technique provides unique benefits for high-quality perovskite crystallisation. Vapour deposition methods are comfortable for substantial area deposition and a scale-up preparation without the use of solvents and annealing steps, allowing the perovskite-on-perovskite deposition to create heterostructures and junctions. To date, the vapour-based methods applied to the synthesis of hybrid perovskites mainly based on vapour-assisted solution processes (VASP) and vacuum evaporation process with few other attempts of ultrasonic spray coating and flash evaporation [131, 132]. These vapour-based methods are classified into 1. Evaporation, which further divided as co-evaporation and subsequent evaporation. 2. vapour-assisted solution process (VASP) which is also organized into atmospheric VASP (AP-VASP) and low-pressure VASP (LP-VASP), and unique deposition techniques (flash evaporation and ultrasonic spray coating (USC)) (Figure 4).

Liu et al. reported a PCE of 15% with a simple planar heterojunction solar cell incorporating a dual-source co-evaporation deposited perovskite as the light-harvesting material, TiO_2 as ETL and Spiro-OMeTAD as HTL. They fabricated a flat $CH_3NH_3PbI_{3-x}Cl_x$ layer by co-evaporating organic (CH_3NH_3I) and inorganic source ($PbCl_2$) in the high vacuum chamber. The X-ray diffraction (XRD) results indicated a similar crystal structure for perovskite films deposited by co-evaporation technique and solution process. The scanning electron microscope (SEM) images in highlight the considerable differences between the film morphologies produced by the two deposition processes the cross-sectional images of the completed devices and h reveal more information about the crystal size [3]. The same group in 2014 achieved 16.5% by studied the optical properties of $CH_3NH_3PbI_3$ and modelled the optical field distribution of the total device [134].

Chen et al. developed a perovskite thin film by using the sequential vacuum deposition. They reported the PCE of 15.4% with the perovskite thin film as an active layer and PEDOT: PSS as HTL and ETL was the C_{60} /Bathophenanthroline (Bphen) [135]. In 2015 Abbas et al. reported the PCE of 13.7% with the sequential vacuum deposited perovskite film and P3HT as HTL [136]. Schematic representation of sequential vapor deposited perovskite film showed in Figure 5.

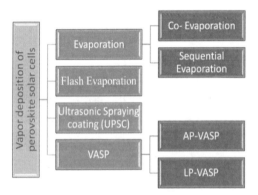

Figure 4. Classification of the vapor-based methods.

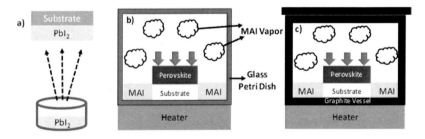

Figure 5. Schematic representation of sequentially vapor deposited perovskite film in two steps: (a) evaporation of PbI_2 followed by vapor assisted growth in (b) glass petri dish or (c) graphite vessel.

Hybrid Organo-Inorganic Perovskite Solar Cells 21

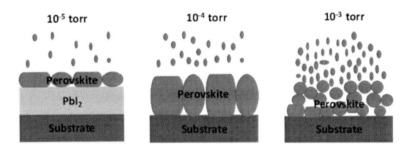

Figure 6. The illustration of perovskite formation under various pressures of organic halide sublimation.

Yang et al. developed precursor layers of lead iodide and MAI by using the alternative sequential vacuum deposition. The device performance resulted in the highest PCE of 16.03% and excellent stability over 62 days with the little degradation of <9% [137]. In 2016 H. J. Bolink studied both the device structure by using the p-i-n and n-i-p, 500 nm layer of the MAPbI$_2$ perovskite absorber, sandwiched in between small molecular weight organic charge transport molecules, using only vacuum-based processes. They achieved the PCE of up to 20% for 0.1 cm^2 and 15% for cell areas close to 1 cm^2 with the combination of intrinsic and molecularly doped charge transport layers [138].

Hsiao et al. reported the record PCE of 17.6% with the short circuit current (J_{sc}) of 22.7 mAcm^{-2}, open circuit voltage of 1.06 V and filled factor of 0.73 by manipulating the partial pressure organic halide reagents during the sequential vapour deposition (Figure 6) [139].

3.2. Buffer Layers of Hybrid Organo-Inorganic Solar Cells

In most of the perovskite SCs mentioned above, at least one of the buffer layers is of an organic nature. Such buffer layers, which have a complex structure and, apparently, are not very stable and radiation-unstable, it is desirable to exclude from the architecture of perovskite solar cells. In this respect, the work [140] on the creation of solar cells with a planar heterojunction of perovskite (CH$_3$NH$_3$)PbI$_3$/CuInS$_2$ formed using low-temperature wet technologies is very interesting. In this work, an Al_2O_3 insulator was used as a scaffold for the perovskite layer. The maximum temperature of the technologies used (annealing of $CuInS_2$ film) was below 250 °C. However, the efficiency of the ITO/$CuInS_2$/Al_2O_3/(CH_3NH_3)PbI_3/Ag photovoltaic cell formed was relatively low and was ~ 5.30%.

It should be noted that the materials of the electron-conducting buffer layers have been fairly well studied, whereas the hole-conducting materials (HTM) require careful theoretical and experimental study. Currently, three categories of HTM are used: inorganic, polymeric and organic materials. Recently, more attention has been paid to inorganic HTM due to their high profitability and stability in the environment. It should also be noted that the number of publications pertaining to this problem is growing every year. In connection with this inorganic HTM, much attention is paid below and above all HTM based on cuprous salts for which good results have recently been obtained.

3.2.1. Electron-Conducting Buffer Layers (Anodes)

The function of anode materials in the PSCs consists of the selective collection of electrons generated in the perovskite absorber with their further transport to the external contacts. ETMs function as blocking of hole transport and retardation of recombination between electrons in the conducting oxide layer and holes in the perovskite. To this end, the ETM buffer layer is formed from an n-type semiconductor material. In view of the fact that in most PSCs architectures systems are realized in which a transparent electrode acts as an anode, the notion of photoanodes is often used. We note that systems with a transparent cathode collecting photoinduced p-type charge carriers are created less often [83]. In this case, we can speak of a photocathode. To date, semiconductor metal oxides have been used as photoanode material. However, there are examples of the use of n-type organic semiconductors, in particular PCBM ([6,6] -Phenyl-C_{61}-butyric acid methyl ester) also used in organic SCs [141, 142].

The most common ETM is titanium dioxide (TiO_2). It can be used in the form of a mesoporous scaffold, or in the form of a thin layer. The hole transport blocking TiO_2 layer is precipitated by spray pyrolysis, atomic layer

Hybrid Organo-Inorganic Perovskite Solar Cells

deposition, or by sintering a spin-coating precursor. The goal is to obtain a thin and homogeneous dense layer [143] free of pinholes [144]. However, it has been shown that PSCs can also work relatively well without [145] or with ultra-thin [146] hole-blocking layers. Pyrolysis spray is a cheap and affordable way of depositing relatively compact and uniform layers in the range from a few nanometers to micrometers. An alternative to spray pyrolysis is atomic layer deposition (ALD) or spin coating of the $TiCl_4$ layer, followed by sintering.

At the top of the blocking layer, a thin mesoporous film made of metal oxide nanoparticles, for example TiO_2, is usually applied. Now the most highly efficient devices contain only a thin-pored TiO_2 film about 100 nm [147]. The role of a thin mesoscopic layer is therefore questioned, especially in view of the high efficiency of planar solar cells. Nevertheless, today the most efficient PSCs still use the mesoscopic scaffold, hence the role of the porous network for electronic transport cannot be ignored.

In the PSCs architecture with a bulk heterojunction, mesoporous metal oxide films are used as the photoanode providing higher rates of charge extraction compared to conventional DSSCs [12] due to their denser perovskite coating. A low-temperature production of solar cells $TiO_2/CH_3N H_3PbI_3$/spiro-OMeTAD was also demonstrated using nanocrystalline rutile TiO_2 deposited by chemical precipitation from the precursor $TiCl_4$. The thickness and morphology of rutile TiO_2 nanoparticles can be regulated by controlling the concentration of the precursor solution. Under optimized conditions, this rutile-based TiO_2 device resulted in an impressive efficiency of 13.7%. This device also exhibits an exceptionally high V_{oc} of 1.11 V. This is explained by the formation of closer contact between the rutile TiO_2 and the perovskite layer, which leads to more efficient extraction of photogenerated electrons [148].

A certain problem in the use of TiO_2 may be its interaction with molecular oxygen and the manifestation of its photocatalytic properties. Photocatalytic reactions are possible on the surface of titanium dioxide, which may affect the stability of some components of the organic photovoltaic system. It is known that TiO_2 contains many oxygen vacancies (or Ti^{3+} sites). The electrons in these regions are approximately 1 eV below the edge of the conduction band and interact with adsorbed molecular oxygen. As a result of charge transfer, a complex ($O_2 - Ti^{4+}$) is formed. When UV excitation of titanium dioxide took place, electron-hole pairs are formed. Holes in the valence band recombine with electrons in the region of oxygen adsorption and molecular oxygen is desorbed. This leaves a free electron in the conduction band and a positively charged unfilled oxygen vacancy on the TiO_2 surface. Since the hole conductor has an excessive number of holes, they will easily recombine with free electrons from the conduction band [149]. This problem is solved by applying a mesoporous scaffold in an inert atmosphere or by using filters that cut off UV radiation.

In addition to TiO_2, other n-type metal oxides (for example, SnO_2 and ZnO) can also be included in planar PSCs. It has been shown that SnO_2-blocking layers provide excellent results [15, 16]. PSCs with these blocking layers had efficiencies of more than 21%. Liu and Kelly reported using a thin film of ZnO as an ETM in planar devices based on perovskite $CH_3NH_3PbI_3$ (15.7% efficiency) [79]. The higher electron mobility in ZnO compared with TiO_2 makes it the best ETM material. ZnO nanoparticles obtained by hydrolysis of the precursor $Zn(O_2CCH_3)_2$ do not require thermal treatment and can be easily deposited on conductive substrates. Nanoparticles II-VI (for example, CdSe) can also be used as ETM in perovskite solar cells, as demonstrated by Wang [149]. Such a device with the configuration of $CdSe/CH_3NH_3PbI_3/spiro-OMeTAD/Ag$ gave an efficiency of 11.7%.

As noted above, various vertical nanowire (NW) or nanotubular (NT) structures can be used instead of the mesoporous metal oxide scaffold. In 2014, [151] considered the possibility of using as nanotube TiO_2 as an ETM, which allowed to increase the efficiency of PSCs somewhat by reducing the rate of recombination processes. Infiltration of HTM into the porous space of these NW and NT structures does not cause difficulties. In addition, such vertical NW and NT structures are better than the mesoporous composition of the photoanode [152] due to more efficient electron transport and a decrease in the rate of recombination processes. The decrease in the total area of interphase surfaces (heterojunctions) in NW/NT structures as compared to the mesocarp does not cause a significant decrease in the photocurrent due to the unique absorption and charge-transport properties of HOIPs. The use of ZnO NW nanomasses as photoanodes in perovskite solar cells with an efficiency of ~ 5% was demonstrated in [74]. It was shown that the ZnO NW composition improves the electron transfer taking into account shorter electron transport paths in ZnO nanorods compared to mesoporous TiO_2 films of the same thickness. The impregnation of the rutile TiO_2 structure on the NW submicron perovskite layer with organic HTM spiro-MeOTAD resulted in solar cells with an efficiency of 9.4%.

Table 1. Structures of organic ETM [86]

Abbreviation	Full name	Structure
C_{60}	Fullerene	
$PC_{61}BM$	[6,6]-Phenyl C_{61} butyric acid methyl ester	
$PC_{71}BM$	[6,6]-Phenyl C_{71} butyric acid methyl ester	
3TPYMB	Tri[3-(3-pyridyl)mesityl]borane	
ICBA	Indene-C60 bisadduct	

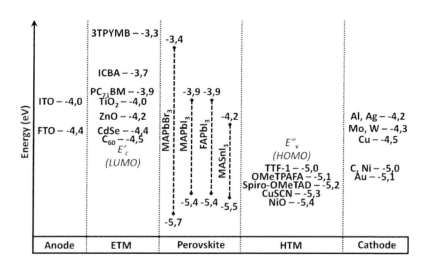

Figure 7. Energy band diagram of the PSC device.

In the case of using the inverted planar (p-i-n) PSCs architecture, it is possible to use n-conducting organic layers. Such typical structures shown in table 1 [86].

When choosing the ETM of the buffer layer, it is important to take into account the mutual arrangement of the electronic levels of perovskite and ETM. In Figure 8 is an energy diagram showing the E'V (LUMO) levels for various ETMs in comparison with the perovskite level of the form $CH_3NH_3PbI_3$ [86].

Fullerene (C_{60}) and its derivatives are the most widely used n-type ETMs in inverted perovskite solar cells. Various fullerene derivatives usually have a solubility in organic solvents and various optoelectronic properties (for example, electron mobility). It is extremely important to optimize the thickness of fullerenes in order to maximize the efficiency of PSCs. It was shown that the $PC_{61}BM$ layer up to 50 nm thick was sufficient to achieve full coverage of the perovskite film, whereas thicker fullerene layers tended to increase the series resistance. In addition to $PC_{61}BM$, other fullerene compounds can be used: C_{60}, ICBA and $PC_{71}BM$. It has been shown that organoboranes (eg, 3TPYMB) can also be suitable ETM for perovskite solar cells [86]. PSCs made using 3TPYMB showed an efficiency of 5.5% compared to 10% for $PC_{61}BM$-based devices.

3.2.2. Hole-Conducting Buffer Layers (Cathodes)

Materials for transporting holes (HTM) are important components in the PSCs architecture. Undoubtedly, the choice of the composition of the buffer HTM greatly influences the performance of PSCs. Currently, 2,2′,7,7′-tetrakis-(N,N-di-4-methoxy-phenylamino)-9,9′-spirobifluorene, better known as spiro-OMeTAD, is the most widely used HTM for producing highly effective SCs. However, this is an extremely expensive material with limited stability and mediocre mobility of holes compared with inorganic HTM. To ensure the large-scale application of technologies based on PSCs, alternative HTMs are being developed. The results obtained with the help of solution technology HTMs are crucial for the development of inexpensive, high-performance and large-format printed PSCs technologies. Recent advances and perspectives in the design of various types of organic HTM predominantly mesoscopic PSCs are presented in the review [153]. Most record PSCs used spiro-MeOTAD. Given the relatively high conductivity of HOIPs and the significantly lowerconductivity (~ 10-5 $\Omega^{-1}cm^{-1}$) of spiro-MeOTAD, a thick layer of this HTM is required, which increases the series resistance and reduces the efficiency of the solar cell.

The effect of spiro-MeOTAD, P3HT, and 4-(diethylamino)-benzaldehyddiphenylhydrozone (DEH) as HTM on the efficacy of PSCs has been studied [154]. The PSCH based on spiro-MeOTAD is more than ten times lower in comparison with P3HT, and is more than 100 times lower than in the device with DEH. Recently, a number of substituted methoxyphenylamine pyrene derivatives have been synthesized and synthesized successfully. PSCs with HTM based on Pyrene derivative. Ru-C demonstrated an efficiency of 12.7% [155]. Its efficiency is slightly lower than PSCs based on HTM spiro-MeOTAD. PEDOT: PSS is also used as an HTM. The substrate structure/ITO/PEDOT:PSS/$CH_3NH_3PbI_{3-x}Cl_x$/PCBM/Al, where PEDOT:PSS and PCBM are used as transport layers of holes and electrons, respectively, made it possible to produce a flexible PSC on PET/ITO with an efficiency of 9.2 [156].

It should be noted that the high cost of all used organic HTMs such as spiro-MeOTAD and P3HT can be an obstacle to their use when commercializing solar cells. It is tempting to use more simple and stable inorganic p-type semiconductors in PSCs. $CH_3NH_3PbI_3$ sensitized PSCs were demonstrated using inorganic CuI as the hole conductor obtained from the solution [78]. In addition, it was shown that an HTM may replace the liquid electrolyte [157].

The use of proton ionic liquids (PIL) as active p-additives in PSC based on triarylamine was demonstrated. An increase in efficiency was observed when replacing a typical p-additive for spiro-MeOTAD lithium salt on PIL [147]. HTM for PSCs should primarily meet the following requirements: (1) good hole mobility; (2) the perovskite-compatible energy level of the valence band, or the next, molecular orbital (HOMO) for effective hole transport; (3) good solubility and film-forming properties, (4) economy and (5) stability under external influences. It should be noted that planar PSCs do not require HTM pore filling, which allows eliminating problems with the pores often found in ssDSS. This expands the choice of materials. Three categories of HTM have been studied in PSC: organic, polymeric and inorganic HTM.

3.2.2.1. Organic and Polymeric P-Type Conducting Materials

The chemical structures of typical organic and polymeric HTM and their corresponding HOMO levels are shown in table 2 [86].

Spiro-OMeTAD is the most commonly used material for transporting holes in perovskite solar cells, mainly because it is well studied in OLEDs and DSSCs. SCs based on HOIPs in combination with spiro-OMeTAD have achieved efficiencies of more than 20% [81]. However, the cost of spiro-OMeTAD remains, unfortunately, high mainly because of its long and complex synthesis. The limitation of the resistance of organic and polymeric HTMs has also proved to be a serious obstacle to the development of highly efficient and cost-effective PSCs. Therefore, it became necessary to develop more economical and more resistant to external influences alternatives to organic and polymeric HTM. In this regard, recently increasing attention is paid to inorganic HTM due to their high profitability and stability in environmental conditions. Inorganic HTMs such as CuI, NiO and CuSCN are well known [77, 158]. In particular, PSCs using CuSCN as a hole-

Hybrid Organo-Inorganic Perovskite Solar Cells 29

conducting layer achieves an efficiency exceeding 20% and possesses high stability [159]. Therefore, we will discuss these HTMs in more detail below.

Table 2. Structures of organic and polymeric HTMs [86]

Abbreviation	Full name	Structure
Spiro-OMeTAD	2,2',7,7'-Tetrakis[N,N-di(4-methoxyphenyl)amino]-9,9'-spirobifluorene	
H101	2,5- bis(4,4'-bis(methoxyphenyl)aminophen-4''-yl)-3,4-ethylenedioxythiophene	
OMeTPA-FA	Tris{N,N-bis(4- methoxyphenyl)-N-phenyl}amine quinolizino acridine	
DR3TBDTT	(5Z,5'Z)-5,5'-((5'',5'''''-(4,8-bis(5-(2-ethylhexyl)thiophen-2-yl)benzo[1,2-b:4,5-b']di-thiophene-2,6-diyl)bis(3,3''-dioctyl-[2,2':5',2''-terthiophene]-5'',5-diyl))bis(methanylylidene))bis(3-ethyl-2-thioxothiazolidin-4-one)	
PANI	Polyaniline	

Table 2. (Continued)

Abbreviation	Full name	Structure
P3HT	Poly(3-hexylthiophene-2,5-diyl)	
PTAA	Poly-triarylamine	
TTF-1	Tetrathiafulvalene	
T103	2,6,14-Tri(N,N-bis(4-methoxyphenyl)amino)-triptycene	
Py-C	Pyrene arylamine	
where		

3.2.2.2. Inorganic P-Type Conducting Materials

There are some general requirements for inorganic HTMs used in PSCs, such as high transparency in the visible region, good chemical stability, increased hole mobility, and the energy level of the valence band (VB) comparable to the energy level of the perovskite valence band. To date, various inorganic p-type compounds such as CuI, CuSCN, CuO, Cu_2O, NiO, MoO_3 and VO_x are used as HTMs in PSCs. Compared to organic HTMs, p-type inorganic semiconductor materials have the following advantages: high hole mobility, large band gap, low cost and availability, and high stability.

3.2.2.2.1. Hole-Conducting Materials Based on Copper (I) Iodide (Cui) and Copper (I) Thiocyanate (Cuscn)

An important requirement for HTM is its transparency. Because of the large band gap, CuI and CuSCN (3.1 and 3.5 eV, respectively), they are transparent throughout the visible light range. In the near UV region, the thin CuSCN film absorbs at $\lambda \sim 350$ nm, and for a thin CuI film at $\lambda \sim 400$ nm [160]. Absorption of CuSCN films obtained by spin-coating and annealed at 90 and 120°C is absorbed somewhat lower than that of unannealed films [161]. Similar results were obtained for CuI films [162]. As the concentration of the precursor solution from which the film forms increases, an increase in absorption is observed, which may be due to an increase in the film thickness and scattering at the boundaries of larger grains. The absorption coefficient (α) was found to be 106 cm^{-1}. A high value of α confirms the existence of a direct band gap for a given semiconductor.

3.2.2.2.2. Iodide of Copper (I) (CuI)

Iodide copper (I) (CuI) is a p-type semiconductor with a wide forbidden band (~ 3.1 eV). CuI is a good candidate for the role of HTM because of its good optical transparency and higher hole mobility (0.5-2 cm^2 / (B·s)) compared with CuSCN.

CuI exists in several crystalline forms. Copper (I) iodide takes the sphalerite structure below 390°C (γ-CuI), the wurtzite structure between 390 and 440°C (β-CuI) and the structure of the halite is above 440°C (α-CuI).

Since most methods for obtaining thin CuI films are low-temperature, the most common form is γ-CuI [77].

In 1995, Tennakone et al. first demonstrated the possibility of using CuI as a hole-transport layer in Solid State Dye Solar Cell Technology (ssDSC) based on cyanidinic dyes [140]. Christians et al. reported the possibility of using CuI as an HTM in mesostructured n-i-p PSCs [158]. The configuration of the FTO/TiO$_2$/mp-TiO$_2$/CH$_3$NH$_3$PbI$_3$/CuI/Au device had an efficiency of 6%. Although the thick CuI layer exhibits a higher electrical conductivity than spiro-OMeTAD, CuI-based PSCs show a lower idling voltage due to the high rate of charge recombination.

Sepalage et al. [96] also produced planar n-i-p PSC with the FTO/TiO$_2$/CH$_3$NH$_3$PbI$_3$/CuI/graphite structure. The CuI and graphite layers were applied by doctor blading. For this PSC, a higher V_{oc} value was observed compared to previously described meso-structured CuI-based devices and an efficiency of 7.5% was achieved. A higher idling voltage was caused by a reduced thickness of the CuI film (~ 400 nm). However, V_{oc} was still below the expected V_{oc} ~ 1 V due to fast recombination at the perovskite/CuI heterojunction. The rapid injection of holes from the perovskite into the CuI layer prevents the appearance of a capacitive current, which is determined by the accumulation of charges at the perovskite boundary. Therefore, unlike spiro-OMeTAD, devices based on CuI do not show pronounced hysteresis when scanning potential in different directions. Strong photoluminescence quenching (PL) and a relatively fast fall of the open circuit voltage indicate a faster removal of positive charge carriers from the perovskite layer upon contact with CuI as compared to spiro-OMeTAD. In addition, CuI-based PSCs have a faster response, determined by electrochemical impedance spectroscopy (EIS) and open-circuit voltage decays (OCVD), which is associated with the current-voltage hysteresis.

Chen et al. used CuI as HTM in inverted planar p-i-n PSC with FTO/CuI/CH$_3$NH$_3$PbI$_3$/PCBM/Al architecture. Due to the high transmittance and good morphology of the CuI film, such CuI-based PSCs exhibit higher V_{oc} and J_{sc} values than devices using PEDOT:PSS [163].

There are several problems in using CuI as an HTM. In particular, there are problems of morphology control of the CuI surface and difficulties in selecting a solvent to exclude large CuI grains. However, these problems are solved, which makes CuI as an excellent material for p-buffer layers.

Characteristics of PSCs with CuI as HTM are given in Table 3 (M - mesostructured, P - planar architecture of PSCs).

Table 3. Efficiency of PSCs with CuI as HTM

Type SC	Architecture	V_{oc} (V)	J_{sc} (mA/cm^2)	FF (%)	η (%)	Reference
M	FTO/TiO$_2$/mp-TiO$_2$/Psk/CuI/Au	0,55	17,8	0,62	6,0	[159]
P	FTO/TiO$_2$/Psk/CuI/graphite	0,78	16,7	0,57	7,5	[152]
P	FTO/CuI/Psk/PCBM/Al	1,04	21,06	0,62	13,58	[164]

3.2.2.2.3. The Copper (I) Thiocyanate (CuSCN)

Thiocyanate copper (I) serves as a universal material for numerous optoelectronic devices, including transparent thin-film transistors, organic, sensitized dyes and PSCs and organic light-emitting diodes. CuSCN combines good hole-transport properties with a large band gap (> 3.5 eV). Compared with spiro-OMeTAD, CuSCN has excellent optical transparency in the entire visible and near-IR spectrum [96].

The copper (I) thiocyanate exists in two polymorphic forms: α and β-CuSCN. The α phase has an orthorhombic crystal lattice, whereas the β phase can be with a hexagonal or rhombohedral crystal lattice. β-CuSCN is more stable and easily accessible form, it consists of layers of SCN ions separating planes of Cu atoms [165].

CuSCN is inexpensive and there are many different methods for obtaining it. It has a higher hole mobility (0.01-0.1 cm^2/(B·s)) compared with spiro-OMeTAD (1.6 × 10^{-3} cm^2/(B·s)) and good chemical stability [79, 98].

In 1996, O'Regan and Schwartz first demonstrated the possibility of using CuSCN in ssDSCs [166]. Wherein, 2.2% efficiency was achieved. As for the possibility of using CuSCN in PSCs, it should be noted that this material can be used both in cells with n-i-p and with p-i-n heterojunction. CuSCN in mesoporous PSCs with the FTO/TiO$_2$/mp-TiO$_2$/CH$_3$NH$_3$PbI$_3$/CuSCN/Au structure was first used by Ito et al. [167]. Such elements achieved an efficiency of 4.85%. It has also been found that CuSCN films limit the photovoltaic degradation of moisture sensitive perovskite materials. Further increase in the efficiency of mesoporous PSCs with the CuSCN buffer layer was achieved mainly by optimizing the deposition of the perovskite layer [168]. CuSCN in planar PSCs with an n-i-p heterojunction was first used by Chavhan [82]. The maximum efficiency of such an element with the architecture FTO/TiO$_2$/CH$_3$NH$_3$PbI$_{3-x}$Cl$_x$/CuSCN/Au was equal to 6.4%. Subbiah et al. [169] reported the possibility of using CuSCN in planar PSCs with the p-i-n FTO heterojunction/ electrodeposited CuSCN/CH$_3$NH$_3$PbI$_{3-x}$Cl$_x$/PCBM/Ag. Such elements achieved an efficiency of 3.8%. Without optimization of the thickness of the

Table 4. Efficiency of PSCs with CuI as HTM

Type SC	Architecture	V_{oc} (V)	J_{sc} (mA/cm^2)	FF (%)	η (%)	Reference
M	FTO/TiO$_2$/mp-TiO$_2$/Psk/CuSCN/Au	0,63	14,5	0,53	4,85	[167]
M	FTO/TiO$_2$/mp-TiO$_2$/Sb$_2$S$_3$/Psk/CuSCN/Au	0,57	17,23	0,52	5,12	[170]
M	FTO/TiO$_2$/mp-TiO$_2$/Psk/CuSCN/Au	1,025	17,91	0,57	10,51	[171]
M	FTO/TiO$_2$/mp-TiO$_2$/Psk/CuSCN/Au	1,016	19,7	0,62	12,4	[76]
M	FTO/TiO$_2$/mp-TiO$_2$/Psk/CuSCN/Au	0,96	18,23	0,68	11,96	[168]
P	FTO/TiO$_2$/Psk/CuSCN/Au	0,97	18,42	0,40	7,19	[168]
P	FTO/TiO$_2$/Psk/CuSCN /Au	0,727	14,4	0,62	6,4	[82]
P	FTO/CuSCN/Psk/PCBM/Ag	0,677	8,8	0,63	3,8	[169]
P	ITO/CuSCN/Psk/C$_{60}$/BCP/Ag	0,97	21,7	0,742	15,6	[92]
P	ITO/CuSCN/Psk/PC$_{61}$BM/bis-C$_{60}$/Ag	1,07	19,6	0,74	15,6	[91]
P	ITO/CuSCN/Psk/PC$_{60}$BM/LiF/Ag	1,06	13,0	0,73	10,06	[91]
P	FTO/TiO$_2$/Psk/CuSCN/graphene oxide/Au	-	-	-	20	[126]

CuSCN film high resistance of the system was observed which led to low efficiency of PACs. Ye et al., using an electrodeposited CuSCN film as an HTM to produce an inverted planar PSC with the ITO/CuSCN/

$CH_3NH_3PbI_3/C_{60}/BCP/Ag$ architecture. By optimizing the electrochemical deposition of the CuSCN layer, they obtained a maximum efficiency of 16.6% [83]. Using as a photoabsorber $CsFAMAPbI_{3-x}Br_x$ and as an HTM CuSCN, Arora et al. achieve an efficiency of 20%, which is today a record for cells with inorganic buffer layers [161].

All the variants received to date of cells with CuSCN as HTM are given in Table 4, where the type of the mesostructured PSC architecture is designated M, and the planar P.

3.2.2.2.4. Methods for Obtaining CuI and CuSCN Thin Films

There are several methods for synthesizing and applying CuSCN and CuI on the substrate: the spin-coining method, the doctor blade method, the vacuum vapor deposition method, the molecular layering method, also called the Successive Ionic Layer Adsorbtion and Reaction (SILAR) method. The first two are the simplest and most frequently used methods. The following two methods are more difficult to implement but provide better crystallinity, morphology and homogeneity of the film.

Spin-Coating Application

Two main solvents are used to produce thin CuI films by spin-coating: acetonitrile (AN) and di-n-propylsulfide (DPS). When AN is used, the CuI solution is applied to the substrate (glass/FTO/TiO_2/$CH_3NH_3PbI_3$) by spin-coining method (1500 rpm) in an argon atmosphere in a glove box followed by annealing at 100 °C for 20 min [172]. When DPS is used, CuI solution is applied to the substrate by spin-coining method (1600 rpm) in an argon atmosphere in a glovebox, followed by annealing at 120 °C for 10 min [173].

DPS and diethyl sulfide (DES) are also used as the solvent for obtaining thin CuSCN films by spin-coating. CuSCN has a limited solubility in DPS, which makes the deposition of thicker films by means of spin-coining very problematic. DES allows to obtain solutions with concentrations up to 40 mg/ml. This greatly improves the homogeneity of the deposited films, which allows better control of the thickness of the HTM. When DES is used, the solution is deposited on the substrate by spin-coating (1500 rpm) in an argon atmosphere in a glove box followed by annealing at 100 °C for 20 min [174].

The method for obtaining a CuSCN layer using DPS is described in [174]. As CuSCN slowly dissolves in DPS, the mixture with a small CuSCN content is stirred overnight and then cooled for 1 day. Before sedimentation in 2 ml of supernatant, 150 μl of DPS is added. The resulting solution is deposited on the substrate by spin-coining (1500 rpm) in an argon atmosphere in a glove box followed by annealing at 100°C for 20 min.

The method of obtaining the HTM CuSCN film by the doctor blade method was described [76]. A solution of 6 mg CuSCN in 1 ml CuSCN propylsulfide was applied at a temperature of 65 °C. Theoretically, the thickness of the film corresponds to the gap between the substrate and the blade. However, the surface tension of the sol, the wetting and the rheological properties of the sol (dynamic viscosity) affect the thickness of the layer actually applied. This leads to the fact that the practical thickness of the film is only 60-70% of the theoretical.

Vacuum Methods of Application

Compounds of univalent copper decompose at high temperatures and this makes it difficult to obtain thin films. Nevertheless, in [175], CuI films were produced by vacuum evaporation. Films obtained by vacuum methods have a uniform and slightly rough surface. Scanning electron microscopy has shown that annealing leads to the formation of larger crystalline structures [175].

Method of Molecular Layering

By the method of molecular layering (SILAR method) - the method of layer-by-layer deposition of a substance from solutions onto a substrate surface is a cyclic layering. Each cycle consists of the adsorption of anions or cations from the aqueous solution on the substrate, washing, reaction and washing. This method of obtaining CuI films is described in [176]. Copper (II) sulfate ($CuSO_4·5H_2O$) was used as a precursor of cations, sodium thiosulfate ($Na_2S_2O_3$) was used as a complexing and reducing agent, which made it possible to reduce the degree of oxidation of copper to +1. Iodide potassium (KI) acted as an anion precursor.

In [177] was described as the SILAR method for obtaining thin CuSCN films. The precursor of cations was an aqueous complex of copper, which is formed by dissolving $CuSO_4 \cdot 5H_2O$ and $Na_2S_2O_3$ in deionized water. The anion precursor was a solution of sodium thiocyanate (NaSCN).

3.2.2.2.5. Hole-Conducting Materials Based on Copper (I) and (II) Oxide (Cu_2O and CuO)

CuO and Cu_2O are well-known p-type semiconductors. It was found that good crystallinity of Cu_2O and CuO films improves hole transport and increases Jsc. PSCs using Cu_2O and CuO as HTM also show good V_{oc} values, and exhibit efficiencies of 13.35% and 12.16%, respectively) [193].

Compared with PSCs using NiO and Cu:NiO as HTM, cells based on Cu_2O show higher characteristics due to higher hole mobility and favorable morphology of the perovskite surface on the Cu_2O film [179].

The traditional methods for the preparation of the Cu_2O film are thermal oxidation, electrodeposition and organometallic chemical vapor deposition [180]. The Cu_2O film can be obtained by in situ conversion of a CuI film in an aqueous solution of NaOH, and a CuO film is produced by heating a Cu_2O film in the air [178].

3.2.2.2.6. Hole-Conducting Materials Based on Nickel Oxide (NiO)

Nickel oxide is a well-known p-type semiconductor used in DSSCs. NiO is a potential candidate for use in PSCs since it has a comparable valence band with a perovskite with $E_g > 3.50$ eV and accordingly has good optical properties. It also has good chemical stability.

There are many different methods for the production of NiO films: pulsed laser deposition, electrodeposition, spray pyrolysis, spin-coating, sputtering and layer-by-atom deposition [181]. The often used method for the production of NiO is a "wet" method which is realized by preparing a precursor solution by adding monoethanolamine to nickel acetate tetrahydrate in methoxyethanol, followed by stirring for 10 hours. The application of the solution of precursors to the substrate is realized by the spin-coating method [179].

In 2014, the first publication was published, which described the use of nickel oxide as HTM in PSCs. The effectiveness of this element was 7.8% [182]. In 2015, copper-doped nickel oxide (Cu: NiO) was used as HTM in planar PSCs. The effectiveness of this element was 15.4% [183]. Later, an efficiency of 16.4% was achieved for PSCs with an HTM pure NiO film obtained by layer-by-atom deposition (ALD) [181]. At present, the highest efficiency of PSCs using NiO (compound $Li_{0.05}Mg_{0.15}Ni_{0.8}O$) as HTM was 18.4% [181]. A significant disadvantage of using NiO as an HTM is that the filling factor and the voltage of the idle (these parameters strongly affect the performance of the device) for such PSCs are lower than for PSCs using organic HTM, in particular, the lowest efficiency is achieved when NiO is applied by "Wet methods" [181].

3.2.2.2.7. Hole-Conducting Materials Based on Molybdenum (VI) Oxide (MoO_3)

MoO_3 is another potential HTM because of its advantages - non-toxicity and stability in the air. However, PSCs with MoO_3 show a low level of efficiency mainly due to the poor quality of perovskite films deposited on MoO_3 [185]. One solution to this problem is to use a MoO_3 / PEDOT composite film: PSS as an HTM. This combines the advantages of MoO_3 such as stability in environmental conditions and at the same time a good morphology of the surface of the perovskite film deposited on PEDOT:PSS. MoO_3 was prepared by thermal decomposition of a solution of ammonium heptamolybdate at 80 °C. Such cells had the structure ITO/MoO_3/PEDOT:PSS/$CH_3NH_3PbI_3$/C_{60}/Bphen/Ag with an efficiency of 14.87% [186].

3.2.2.2.8. Hole-Conducting Materials Based on Vanadium Oxide (VO_x)

VO_x is a p-type semiconductor material with high stability. It can be easily obtained by low-temperature "wet methods" [149]. Initially, VO_x was used as an HTM in organic solar cells (OSC). PSCs with such HTM achieved a high enough efficiency of 14.23% [187]. However, the poor morphology of the VO_x surface leads to defects in the deposition of perovskite films, which significantly limits the effectiveness of PSCs.

3.3. Selection of Materials of Buffer Layers and Ohmic Contacts

To create highly efficient SCs based on HOIPs, it is necessary to develop heterostructures that optimally combine perovskite materials with other known or newly synthesized materials. The transfer of electrons and holes in heterojunctions and then to SCs contacts should occur with minimal fundamental voltage losses, which is important for the relative positions of electron energy levels in the HTM/perovskite/ETM structure (Figure 8). On the one hand, the level shift at each heterojunction must be large enough to ensure intensive injection of free charges and allow the bound pairs (excitons) to dissociate in a strong electric field of heterojunctions.

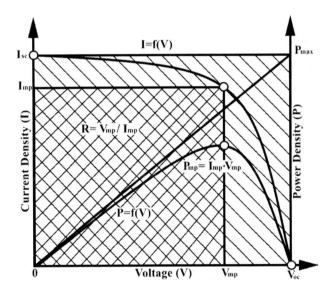

Figure 8. Volt-ampere characteristic and the dependence of the power density on the voltage of the illuminated solar cell (I_{SC} – short-circuit current density, V_{oc} - open-circuit voltage, P_{max} - maximum theoretical power density, P_{mp} - maximum power density, R= $V_{MP}/I_{MP} \sim V_{oc}/I_{SC}$ - characteristic resistance of the solar cell - the inverse of the slope of the line 0 - I_{sc}, V_{oc}).

The competing requirement is that the level shift must be minimized to achieve a high open-circuit voltage V_{oc}. In addition, the transport properties of the heterojunction in the entire volume of the device are also dictated by the morphology of the perovskite layer and the quality of contacts at the interphase boundaries with buffer materials. Dependence of these structural characteristics of PSCs on the choice of the architecture of PSCs, buffer materials and electrodes, and the procedure for the formation of the device as a whole seems to be the central problem of current research. For an effective collection of charges and the formation of effective ohmic contacts with electrodes, the Fermi levels of SCs should be consistent with the transport energy levels of holes and electrons in the buffer layers. The Fermi level of the anode collecting the holes should be close to the HOMO level of the HTM, and the level of the cathode collecting the electrons to the LUMO level of the ETM.

Therefore, for the successful operation of SCs, the following conditions must be met: the optimum difference between the energy levels E "V (HOMO) and E'C (LUMO), respectively, of the buffer HTM and ETM, sufficient mobility of the injected charge carriers and good ohmic contacts of the HTM with the cathode and ETM with anode.

The choice of materials of ohmic contacts is based on the efficiency of collection of charges with the necessary material work function [86]. The following are the most frequently used ohmic contacts in PSCs. The work function of counter-electrodes (cathodes) are: gold -5.1 eV; silver -4.26 eV; aluminum -4.28 eV. The work function of transparent conductive face electrodes (anodes) are: fluorine doped (FTO) oxide -4.4 eV; tin oxide doped with indium (ITO) -4.0 eV.

Most often perovskite solar cells have metal rear contacts. Preference is given to gold, less often silver. However, silver in contact with $CH_3NH_3PbI_3$ oxidizes to AgI. Semi-transparent contacts using silver nanowires are known [188, 189]. Also, counter-electrodes of indium oxide, deposited on MoO_x, [117, 190] have also been successfully used as counter electrodes of tandem SC $Si/CH_3NH_3PbI_3$.

4. ELECTRONIC PROCESSES, PHOTOELECTRIC CHARACTERISTICS AND STABILITY OF HYBRID ORGANO-INORGANIC PEROVSKITE SOLAR CELLS

4.1. Parameters and Characteristics of Solar Cells

To characterize solar cells, a set of special parameters and characteristics is used, which allows a comparative assessment of various types of solar cells [191, 192]. The characteristics of a solar cell include spectral response, current-voltage characteristic (I_{VC}), fill factor (FF), open circuit-voltage (V_{oc}), short-circuit current (I_{SC}) (or short-circuit current density J_{SC}) and PCE efficiency. The spectral characteristic is the dependence of the quantum efficiency on the wavelength of the incident radiation. The short-circuit current can be considered the maximum current that a solar cell is capable of creating. The short-circuit current density J_{SC} depends on the power of the incident radiation and the spectrum of the incident radiation. V_{oc} is the maximum voltage created by the SC when there is no load (at zero current).

The SC current-voltage characteristic shows the dependence of the current on the voltage at the SC terminals (Figure 9). For an ideal cell in which the shunt voltage R_{SH} is infinite and the series resistance R_S is zero, the current-voltage characteristic is described by the equation:

$$I = I_L - I_0 \left[\exp\left(\frac{qV}{nkT}\right) - 1\right] \quad (1)$$

where I_o - dark current, I_L - photogenerated current, T - the temperature, V - the voltage, g - elementary charge, n - ideal cell coefficient of the solar cell. The magnitude of the ideality coefficient indicates the type of recombination in the solar cell. With the usual recombination mechanisms, n-factor is 1.

Measurements of the current – voltage characteristics and other characteristics of CS are carried out under standard conditions, usually at AM 1.5 and at a temperature of 25°C.

The fill factor FF determines the maximum power of the solar cell. FF is equal to the ratio of the experimentally determined maximum power density P_{mp} (Figure 8) to the theoretically maximum power density P_{max}:

$$FF = P_{mp} / P_{max} = (I_{mp} \cdot V_{mp}) / (I_{sc} \cdot V_{oc})$$

The efficiency of photoelectric converters is largely determined by the mobility of electrons and holes that are prone to recombination. The efficiency of charge transfer in semiconductors is characterized by the diffusion length the distance at which the diffusion flux of nonequilibrium charge carriers in the absence of an electric field decreases by a factor of e as a result of the recombination of electrons and holes. The length of the diffusion of charges is determined by the diffusion coefficient D and the lifetime τ of the charges:

$$L = \sqrt{D\tau}.$$

4.2. Electronic Processes in Hybrid Organic-Inorganic Perovskite Solar Cells

4.2.1. Principles of Photoelectric Conversion in Semiconductor Heterostructures and Analysis of Real Achievable Photoelectric Conversion Results in Perovskite Systems

Carrier generation in HOIPs occurs under illumination when photons are absorbed by the semiconductor and electrons are in the conduction band and holes are formed in the valence band. This corresponds to splitting the Fermi level into two quasi-Fermi levels (one for electrons, one for a hole), which are non-equilibrium.

The characteristics of SCs are primarily determined by the structure and physical properties of the three functionally different materials constituting the heterostructural photovoltaic device. In the SCs considered here, the main component - a HOIP photoabsorber in the form of a thin layer (or nanoparticle conglomerate) is placed between materials of different chemical composition - buffers that form two different interfaces (heterojunction) with the absorber. An interface with an electron-conducting material (ETM) transports photo-induced electrons and blocks the movement of photo-hole carriers, the second interface with a hole-conducting material (HTM) performs the opposite function. In turn, the buffer layers are associated with external current-carrying contacts (electrodes). The efficiency of charge collection is determined by the ratio of the collection current and the recombinant current.

The transfer of electrons and holes from the HOIP layer to the SC electrodes through the buffer layers should occur with minimal fundamental voltage losses, so the choice of buffer ETM and HTM is based on the relative position of the electronic levels (relative to the vacuum level) of all three of these materials of the SC components, namely, conduction band edges E_C and valence bands E_V of perovskite, and pairs of relevant levels, E'_C, E'_V for ETM and E''_C, E''_V for HTM. For organic ETM and HTM, the states at the E'_C and E''_C levels are referred to as the lowest vacant molecular orbitals (LUMO), and at the E'_V and E''_V, as the highest occupied molecular orbitals (HOMO). Selection and removal of photoinduced electron and hole carriers to opposite contacts in SC, as well as blocking their reverse movement, dictates a certain order of displacement of the initial levels. For example, in the simplest case, bias conditions can be represented as $E'_V < E_V < E''_V$ and $E'_C < E_C < E''_C$. On the one hand, the level shift should be large enough to provide an intensive injection of free charges. The competing requirement is that the level offset should be minimized in order to achieve a high V_{oc} open circuit voltage [1]. In terms of the geometric architecture of these three components in perovskite SC, one should distinguish between meso- (or nano-) structured bulk and thin-film planar architecture, which, for brevity, can be referred to as bulk and planar heterostructural SC, respectively. The choice of bulk (non-planar) architecture is dictated, first of all, by the requirement to increase the surface of at least one of the heterojunctions in order to enhance the selection process of charge carriers, which are photoinduced in perovskite. In order to emphasize the distinctive features of the SC operation with a bulk heterostructure, the term "pseudo p-i-n" transition [14] is used. The operation of an SC with a planar

HTM/perovskite/ETM heterostructure is usually considered within the framework of the p-i-n transition concept (Figure 10).

The most important physical characteristics of hybrid perovskites include, first of all, the band gap $E_g = E_C - E_V$, the diffusion lengths (the average distance that the carrier can travel from the place of its appearance to the place where it recombines) of electrons and holes L_e and L_h, as well as the maximum binding energy E_{ex} of exciton excitation. In the most commonly used perovskite lead methyl ammonium trihalide, $CH_3NH_3PbX_3$, the value of E_g varies from ~ 1.5 eV to ~ 3.1 eV when replacing the halogen in the series X = I, Br, Cl [167], or by replacing [193] methylammonium $[CH_3NH_3]^+$ to another organic cation (for example, to a cation of formamidinium, $[NH_2CH = NH_2]^+$) with a close effective ionic radius. We also note the ambipolar transport, $L_e \sim L_h$, in hybrid perovskites, where the diffusion lengths range from ~ 100 nm to more than one micron [194]. The long diffusion length allows the formation of effective SCs in a planar architecture. It has been established that the transport of photocarriers in hybrid perovskites occurs mainly through free electrons and holes, which is explained by the low exciton dissociation energy, E_{ex} <50 meV, comparable to thermal energy [195].

Efficiency - Power conversion efficiency (PCE) SCs defined as follows: PCE = J_{SC} V_{oc} FF / P_0, where J_{SC} - short-circuit current, V_{oc} - the open circuit voltage, FF - fill factor and P_0 - incident radiation power. An analysis of the realistically achievable efficiency values for the optimized composition of perovskite solar cells was proposed by Park [14] in 2013, based on empirical data known at that time. Thus, according to [196], the maximum photocurrent density of the J_{sc} = 28 mA/cm^2 can be achieved with full conversion of solar radiation in the wavelength range of 280-820 nm, where 820 nm corresponds to the minimum band gap of 1.5 eV. Taking into account the reflection of ~ 20% of the solar radiation from a standard upper transparent conductive contact deposited on a transparent substrate, the current density decreases to 22 mA / cm^2. Fundamental voltage losses of ~ 0.4 eV caused by a shift of energy levels of ~ 0.2 eV at each of the ETM/absorber and HTM/absorber heterojunctions, with a gap width of 1.5 eV, limit the maximum possible V_{oc} value to 1.1 eV.

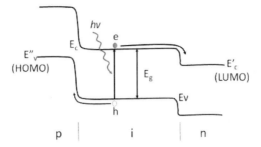

Figure 9. Energy diagram of a solar cell with p-i-n junction [14]. Here p is a hole-conducting material (HTM), i is a perovskite, n is an electron-conducting material (ETM). The directions of motion of the photo-induced carriers in the perovskite layer are also shown.

According to [14], the PCE up to 20% can be achieved by optimizing the following factors: an increase in the filling factor due to an increase in shunt resistance and a decrease in series resistance in the entire device; a decrease in fundamental voltage losses at heterojunctions due to a better choice of ETM and HTM, or the elimination of one of the heterojunctions; the use of antireflection or plasmon technology for the enlightenment of the upper conductive contact. The ability to create highly efficient samples with a simple planar thin-film architecture of PSCs is primarily associated with large and comparable values of the optical absorption length and the carrier diffusion length.

A pressing issue for PSCs is their instability. HOIPs can degrade under the influence of moisture and oxygen under ambient conditions [197]. UV-induced degradation associated with the interaction of photogenerated holes inside TiO_2 and oxygen radicals on the surface of TiO_2 may also be an obstacle to the long-term stability of PSCs with mesoporous architecture [186].

When considering the stability problem of PSCs, it is also necessary to take into account the structural phase transition in perovskite. The phase transition from cubic to a tetragonal structure in a bulk sample of $CH_3NH_3PbI_3$ occurs at a temperature of 54-55 °C. It should be expected that the crystal structure of the perovskite absorber, obtained at temperatures above this phase transition, at lower operating temperatures PSCs undergoes a structural transition with a possible decrease in the efficiency and structural stability of PSCs.

The problem of instability of PSCs is adjacent to another serious problem related to the measurement of their current-voltage characteristics [199, 200]. Unlike other types of solar cells, the JV-curves of perovskite solar cells exhibits hysteresis when the scanning direction changes [198]. Therefore, the reported effectiveness in the literature, based only on the results of express JV scans, should be considered as insufficiently unreliable, i.e., not reflecting the average long-term performance of the solar cell. Various causes of hysteresis have been proposed, such as ion motion, polarization, ferroelectric effects, trap filling (trapped states) [201].

4.3. Efficiency of Hybrid Organic-Inorganic Perovskite Solar Cells

The effectiveness of PSCs is largely determined by their architecture, which in turn dictates the choice of materials, methods of applying materials and compatibility between the various components in the device. The fundamental limit of the effectiveness of the Shockley-Queisser (SQ) for PSCs with a band gap of 1.55 eV is about 31% under standard conditions AM1.5G [129]. To calculate the effectiveness of SCs based on the detailed balance model, the Matlab program [202] was developed.

The drift diffusion model also successfully predicts the efficiency limit of PSCs [203]. This allows you to more deeply understand the physics of processes. Drift-diffusion model using two procedures gives an accurate prediction of the efficiency limit and the assessment of the degradation of the efficiency of PSCs [203]. Two points should be taken into account. First, it is advisable to investigate the intrinsic radiative recombination after adapting optical structures that significantly affect the open-circuit voltage at the Shockley-Queisser limit. Secondly, the contact characteristics of the electrodes must be carefully selected to eliminate charge accumulation and surface recombination on the electrodes.

A series of first principle studies was conducted to numerically determine the characteristics of HOIPs, which take into consideration the band gap, the effective mass and the level of HOIPs defects, etc. [204-207]. Agrawal et al. [208, 209] proposed a numerical simulation approach based on an analysis of almost perfect efficiency, considering the importance of the interfaces of perovskite and hole / electronic transport layers [210]. At the same time, Sun et al. [211] obtained a more compact model of PSCs based on experimental transport data.

4.4. Stability and Photoelectric Hysteresis

For the production of PSCs, it is necessary to solve the problems of the long-term internal and external stability of the HOIP layer, which can be sensitive to heat, moisture and oxygen. Until 2015, the size of PSCs did not exceed 1 cm^2 and degraded in a humid environment [212]. The instability of PSCs to a large extent overshadow the success in increasing efficiency [213, 214]. Under actual operating conditions at elevated PSCs temperatures, for example, at 60°C, problems such as stoichiometric polarization and phase separation and the associated changes in SCs properties can be expected. An ageing study showed that even when the SCs were under argon, trace amounts of water destroy the perovskite layer within a few days. The addition of hydrophobic molecules (PCBM and PCBM derivatives of α-bis-PCBM) to the perovskite layer slows down the ingress of water and the hydration of the photoactive layer [215]. The problem of degradation is gradually being solved and in 2017 perovskite solar modules have already been demonstrated without visible degradation for one year [216]. The challenge now is to develop fully printed PSCs with an efficiency of 22% and with 90% of the initial efficiency [217]. The laboratory research technique was developed to simulate ageing under actual operating conditions and an algorithm was created that can track the optimal power of cells based on perovskite [102].

Another problem associated with the practical use of PSCs is the manifestation of hysteresis when a voltage is deployed in different directions

[64, 198]. The power conversion efficiency of the PSCs, measured with a voltage change from V_{oc} to J_{SC}, is usually higher than with a reverse stroke from J_{sc} to V_{oc}. Volt-ampere hysteresis makes an ambiguous determination of the effectiveness of PSCs when standard conditions are not met (scanning direction, scanning speed, light attenuation, displacement, etc.) [200, 218]. This has affected the certification process conducted by accredited laboratories, in particular NREL. To compare the results of different laboratories, a standard protocol for measuring the characteristics of PSCs, including the Matlab code, was developed [219].

As a measure of the effectiveness of PSCs, Snaith et al. [218] proposed "stabilized power." This value is determined by maintaining the voltage near the maximum power point and tracking the output power until it becomes constant. It was shown that both methods give lower values of efficiency compared to the efficiency determined by fast scanning [200, 218].

It is assumed that the hysteresis is not an exclusively capacitive effect of charge-discharge, and the efficiency of charge collection in the cell at least temporarily increases after switching to forward bias [220]. According to Unger et al. [200] hysteresis can be caused by polarization, ion motion, ferroelectric effects, and trap filling. It is shown that in the current-voltage scanning it is important to take into account the time required for the perovskite to reach the electronic steady state. The large chemical capacitance of PSCs [221] is a problem for fast power tracking. A very slow voltage scan allows the system to be in stationary conditions at each measurement point. This can prevent inconsistencies between the forward and reverse scan curves. It was shown that the surface passivation of the perovskite absorber contributes to the stabilization of efficiency near the efficiency obtained with fast scanning [222, 223]. It is assumed that in the «inverted architecture» with a transparent cathode there is practically no hysteresis [224]. This indicates that interfaces can play a decisive role in hysteresis since the main difference between an inverted architecture and an ordinary architecture is the replacement of ETM from metal oxide with an organic one.

In the long term, the solution to problems of stability and hysteresis will decide the fate of $CH_3NH_3PbI_3$ in solar energy [102]. Therefore, it is important to consider these two issues together and understand the relationship between stability and efficiency.

CONCLUSION

Hybrid organo-inorganic perovskite solar cells technology fast evolution shows impressive progress and high competitiveness as compared to the other photovoltaic technologies. This new generation of SC has the advantage because of the use of widespread in natural elements, technology simplicity and overall low cost. HOIP have another unexpected and very important property: very long current carriers diffusion length L ~1 μm. It is two orders of magnitude higher than L for the other semiconductor thin films grown from the solution at low temperatures. The high light conversion efficiency of perovskite SC is also due to the comparable values of the optic absorption and diffusion lengths. A large variety of possible electron and holes conducting materials allows to find semiconductor pairs providing low dark current and so high open circuit voltages $V_{OC} > 1.1$ V. This value is much larger than V_{OC} for the other traditional organic and chalcogenide thin film solar cells and is about the V_{OC} of silicon single crystal SC.

Targeted modification of perovskite photoabsorber chemical composition along with the optimal choice of electron and hole conducting materials and quality control opens wide opportunities to tailor optical and transport properties of PSC and to increase their PCE. It is also very important that perovskite technology is compatible with technologies of the first and second generation SCs and this can support quickly solution of the PSC stability problem and their wide commercial production.

Up to now the architecture of PSC was simplified from the initial architecture with volume heterostructure specific for the die sensitize SC to classic architecture with planar p-i-n heterodiode. The advantage of this approach consists of wide use of «wet» chemistry methods which can ensure the growth of heterostructure layers from the solutions and another low-temperature synthesis methods.

But in spite of all these advantages, we must overcome some drawbacks to use HOIP widely. There is a lack of stability for commercial use because HOIPs are very sensitive to oxygen and water vapor. It requires to grow perovskite thin films in the inert atmosphere. It is also necessary to investigate the long term perovskite SC stability paying attention on the structural phase transitions. We have to scale lab PSC technologies to the industrial technologies of continues thin films of a large area including «roll-to-roll» technology.

To increase the economical efficiency of PSC we must use cheaper adjacent to perovskite absorber materials. We must pay special attention to the quality increase of the contact spacer layers by using the advanced methods (spray deposition, thermal vaporization). Apparently p-i-n structure PSC will compete seriously with traditional silicon photovoltaics. There is a great potential of PCE increase on the way of tandem SC which combines perovskite and crystalline silicon SCs or the use of advanced thin film SC. This can allow the increase of PSC PCE up to 30%. As a result perovskite technology can allow to organize mass production of cheap SC with high efficiency.

Although it is not completely clear, when HOIP will be widely used, there is a tremendous effort invested in the development of complex PSCs' recycling technologies [210]. Lead is the most unwanted element from the ecological point of view. However, the last investigations show that the environmental impact of lead is not determinative [135, 211-214].

REFERENCES

[1] Kojima, A., Teshima, K., Shirai, Y., Miyasaka, T. (2009). Organometal halide perovskites as visible-light sensitizers for photovoltaic cells. *J. Am. Chem. Soc.*, 131: 6050–6051.

[2] Grätzel, M. (2003). Dye-sensitized solar cells. *Journal of Photochemistry and Photobiology C: Photochemistry Reviews*, 4: 145–153.

[3] Sauvage, F. (2014). A Review on Current Status of Stability and Knowledge on Liquid Electrolyte-Based Dye-Sensitized Solar Cells. *Advances in Chemistry*, 1–23.

[4] Kim, H., Lee, C., Im, J. H., Lee, K. B., Moehl, T., Marchioro, A., Moon, S. J., Humphry-Baker, R., Yum, J. H., Moser, J. E., Grätzel, M., Park, N. G. (2012). Lead Iodide Perovskite Sensitized All-Solid-State Submicron Thin Film Mesoscopic Solar Cell with Efficiency Exceeding 9%. *Scientific Reports*, 2: 591.

[5] Krüger, J., Plass, R., Cevey, L., Piccirelli, M., Gatzel, M. (2001). High efficiency solid-state photovoltaic device due to inhibition of interface charge recombination. *Appl. Phys. Lett.*, 79: 2085–2087.

[6] Eperon, G. E., Burlakov, V. M., Docampo, P., Snaith, H. J. (2014). Morphological Control for High Performance, Solution-Processed Planar Heterojunction Perovskite Solar Cells. *Advanced Functional Materials*, 24 (1): 151–157.

[7] Burschka, J., Pellet., N, Moon, S. J., Humphry-Baker, R., Gao, P., Nazeeruddin, M. K., Gräetzel M. (2013). Sequential deposition as a route to high-performance perovskite-sensitized solar cells. *Nature*, 499: 316–319.

[8] Cartwright, J. (2013). A Flat-Out Major Advance for an Emerging Solar Cell Technology. *Science*, 362. URL: https://www.sciencemag.org/news/2013/09/flat-out-major-advance-emerging-solar-cell-technology.

[9] Lotsch, B. V. (2004). New Light on an Old Story: Perovskites Go Solar. *Angewadte Chemie Int. Ed.*, 53 (3): 635–637.

[10] Lee, M. M., Teuscher J., Miyasaka T., Murakami T. N. and Snaith H. J. (2012). Efficient hybrid solar cells based on meso-superstructured organometal halide perovskites. *Science*, 338 (6107): 643–647.

[11] Xing, G., Mathews N., Sun, S., Lim, S. S., Lam, Y.M., Grätzel, M., Mhaisalkar, S. and Sum, T. C. (2013). Long-range balanced electron- and hole-transport lengths in organic-inorganic $CH_3NH_3PbI_3$. *Science*, 342 (6156): 344–347.

[12] Stranks, S. D., Eperon, G. E., Grancini, G., Menelaou, C., Alcocer, M. J. P., Leijtens, T., Herz, L. M., Petrozza, A. and Snaith, H. J. (2013).

Electron-hole diffusion lengths exceeding 1 micrometer in an organometal trihalide perovskite absorber. *Science*, 342 (6156): 341–344.

[13] Im, J. H., Kim, H. S., Park, N. G. (2014). Morphology-photovoltaic property correlation in perovskite solar cells: One-step versus two-step deposition of $CH_3NH_3PbI_3$. *APL Materials*, 2: 081510.

[14] Park, N. G. (2013). Organometal perovskite light absorbers toward a 20% efficiency low-cost solid-state mesoscopic solar cell. *J. Phys. Chem. Lett.*, 4: 2423–2430.

[15] Anaraki, E. H., Kermanpur, A., Steier, L., Domanski, K., Matsui, T., Tress, W., Saliba, M., Abate, A., Grätzel, M., Hagfeldt, A. and Correa-Baena, J. P. (2016). Highly efficient and stable planar perovskite solar cells by solution-processed tin oxide. *Energy Environ. Sci.*, 9 (10): 3128–3134.

[16] Correa-Baena, J. P., Steier, L., Tress, W., Saliba, M., Neutzner, S., Matsui, T., Giordano, F., Jacobsson, T. J., Kandada, A. R. S., Zakeeruddin, S. M., Petrozza, A., Abate, A., Nazeeruddin, M. K., Grätzel, M. and Hagfeldt, A. (2015). Highly efficient planar perovskite solar cells through band alignment engineering. *Energy Environ. Sci.*, 8 (10): 2928–2934.

[17] Yang, W. S. et al. (2015). High-performance photovoltaic perovskite layers fabricated through intramolecular exchange. *Science*, 348: 1234–1237.

[18] Tan, H. et al. (2017). Efficient and stable solution-processed planar perovskite solar cells via contact passivation. *Science*, 355: 722–726.

[19] Yang, W. S., Park, B. W., Jung, E. H., Jeon, N. J., Kim, Y. C., Lee, D. U., Shin, S. S., Seo J., Kim, E. K., Noh, J. H. (2017). Iodide management in formamidinium-lead-halide-based perovskite layers for efficient solar cells. *Science*, 356: 1376–1379.

[20] NREL *Best Research-Cell Efficiencies.* (2017) URL: https://www.nrel.gov/pv/assets/pdfs/pv-efficiencies-07-17-2018.pdf.

[21] Zhao, D., Wang, Ch., Song, Zh., Yu, Y., Chen, C., Zhao, X., Zhu, K., Yan, Y. (2018). Four-Terminal All-Perovskite Tandem Solar Cells

Achieving Power Conversion Efficiencies Exceeding 23%. *ACS Energy Lett.,* 3: 305–306.

[22] He, M., Zheng, D. G., Wang, M. Y., Lin, C. J., and Lin, Z. Q. (2014). High efficiency perovskite solar cells: from complex nanostructure to planar heterojunction. *J. Mater. Chem.,* 2: 5994–6003.

[23] Snaith, H. J. (2013). Perovskites: the emergence of a new era for low-cost, high-efficiency solar cells. *J. Phys. Chem. Lett.,* 4: 3623–3630.

[24] Гладышев, П.П., Юшанхай, В.Ю., Сюракшина, Л.А. (2015). Гибридные органо-неорганические перовскитные структуры и фотоэлектрические преобразователи на их основе: физические и химические проблемы. *Органические и гибридные наноматериалы: получение и перспективы применения: монография. Изд. Ивановского гос. ун-та.* 426–556. [Gladyshev, P. P., Yushanhaj, V. Y., Syurakshina, L. A. (2015). Hybrid organic-inorganic perovskite structures and photoelectric converters based on them: physical and chemical problems. *Organic and hybrid nanomaterials: production and application prospects: monograph. Publishing House of Ivanovo State University,* 426–556.]

[25] Seigo, I. (2016). Research Update: Overview of progress about efficiency and stability on perovskitesolar cells. *APL Materials,* 4 (9): 091504.

[26] Di Giacomo, F., Fakharuddin, A., Jose, R. and Brown, T. M. (2016) Progress, challenges and perspectives in flexible perovskite solar cells. *Energy & Environmental Science,* 9 (10): 3007–3035.

[27] Aurélien, M. A. L., Azarhoosh, P., Alonso, M. I., Campoy-Quiles, M., Weber, O. J., Yao, J., Bryant, D., Weller, M. T., Nelson, J., Walsh, A., van Schilfgaarde, M., Barnes, P. R. F. (2016). Experimental and theoretical optical properties of methylammonium lead halide perovskites. *Nanoscale,* 8: 6317–6327.

[28] Wang, D., Wright, M., Elumalai, N. K., Uddin, A. (2016). Stability of perovskite solar cells. *Solar Energy Materials & Solar Cells,* 147: 255–275.

[29] Tang, H., He S., Peng, C. (2017). A Short Progress Report on High-Efficiency Perovskite Solar Cells. *Nanoscale Research Letters,* 12: 410.

[30] Asghar, M. I., Zhang, J., Wang, H., Lunda, P. D. (2017). Device stability of perovskite solar cells. *Renewable and Sustainable Energy Reviews,* 77: 131–146.

[31] Ibn- Mohammeda, T., Koha, S. C. L., Reaney, I. M., Acquaye, A., Schileo, G., Mustapha, K. B., Greenough, R. (2017). Perovskite solar cells: An integrated hybrid lifecycle assessment and review in comparison with other photovoltaic technologies. *Renewable and Sustainable Energy Reviews,* 80: 1321–1344.

[32] Zhang, H. & Toudert, J. (2018). Optical management for efficiency enhancement in hybrid organic-inorganic lead halide perovskite solar cells. *Science and Technology of Advanced Materials,* 19 (1): 411−424.

[33] Shi, Z. and Jayatissa, A. H. (2018). Perovskites-Based Solar Cells: A Review of Recent Progress, Materials and Processing. *Methods Materials,* 11 (729): 1−34.

[34] Yun, S., Qin, Y., Uhl, A. R., Vlachopoulos, N., Yin, M., Li, D., Han, X., Hagfeldt, A. (2018). New-generation integrated devices based on dye-sensitized and perovskite solar cells. *Energy Environ. Sci.,* 11: 476−526.

[35] Qiu, L, Ono, L. K., Qi, Y. (2018). Advances and challenges to the commercialization of organic-inorganic halide perovskite solar cell technology. *Materials Today Energy,* 7: 169−189.

[36] Assadi, M. K., Bakhodaa, S., Saidur, R., Hanaei, H. (2018). Recent progress in perovskite solar cells. *Renewable and Sustainable Energy Reviews,* 81: 2812–2822.

[37] Yin, W., Yang, J., Kang, J., Yan, Y. and Wei, S. (2014). Halide Perovskite Materials for Solar Cells: A Theoretical Review. *J. Mater. Chem. A.,* 1−18.

[38] Park, N. G., Grätzel, M., Tsutomu, M., (Eds.) (2016). *Organic-Inorganic Halide Perovskite Photovoltaics From Fundamentals to Device Architectures* Editors. Springer, 8: 366.

[39] 1 Giorgi, G., Yamashita, K. (2017). *Theoretical Modeling of Organohalide Perovskites for Photovoltaic Applications* Editors. Taylor and Francis Group, 244.

[40] Wei-guang, D. E., Chao-yu, C. P. (2017). *Perovskite Solar Cells: Principle, Materials and Devices* Editors. World Scientific Publishing Co, 244.

[41] Yusoff, R. B. N. (2019). Perovskite Solar Cells - Fabrication and Characterization. *Wiley-Vch.*

[42] Bisquert, J. (2017). The Physics of Solar Cells: Perovskites, Organics, and Photovoltaic Fundamentals. *CRC Press*, 226.

[43] Ming, W., Yang, D., Li T., Zhang, L., Du, M.-H. (2017). Formation and Diffusion of Metal Impurities in Perovskite Solar Cell Material $CH_3NH_3PbI_3$: Implications on Solar Cell Degradation and Choice of Electrode. *Advanced Science*, 5 (2): 1700662 1−10.

[44] Fu, K., Ho-Baillie, A., Trang, P. T. T., Mulmudi, H. K. (2018). Perovskite Solar Cells: Technology and Practices. *Taylor & Francis Group*, 299.

[45] Ma, J., Zheng, X., Lei, H., Ke, W., Chen, C., Chen, Z., Fang, G. (2017). Highly Efficient and Stable Planar Perovskite Solar Cells With Large-Scale Manufacture of E-Beam Evaporated SnO_2 Toward Commercialization. *Solar RRL*, 1 (10): 1700118.

[46] Lioz, E. (2016) Hole Conductor Free Perovskite-based Solar Cells. *Springer*, 8: 59.

[47] Sankir, N. D., Sankir M. (2017). Printable Solar Cells. *John Wiley & Sons*, 576.

[48] Nanduri, S. N. R. (2017). Numerical Simulation and Performance Optimization of Perovskite Solar Cell. *A Thesis In Electrical Engineering*, 77.

[49] Zhao, W., Yao, Z., Yu, F., Yang, D., & Liu, S. F. (2017). Alkali Metal Doping for Improved $CH_3NH_3PbI_3$ Perovskite Solar Cells. *Advanced Science*, 5 (2): 1700131.

[50] Gratia, P., Nazeeruddin, M. K., Graetzel, M. (2018). Compositional Characterization of Organo-Lead Tri-Halide Perovskite Solar Cells. *Ecole Polytechnique Fédérale de Lausanne*, 168.

[51] Li, J., Li, W., Dong, H., Li, N., Guo, X., & Wang, L. (2015). Enhanced performance in hybrid perovskite solar cell by modification with spinel lithium titanate. *Journal of Materials Chemistry A,* 3 (16): 8882–8889.

[52] Wali, Q., Elumalai, N. K., Iqbal, Y., Uddin, A., & Jose, R. (2018). Tandem perovskite solar cells. *Renewable and Sustainable Energy Reviews,* 84: 89–110.

[53] Dunbar, R. B., Duck, B. C., Moriarty, T., Anderson, K. F., Duffy, N. W., Fell, C. J., Wilson, G. J. (2017). How reliable are efficiency measurements of perovskite solar cells? The first inter-comparison, between two accredited and eight non-accredited laboratories. *Journal of Materials Chemistry A,* 5 (43): 22542–22558.

[54] Erin, M. (2017). Sanehira Efficient, Stable Perovskite Solar Cells Enabled by Electrode Interface Engineering and Nanoscale Phase. *Stabilization University of Washington Libraries,* 135 p.

[55] Wang, Q., Lin, F., Chueh, C. C., Zhao, T., Eslamian, M., & Jen, A. K. Y. (2018). Enhancing efficiency of perovskite solar cells by reducing defects through imidazolium cation incorporation. *Materials Today Energy,* 7: 161–168.

[56] Li, W., Yang, J., Jiang, Q., Li, R., & Zhao, L. (2018). Electrochemical deposition of PbI$_2$ for perovskite solar cells. *Solar Energy,* 159: 300–305.

[57] Schmidt-Mende L., Weickert J. (2016). Organic and Hybrid Solar Cells: An Introduction. *Walter de Gruyter GmbH & Co KG,* 304 p.

[58] Yu, Z., & Sun, L. (2017). Inorganic Hole-Transporting Materials for Perovskite Solar Cells. *Small Methods,* 2 (2): 1700280.

[59] Kapoor, V., Bashir, A., Haur, L. J., Bruno, A., Shukla, S., Priyadarshi, A., Mhaisalkar, S. (2017). Effect of Excess PbI$_2$ in Fully Printable Carbon-based Perovskite Solar Cells. *Energy Technology,* 5 (10): 1880–1886.

[60] Zhang, X., Liu, J., Kou, D., Zhou, W., Zhou, Z., Tian, Q., Ouyang, C. (2017). Performances Enhancement in Perovskite Solar Cells by Incorporating Plasmonic Au NRs@SiO$_2$ at Absorber/HTL Interface. *Solar RRL,* 1 (12): 1700151.

[61] Gujar, T. P., & Thelakkat, M. (2016). Highly Reproducible and Efficient Perovskite Solar Cells with Extraordinary Stability from Robust CH$_3$NH$_3$PbI$_3$: Towards Large-Area Devices. *Energy Technology,* 4 (3): 449–457.

[62] Zhang, Y., Wu, Z., Li, P., Ono, L. K., Qi, Y., Zhou, J., Zheng, Z. (2017). Fully Solution-Processed TCO-Free Semitransparent Perovskite Solar Cells for Tandem and Flexible Applications. *Advanced Energy Materials,* 8 (1): 1701569.

[63] Duan, M., Hu, Y., Mei, A., Rong, Y., & Han, H. (2018). Printable carbon-based hole-conductor-free mesoscopic perovskite solar cells: From lab to market. *Materials Today Energy,* 7: 221–231.

[64] Zhao, P., Kim, B. J., & Jung, H. S. (2018). Passivation in perovskite solar cells: A review. *Materials Today Energy,* 7: 267–286.

[65] Rahul, Singh, P. K., Singh, R., Singh, V., Bhattacharya, B., & Khan, Z. H. (2018). New class of lead free perovskite material for low-cost solar cell application. *Materials Research Bulletin,* 97: 572–577.

[66] Chen, W., Liu, F. Z., Feng, X. Y., Djurišić, A. B., Chan, W. K., & He, Z. B. (2017). Cesium Doped NiO$_x$ as an Efficient Hole Extraction Layer for Inverted Planar Perovskite Solar Cells. *Advanced Energy Materials,* 7 (19): 1700722.

[67] Raga, S. R., Jiang, Y., Ono, L. K., & Qi, Y. (2017). Application of Methylamine Gas in Fabricating Organic-Inorganic Hybrid Perovskite Solar Cells. *Energy Technology,* 5 (10): 1750–1761.

[68] Zhang, J, Chen, R., Wu, Y., Shang, M., Zeng, Z., Zhang, Y., Han, L. (2017). Extrinsic Movable Ions in MAPbI$_3$ Modulate Energy Band Alignment in Perovskite Solar Cells. *Advanced Energy Materials,* 8 (5): 1701981.

[69] Richard, J. D. Tilley, D. Sc. (2016). Perovskites: Structure-Property Relationships. *John Wiley & Sons,* 328.

[70] Jiang, Y., Leyden, M. R., Qiu, L., Wang, S., Ono, L. K., Wu, Z., Qi, Y. (2017). Combination of Hybrid CVD and Cation Exchange for Upscaling Cs-Substituted Mixed Cation Perovskite Solar Cells with High Efficiency and Stability. *Advanced Functional Materials,* 28 (1): 1703835.

[71] Ummadisingu, A. (2018). Fundamentals of Perovskite Formation for Photovoltaics. *Ecole Polytechnique Fédérale de Lausanne* [Federal Institute of Technology in Lausanne], 97.

[72] Wang, Y., Steigert, A., Yin, G., Parvan, V., Klenk, R., Schlatmann, R., Lauermann, I. (2017). Cu$_2$O as a Potential Intermediate Transparent Conducting Oxide Layer for Monolithic Perovskite-CIGSe Tandem Solar Cells. *Phys. Status Solidi C*, 14: 1700164.

[73] Im, J. H., Lee, C. R., Lee, J. W., Park, S. W., and Park, N. G. (2011). 6.5% efficient perovskite quantum-dot-sensitized solar cell. *Nanoscale*, 3: 4088-4093.

[74] Fan, J., Jia, B., & Gu, M. (2014). Perovskite-based low-cost and high-efficiency hybrid halide solar cells. *Photonics Research*, 2 (5): 111.

[75] Liu, M., Johnston, M. B., Snaith, H. J. (2013). Efficient planar heterojunction perovskite solar cells by vapour deposition. *Nature*, 501: 395–398.

[76] Qin, P., Tanaka, S., Ito, S., Tetreault, N., Manabe, K., Nishino, H., Nazeeruddin, M. K., Grätzel, M. (2014). Inorganic hole conductor-based lead halide perovskite solar cells with 12.4% conversion efficiency. *Nat. Commun.*, 5: 3834.

[77] González-Vázquez, J. P., Morales-Florez, V., Anta, J. A. (2012). How important is working with an ordered electrode to improve the charge collection efficiency in nanostructured solar cells? *J. Phys. Chem. Lett.*, 3: 386–393.

[78] Ball, J. M., Lee, M. M., Hey, A., Snaith, H. J. (2013). Low-temperature processed meso-superstructured to thin-film perovskite solar cells. *Energy & Environmental Science*, 6: 1739-1743.

[79] O'Regan, B., Schwartz, D. T. (1996). Efficient dye-sensitized charge separation in a wide-band-gap p-n heterojunction. *J. Appl. Phys.*, 80: 4749-4754.

[80] Chen, Q., Zhou, H., Hong, Z., Luo, S., Duan, H.-S., Wang, H.-H., Liu, Y., Li, G., and Yang, Y. (2014). Planar Heterojunction Perovskite Solar Cells via Vapor-Assisted Solution Process. *J. Am. Chem. Soc.*, 136: 622–625.

[81] Liu, D. Y. and Kelly, L. (2014). Perovskite solar cells with a planar heterojunction structure prepared using room-temperature solution processing techniques. *Nat. Photonics,* 8:133–138.

[82] Docampo, P., Ball, J. M., Darwich, M., Eperon, G. E., & Snaith, H. J. (2013). Efficient organometal trihalide perovskite planar-heterojunction solar cells on flexible polymer substrates. *Nature Communications,* 4 (1): 4:2761.

[83] Xiao, Z., Bi, C., Shao, Y., Dong, Q., Wang, Q., Yuan, Y., Huang, J. (2014). Efficient, high yield perovskite photovoltaic devices grown by interdiffusion of solution-processed precursor stacking layers. *Energy Environ. Sci.,* 7 (8): 2619–2623.

[84] Zhou, H., Chen, Q., Li, G., Luo, S., Song, T. B., Duan, H. S., Hong, Z., You, J., Liu, Y., Yang, Y. (2014). Interface engineering of highly efficient perovskite solar cells. *Science,* 345 (6196): 542–546.

[85] Perovskite solar cell. URL: http://en.wikipedia.org/wiki/Perovskite_solar_cell.

[86] Salim, T., Sun, S. (2015). Perovskite-based solar cells: impact of morphology and device architecture on device performance. *J. Mater. Chem. A.,* 3: 8943−8969.

[87] Li, Y., Ye, S., Sun, W., Yan, W., Li, Y., Bian, Z., Huang, C. (2015). Hole-conductor-free planar perovskite solar cells with 16.0% efficiency. *Journal of Materials Chemistry A,* 3 (36): 18389–18394.

[88] Fakharuddin, A., De Rossi, F., Watson, T. M., Schmidt-Mende, L., & Jose, R. (2016). Research Update: Behind the high efficiency of hybrid perovskite solar cells. *APL Materials,* 4 (9): 091505.

[89] Qin, P., Tanaka, S., Ito, S., Tetreault, N., Manabe, K., Nishino, H., Grätzel, M. (2014). Inorganic hole conductor-based lead halide perovskite solar cells with 12.4% conversion efficiency. *Nature Communications,* 5 (1): 5:3834.

[90] Chavhan, S., Miguel, O., Grande, H. J., Gonzalez-Pedro, V., Sánchez, R. S., Barea, E. M., Tena-Zaera, R. (2014). Organo-metal halide perovskite-based solar cells with CuSCN as the inorganic hole selective contact. *J. Mater. Chem. A,* 2 (32): 12754–12760.

[91] Jung, J. W., Chueh, C. C., & Jen, A. K. Y. (2015). High-Performance Semitransparent Perovskite Solar Cells with 10% Power Conversion Efficiency and 25% Average Visible Transmittance Based on Transparent CuSCN as the Hole-Transporting Material. *Advanced Energy Materials*, 5 (17): 1500486.

[92] Ye, S., Sun, W., Li, Y., Yan, W., Peng, H., Bian, Z., Huang, C. (2015). CuSCN-Based Inverted Planar Perovskite Solar Cell with an Average PCE of 15.6%. *Nano Letters*, 15 (6): 3723–3728.

[93] Heo, J. H., Im, S. H., Noh, J. H., Mandal, T. N., Lim, C. S., Chang, J. A., nSeok, S. I. (2013). Efficient inorganic–organic hybrid heterojunction solar cells containing perovskite compound and polymeric hole conductors. *Nature Photonics*, 7 (6): 486–491.

[94] Ryu, S., Noh, J. H., Jeon, N. J., Kim, Y. C., Yang W. S., Seo J., Seok S. I. (2014). Voltage output of efficient perovskite solar cells with high open-circuit voltage and fill factor. *Energy & Environmental Science*, 7: 2614.

[95] Jeon, N. J., Noh, J. H., Kim, Y. C., Yang, W. S., Ryu, S., & Seok, S. I. (2014). Solvent engineering for high-performance inorganic–organic hybrid perovskite solar cells. *Nature Materials*, 13 (9): 897–903.

[96] Sepalage, G. A., Meyer, S., Pascoe, A., Scully, A. D., Huang, F., Bach, U., Cheng, Y.-B., Spiccia, L. (2015). Copper(I) iodide as hole-conductor in planar perovskite solar cells: Probing the origin of J-V hysteresis. *Adv. Funct. Mater.*, 25: 5650–5661.

[97] Smith, D. L., Saunders, V. I. (1981). The structure and polytypism of the beta-modification of copper(I) thiocyanate. *Acta. Crystallogr. B*, 37: 1807-1812.

[98] Nilushi, W., Thomas, D. A. (2015). Copper(I) thiocyanate (CuSCN) as a hole-transport material for large-area opto/electronics. *Semicond. Sci. Technol.*, 30: 104002.

[99] Hu, H.; Dong, B.; Hu, H.; Chen, F.; Kong, M.; Zhang, Q.; Luo, T.; Zhao, L.; Guo, Z.; Li, J.; et al. (2016). Atomic layer deposition of TiO_2 for a high-efficiency hole-blocking layer in hole-conductor-free PSCs

processed in ambient air. *ACS Appl. Mater. Interfaces,* 8: 17999–18007.

[100] Mei, A., Li, X., Liu, L., Ku, Z., Liu, T., Rong, Y., Han, H. (2014). A hole-conductor-free, fully printable mesoscopic perovskite solar cell with high stability. *Science,* 345 (6194): 295–298.

[101] Xiong, Y., Liu, Y., Lan, K., Mei, A., Sheng, Y., Zhao, D., & Han, H. (2018). Fully printable hole-conductor-free mesoscopic perovskite solar cells based on mesoporous anatase single crystals. *New Journal of Chemistry,* 42 (4): 2669–2674.

[102] Norman, P. (2017). Investigations on hybrid organic-inorganic perovskites for high performance solar cells. *Lausanne, EPFL.*

[103] Sha, W. E. I., Ren, X., Chen, L. & Choy, W. C. H. (2015). The efficiency limit of $CH_3NH_3PbI_3$ perovskite solar cells. *Applied Physics Letters,* 106 (22): 221104.

[104] Rühle, S. (2016). Tabulated values of the Shockley–Queisser limit for single junction solar cells. *Solar Energy,* 130: 139–147.

[105] Rühle, S. (2017). The detailed balance limit of perovskite/silicon and perovskite/CdTe tandem solar cells. *Physica Status Solidi (a),* 214 (5): 1600955.

[106] Werner, J., Niesen, B., & Ballif, C. (2017). Perovskite/Silicon Tandem Solar Cells: Marriage of Convenience or True Love Story? - An Overview. *Advanced Materials Interfaces,* 5 (1): 1700731.

[107] Chen, B., Zheng, X., Bai, Y., Padture, N. P., & Huang, J. (2017). Progress in Tandem Solar Cells Based on Hybrid Organic-Inorganic Perovskites. *Advanced Energy Materials,* 7 (14): 1602400.

[108] Lal, Niraj N.; Dkhissi, Lal, N. N., Dkhissi, Y., Li, W., Hou, Q., Cheng, Y. B., & Bach, U. (2017). Perovskite Tandem Solar Cells. *Advanced Energy Materials,* 7 (18): 1602761.

[109] Werner, J., Dubuis, G., Walter, A., Löper, P., Moon, S. J., Nicolay, S., Ballif, C. (2015). Sputtered rear electrode with broadband transparency for perovskite solar cells. *Solar Energy Materials and Solar Cells,* 141: 407–413.

[110] Duong, T., Lal, N., Grant, D., Jacobs, D., Zheng, P., Rahman, S., Catchpole, K. R. (2016). Semitransparent Perovskite Solar Cell with

Sputtered Front and Rear Electrodes for a Four-Terminal Tandem. *IEEE Journal of Photovoltaics,* 6 (3): 679–687.
[111] Werner, J., Barraud, L., Walter, A., Bräuninger, M., Sahli, F., Sacchetto, D., Ballif, C. (2016). Efficient Near-Infrared-Transparent Perovskite Solar Cells Enabling Direct Comparison of 4-Terminal and Monolithic Perovskite/Silicon Tandem Cells. *ACS Energy Letters,* 1 (2): 474–480.
[112] Duong, T., Wu, Y., Shen, H., Peng, J., Fu, X., Jacobs, D., Catchpole, K. (2017). Rubidium Multication Perovskite with Optimized Bandgap for Perovskite-Silicon Tandem with over 26% Efficiency. *Advanced Energy Materials,* 7 (14): 1700228.
[113] Ramírez, Q., César, O., Shen, Y., Salvador, M., Forberich, K., Schrenker, N., Spyropoulos, G. D., Heumüller, T., Wilkinson, B., Kirchartz, T., Spiecker, E., Verlinden, P. J., Zhang, X., Green, M. A., Ho-Baillie, A., Brabec, C. J., (2018). Balancing electrical and optical losses for efficient 4-terminal Si–perovskite solar cells with solution processed percolation electrodes. *Journal of Materials Chemistry A,* 6 (8): 3583–3592.
[114] Shen, H., Duong, T., Peng, J., Jacobs, D., Wu, N., Gong, J., Wu, Y., Karuturi, S. K., Fu, X., Weber, K., Xiao, X., White, T. P., Catchpole, K. (2018). Mechanically-stacked perovskite/CIGS tandem solar cells with efficiency of 23.9% and reduced oxygen sensitivity. *Energy & Environmental Science,* 11 (2): 394–406.
[115] Han, Q., Hsieh, Y. T., Meng, L., Wu, J. L., Sun, P., Yao, E. P., Chang, S. Y., Ba, S. H., Kato, T., Bermudez, V., Yang, Y. (2018). High-performance perovskite/Cu(In,Ga)Se$_2$ monolithic tandem solar cells. *Science,* 361: 904–908.
[116] Mailoa, J. P., Bailie, C. D., Johlin, E. C., Hoke, E. T., Akey, A. J., Nguyen, W. H., McGehee, M. D., Buonassisi, T. (2015). A 2-terminal perovskite/silicon multijunction solar cell enabled by a silicon tunnel junction. *Applied Physics Letters,* 106 (12): 121105.
[117] Albrecht, S., Saliba, M., Correa-Baena, J. P., Lang, F., Kegelmann, L., Mews, M., Steier, L., Abate, A., Rappich, J., Korte, L., Schlatmann, R., Nazeeruddin, M. K., Hagfeldt, A., Grätzel, M., Rech,

B. (2016). Monolithic perovskite/silicon-heterojunction tandem solar cells processed at low temperature. *Energy & Environmental Science,* 9 (1): 81–88.

[118] Werner, J., Weng, C. H., Walter, A. Fesquet, L., Seif, J. P., De Wolf, S., Niesen, B., Ballif, C. (2015). Efficient Monolithic Perovskite/Silicon Tandem Solar Cell with Cell Area >1 cm. *The Journal of Physical Chemistry Letters,* 7 (1): 161–166.

[119] Bush, K. A., Palmstrom, A. F., Yu, Z. J., Boccard, M., Cheacharoen, R., Mailoa, J. P., McMeekin, D. P., Hoye, R. L. Z., Bailie, C. D. (2017). 23.6%-efficient monolithic perovskite/silicon tandem solar cells with improved stability. *Nature Energy,* 2 (4): 17009.

[120] Sahli, F., Werner, J., Kamino, B. A., Bräuninger, M., Monnard, R., Paviet-Salomon, B., Barraud, L., Ding, L., Diaz L., Juan J., Sacchetto, D., Cattaneo, G., Despeisse, M., Boccard, M., Nicolay, S., Jeangros, Q., Niesen, B., Ballif, C. (2018). Fully textured monolithic perovskite/silicon tandem solar cells with 25.2% power conversion efficiency. *Nature Materials,* 17: 820–826.

[121] Osborne M. (2018). Oxford PV takes record perovskite tandem solar cell to 27.3% conversion efficiency. *PV-Tech,* URL: https://www.pv-tech.org/news/oxford-pv-takes-record-perovskite-tandem-solar-cell-to-27.3-conversion-effi.

[122] Eperon, G. E., Leijtens, T., Bush, K. A., Prasanna, R., Green, T., Wang, J. T.-W., McMeekin, D. P., Volonakis, G., Milot, R. L., (2016). Perovskite-perovskite tandem photovoltaics with optimized band gaps. *Science,* 354 (6314): 861–865.

[123] Zhao, D., Yu, Y., Wang, C., Liao, W., Shrestha, N., Grice, C. R., Cimaroli, A. J., Guan, L., Ellingson, R. J., (2017). Low-bandgap mixed tin–lead iodide perovskite absorbers with long carrier lifetimes for all-perovskite tandem solar cells. *Nature Energy,* 2 (4): 17018.

[124] Мельников, М. Я. (2004) Экспериментальные методы химической кинетики. Фотохимия. *Изд-во Моск. ун-та.* 125. [Melnikov, M. J. (2004). Experimental methods of chemical kinetics. Photochemistry. *Publishing House of Moscow University,* 125.

[125] Service, R. F. (2013). Turning up the light. *Science,* 342: 794–797.

[126] Abrusci, A., Stranks, S. D., Docampo, P., Yip, H. L, Jen, A., Snaith, H.J. (2013). High-performance perovskite-polymer hybrid solar cells via electronic coupling with fullerene monolayers. *Nano Lett.,* 13: 3124–3128.

[127] Chen, C., Li, C., Li, F., Wu, F., Tan, F., Zhai, Y., Zhang, W. (2014). Efficient perovskite solar cells based on low-temperature solution-processed (CH_3NH_3)PbI_3 perovskite/$CuInS_2$ planar heterojunctions. *Nanoscale Research Letters,* 9: 457.

[128] Eperon, G. E., Burlakov, V. M., Docampo, P., Goriely, A., Snaith, H. J. (2014). Morphological control for high performance, solution-processed planar heterojunction perovskite solar cells. *Advanced Functional Materials,* 24 (1): 151–157.

[129] Dualeh, A., Tétreault, N., Moehl, T., Gao, P., Nazeeruddin, M. K., Grätzel, M. (2014). Effect of annealing temperature on film morphology of organic-inorganic hybrid pervoskite solid-state solar cells. *Advanced Functional Materials,* 24 (21): 3250–3258.

[130] Liu, M., Johnston, M. B., Snaith, H. J. (2013). Efficient planar heterojunction perovskite solar cells by vapour deposition. *Nature,* 501 (7467): 395–398.

[131] Li, Z., Klein, T. R., Kim, D. H., Yang, M., Berry, J. J., van Hest, M. F., Zhu, K. (2018). Scalable fabrication of perovskite solar cells. *Nature Reviews Material,* 3 (4): 18017.

[132] Ono, L. K., Leyden, M. R., Wang, S., Qi, Y. (2016). Organometal halide perovskite thin films and solar cells by vapor deposition. *Journal of Materials Chemistry A,* 4 (18): 6693–6713.

[133] Shen, P.-S., Chiang, Y. H., Li, M. H., Guo, T. F., Chen, P. (2016). Research Update: Hybrid organic-inorganic perovskite (HOIP) thin films and solar cells by vapor phase reaction. *APL Materials,* 4 (9): 091509.

[134] Lin, Q., Armin, A., Nagiri, R. C. R., Burn, P. L., Meredith, P., (2015). Electro-optics of perovskite solar cells. *Nature Photonics,* 9 (2): 106.

[135] Chen, C. W., Kang, H. W., Hsiao, S. Y., Yang, P. F., Chiang, K. M., Lin, H. W. (2014). Efficient and uniform planar-type perovskite solar

cells by simple sequential vacuum deposition. *Advanced Materials,* 26 (38): 6647–6652.

[136] Abbas, H. A., Kottokkaran, R., Ganapathy, B., Samiee, M., Zhang, L., Kitahara, A., Noack, M., Dalal, V. L. (2015) High efficiency sequentially vapor grown nip $CH_3NH_3PbI_3$ perovskite solar cells with undoped P3HT as p-type heterojunction layer. *APL Materials,* 3 (1): 016105.

[137] Yang, D., Yang, Z., Qin, W., Zhang, Y., Liu, S. F., Li, C. (2015). Alternating precursor layer deposition for highly stable perovskite films towards efficient solar cells using vacuum deposition. *Journal of Materials Chemistry A,* 3 (18): 9401–9405.

[138] Momblona, C., Gil-Escrig, L., Bandiello, E., Hutter, E. M., Sessolo, M., Lederer, K., Blochwitz-Nimoth, J., Bolink, H. J. (2016). Efficient vacuum deposited pin and nip perovskite solar cells employing doped charge transport layers. *Energy & Environmental Science,* 9 (11): 3456–3463.

[139] Hsiao, S. Y., Lin, H. L., Lee, W. H., Tsai, W. L., Chiang, K. M., Liao, W. Y., Ren-Wu, C. Z., Chen, C. Y., Lin, H. W. (2016). Efficient All-Vacuum Deposited Perovskite Solar Cells by Controlling Reagent Partial Pressure in High Vacuum. *Advanced Materials,* 28 (32): 7013-7019.

[140] Tennakone, K., Kumara, G. R. R. A., Kumarasinghe, A. R., Wijayantha, K. G. U., & Sirimanne, P. M. (1995). A dye-sensitized nano-porous solid-state photovoltaic cell. *Semiconductor Science and Technology,* 10 (12): 1689–1693.

[141] You, J. B., Hong, Z. R., Yang, Y., Chen, Q. and Cai, M. (2014). Low temperature solution-processed perovskite solar cells with high efficiency and flexibility. *ACS Nano,* 8: 1674–1680.

[142] Wu, Y., Yang, X., Chen, H., Zhang, K., Qin, C., Liu, J., Peng, W., Islam, A., Bi, E., Ye, F., Yin, M., Zhang, P. and Han, L. (2014). Highly compact TiO_2 layer for efficient hole-blocking in perovskite solar cells. *Applied Physics Express,* 7: 052301.

[143] Kavan, L., Tétreault, N., Moehl, T. and Grätzel, M. (2014) Electrochemical Characterization of TiO_2 Blocking Layers for Dye-

Sensitized Solar Cells. *The Journal of Physical Chemistry C,* 118: 16408–16418.

[144] Ke, W., Fang, G., Wan, J., Tao, H., Liu, Q., Xiong, L., Qin, P., Wang, J., Lei, H., Yang, G., Qin, M., Zhao, X. and Yan, Y. (2015). Efficient hole-blocking layer-free planar halide perovskite thin-film solar cells. *Nature Communications,* 6: 6700.

[145] Chandiran, A. K., Yella, A., Mayer, M. T., Gao, P., Nazeeruddin, M. K. and Grätzel, M. (2014). SubNanometer Conformal TiO$_2$ Blocking Layer for High Efficiency Solid-State Perovskite Absorber Solar Cells. *Advanced Materials,* 26: 4309–4312.

[146] Bi, D., Tress, W. M., Dar, I., Gao, P., Luo, J., Renevier, C., Schenk, K., Abate, A., Giordano, F., Correa-Baena, J. P., Decoppet, J. D., Zakeeruddin, S. M., Nazeeruddin, M. K., Graetzel, M. and Hagfeldt, A. (2016). Efficient luminescent solar cells based on tailoredmixed-cation perovskites. *Science Advances,* 2: e1501170–e1501170.

[147] Conings, B., Baeten, L., Jacobs, T., Dera, R., D'Haen, J., Manca, J., Boyen, H.G. (2014). An easy-to-fabricate low-temperature TiO$_2$ electron collection layer for high efficiency planar heterojunction perovskite solar cells. *APL Mater.,* 2: 081505.

[148] Löper, P., Moon, S. J., De Nicolas, S. M., Niesen, B., Ledinsky, M., Nicolay, S., Bailat, J., Yum, J. H., De Wolf, S. and Ballif, C. (2015). Organic–inorganic halide perovskite/crystalline silicon four-terminal tandemsolar cells. *Phys. Chem. Chem. Phys.,* 17 (3): 1619–1629.

[149] Wang, L., Fu, W., Gu, Z., Fan, C., Yang, X., Li H. and Chen, H. (2014). Low Temperature Solution Processed Planar Heterojunction Perovskite Solar Cells with CdSe Nanocrystal as Electron Transport/Extraction Layer. *J. Mater. Chem. C,* 2: 9087.

[150] Gao, X., Li, J., Baker, J., Hou, Y., Guan, D., Chen, J., Yuan, C. (2014). Enhanced photovoltaic performance of perovskite CH$_3$NH$_3$PbI$_3$ solar cells with freestanding TiO$_2$ nanotube array films. *Chemical Communications,* 50: 6368–6371.

[151] Zhu, K., Neale, N. R., Miedaner, A., Frank, A. J. (2007). Enhanced charge-collection efficiencies and light scattering in dye sensitized solar cells using oriented TiO$_2$ nanotubes arrays. *Nano Lett.,* 7: 69–74.

[152] Malinkiewicz, O., Roldán-Carmona, C., Soriano, A., Bandiello, E., Camacho, L., Nazeeruddin, M. K., Bolink, H. J. (2014). Metal-oxide-free methylammonium lead iodide perovskite-based solar cells: the influence of organic charge transport layers. *Adv Energy Mater.*, 4: 1400345.

[153] Vivo, P., Salunke, J. K. and Priimagi, A. (2017). Hole-Transporting Materials for Printable Perovskite Solar Cells. *Materials*, 10 (9): 1087.

[154] Bi, D. Q., Yang, L., Boschloo, G., Hagfeldt, A., and Johansson E. M. J. (2013). Effect of different hole transport materials on recombination in $CH_3NH_3PbI_3$ perovskite-sensitized mesoscopic solar cells. *J. Phys. Chem. Lett.*, 4: 1532–1536.

[155] Jeon, N. J., Lee, J., Noh, J. H., Nazeeruddin, M. K. and Gratzel, M. (2013). Efficient inorganic–organic hybrid perovskite solar cells based on pyrenearylamine derivatives as hole-transporting Materials. *J. Am. Chem. Soc.*, 135: 19087–19090.

[156] Etgar, L., Gao, P., Xue, Z. S., Peng, Q., Chandiran, A. K., Liu, B., Nazeeruddin M. K., Grätzel M. (2012). Mesoscopic $CH_3NH_3PbI_3$/TiO_2 heterojunction solar cells. *J. Am. Chem. Soc.*, 134: 17396–17399.

[157] Abate A., D. J. Hollman D. J., Teuscher J., Pathak S., and Avolio R. (2013). Protic ionic liquids as p-dopant for organic hole transporting materials and their application in high efficiency hybrid solar cells. *J. Am. Chem. Soc.*, 135: 13538–13548.

[158] Christians, J., Fung, R., Kamat, P. (2014). An Inorganic Hole Conductor for Organo-Lead Halide Perovskite Solar Cells. Improved Hole Conductivity with Copper Iodide. *J. Am. Chem. Soc.*, 136: 758–764.

[159] Arora, N., Dar, M. I., Hinderhofer, A., Pellet, N., Schreiber, F., Zakeeruddin, S. M., Grätzel, M. (2017). Perovskite solar cells with CuSCN hole extraction layers yield stabilized efficiencies greater than 20. *American Association for the Advancement of Science*, 358: 768.

[160] Chaudhary, N., Chaudhary, R., Kesari, J. P., Patra, A., & Chand, S. (2015). Copper thiocyanate (CuSCN): an efficient solution-

processable hole transporting layer in organic solar cells. *Journal of Materials Chemistry C,* 3 (45): 11886–11892.

[161] Amalina, M. N., Azman, M. A., & Rusop M. M. (2011). Effect of the Precursor Solution Concentration of CuI Thin Film Deposited by Spin Coating Method. *Advanced Materials Research,* 364: 417–421.

[162] Agarwal, S., Nair, P. R. (2014). Performance optimization for Perovskite based solar cells. *Photovoltaic Specialist Conference (PVSC),* 40: 1515–1518.

[163] Chen, W. Y., Deng, L. L., Dai, S. M., Wang, X., Tian, C. B., Zhan, X. X., Xie, S. Y., Huang, R. B., Zheng, L. S. (2015). Low-cost solution-processed copper iodide as an alternative to PEDOT:PSS hole transport layer for efficient and stable inverted planar heterojunction perovskite solar cells. *J. Mater. Chem. A,* 3: 19353–19359.

[164] Ezealigo, B. N., Nwanya, A. C., Simo, A. (2017). A study on solution deposited CuSCN thin films: Structural, electrochemical, optical properties. *Arabian Journal of Chemistry,* 1–11.

[165] Ito, S., Tanaka, S., Vahlman, H., Nishino, H., Manabe, K., Lund, P. (2014). Carbon-double-bond-free printed solar cells from TiO_2/$CH_3NH_3PbI_3$/CuSCN/Au: Structural control and photoaging effects. *Chem Phys Chem,* 15: 1194–1200.

[166] Ito, S., Tanaka, S., Nishino, H. (2015). Lead-halide perovskite solar cells by $CH_3NH_3PbI_3$ dripping on PbI_2- CH_3NH_3I-DMSO precursor layer for planar and porous structures using CuSCN hole-transporting material. *J. Phys. Chem. Lett.,* 6: 881–886.

[167] Subbiah, A. S., Halder, A., Ghosh, S., Mahuli, N., Hodes, G., Sarkar, S. K. (2014). Inorganic hole conducting layers for perovskite-based solar cells. *J. Phys. Chem. Lett.,* 5: 1748–1753.

[168] Ito, S., Tanaka, S., Manabe, K., Nishino, H. (2014). Effects of surface blocking layer of Sb_2S_3 on nanocrystalline TiO_2 for $CH_3NH_3PbI_3$ perovskite solar cells. *J. Phys. Chem. C,* 118: 16995–17000.

[169] Ito, S., Tanaka, S., Nishino, H. (2015). Substrate-preheating effects on PbI2 spin coating for perovskite solar cells via sequential deposition. *Chem. Lett.,* 44: 849–851.

[170] Wu, Z. W., Bai, S., Xiang, J., Yuan, Z. C., Yang, Y. G., Cui, W., Gao, X. Y., Liu, Z., Jin, Y. Z., Sun, B. Q. (2014). Efficient planar heterojunction perovskite solar cell employing graphene oxide as hole conductor. *Nanoscale,* 6: 10505–10510.

[171] Zhi, Y., Minqiang, W., Sudhanshu, S., Yue, Z., Jianping, D., Hu, G., Xingzhi, W. (2015). Developing Seedless Growth of ZnO Micro/Nanowire Arrays towards ZnO/FeS$_2$/CuI P-I-N Photodiode Application. *Scientific Reports,* 5: 1–11.

[172] Brian, C., Lenzmann, F. (2004). Charge Transport and Recombination in a Nanoscale Interpenetrating Network of n-Type and p-Type Semiconductors: Transient Photocurrent and Photovoltage Studies of TiO$_2$/Dye/CuSCN Photovoltaic Cells. *J. Phys. Chem. B,* 108: 4342–4350.

[173] Xia, M., Gu, M., Liu, X., Liu, B., Huang, S., & Ni, C. (2015). Luminescence characteristics of CuI film by iodine annealing. *J Mater Sci: Mater Electron,* 26 (7): 5092–5096.

[174] Ezealigo, B. N., Nwanya, A. C. (2017). Optical and electrochemical capacitive properties of copper (I) iodide thin film deposited by SILAR method. *Arabian Journal of Chemistry,* 1–12.

[175] Gao, X. D., Li, X. M., Yu, W. D., Qiu, J. J., Gan, X. Y. (2008). Room-temperature deposition of nanocrystalline CuSCN film by the modified successive ionic layer adsorption and reaction method. *Thin Solid Films,* 517: 554–559.

[176] Zuo, C., Ding, L. (2015). Solution-processed Cu$_2$O and CuO as hole transport materials for efficient perovskite solar cells. *Small,* 11: 5528–5532.

[177] Chen, L. C. (2013). Review of preparation and optoelectronic characteristics of Cu2O-based solar cells with nanostructure. *Materials Science in Semiconductor Processing,* 16: 1172–1185.

[178] Jeng, J. Y., Chen, K. C., Chiang, T. Y. et al. (2014). Nickel oxide electrode interlayer in CH$_3$NH$_3$PbI$_3$ perovskite/PCBM planar-heterojunction hybrid solar cells. *Advanced Materials,* 26: 4107–4133.

[179] Seo, S., Park, I.J., Kim, M. et al. (2016). An ultra-thin, un-doped NiO hole transporting layer of highly efficient (16.4%) organic–inorganic hybrid perovskite solar cells. *Nanoscale*, 8: 11403–11412.

[180] Hu, L., Peng, J., Wang, W. et al. (2014). Sequential deposition of $CH_3NH_3PbI_3$ on planar NiO film for efficient planar perovskite solar cells. *ACS Photonics*, 1: 547–553.

[181] Kim, J. H., Liang, P.-W., Williams, S. T. et al. (2015). High-performance and environmentally stable planar heterojunction perovskite solar cells based on a solution-processed copper-doped nickel oxide hole-transporting layer. *Advanced Materials*, 27: 695–701.

[182] Chen, W., Wu, Y., Yue, Y. et al. (2015). Efficient and stable large-area perovskite solar cells with inorganic charge extraction layers. *Science*, 350: 944–948.

[183] Tseng, Z. L., Chen, L. C., Chiang, C. H., Chang, S. H., Chen, C. C., Wu, C. G. (2016). Efficient inverted-type perovskite solar cells using UV-ozone treated MoO_x and WO_x as hole transporting layers. *Solar Energy*, 139: 484–488.

[184] Hou, F., Su, Z., Jin, F. et al. (2015). Efficient and stable planar heterojunction perovskite solar cells with an MoO_3/PEDOT:PSS hole transporting layer. *Nanoscale*, 7: 9427–9432.

[185] Li, P., Liang, C., Zhang, Y., Li, F., Song, Y., Shao, G. (2016). Polyethyleneimine High-Energy Hydrophilic Surface Interfacial Treatment toward Efficient and Stable Perovskite Solar Cells. *ACS Applied Materials & Interfaces*, 8: 32574–32580.

[186] Sun, H., Hou, X., Wei, Q. et al. (2016). Low-temperature solution-processed p-type vanadium oxide for perovskite solar cells. *Chemical Communications*, 52: 8099–8102.

[187] Bailie, C. D., Christoforo, M. G., Mailoa, J. P., Bowring, A. R., Unger, E. L., Nguyen, W. H., Burschka, J., Pellet, N., Lee, J. Z., Grätzel, M., Noufi, R., Buonassisi, T., Salleo, A. and McGehee, M. D. (2015). Semi-transparent perovskite solar cells for tandems with silicon and CIGS. *Energy Environ. Sci.*, 8 (3): 956–963.

[188] Jang, J., Im, H. G., Jin, J., Lee, J., Lee, J. Y. and Bae, B. S. (2016). A Flexible and Robust Transparent Conducting Electrode Platform Using an Electroplated Silver Grid/Surface-Embedded Silver Nanowire Hybrid Structure. *ACS Applied Materials & Interfaces,* 8: 27035–27043.

[189] Green, M. A. (1982). Accuracy of Analytical Expressions for Solar Cell Fill Factors. *Solar Cells,* 7: 337–340.

[190] Luque, A., Hegedus, S. (2003). Handbook of Photovoltaic Science and Engineering. *Wiley,* 1168.

[191] Eperon, G. E., Stranks, S. D., Menelaou, C., Johnston, M. B., Herz, L. M., Snaith, H. J. (2014). Formamidinium lead trihalide: a broadly tunable perovskite for efficient planar heterojunction solar cells. *Energy Environ. Sci.,* 7: 982–988.

[192] D'Innocenzo, V., Grancini, G., Alcocer, M. J. P., Kandada, A. R. S., Stranks, S. D., Lee, M. M., Lanzani, G., Snaith, H. J. et al. (2014). Excitons versus free charges in organo-lead tri-halide perovskites. *Nature Communications,* 5 (1): 3586 1–6.

[193] Smestad, G. P., Krebs, F. C., Lampert, C. M., Granqvist, C. G., Chopra, K. L., Mathew, X., Takakura, H. (2008). Reporting Solar Cell Efficiencies in Solar Energy Materials and Solar Cells. *Solar Energy Mat. Solar Cells,* 92: 371–373.

[194] Habisreutinger, S. N., Leijtens, T., Eperon, G. E., Stranks, S. D., Nicholas, R. J., Snaith, H. J. (2014). Carbon Nanotube/Polymer Composites as a Highly Stable Hole Extraction Layer in Perovskite Solar Cells. *Nano Letters,* 14 (10): 5561–5568.

[195] Leijtens, T., Eperon, G. E., Pathak, S., Abate, A., Lee, M. M., Snaith, H. J. (2013). Overcoming ultraviolet light instability of sensitized TiO_2 with meso-superstructured organometal tri-halide perovskite solar cells. *Nature Communications,* 6: 2885.

[196] Snaith, H. J., Abate, A., Ball, J. M., Eperon, G. E., Leijtens, T., Noel, N. K., Wang, J. T. W., Wojciechowski, K., Zhang, W., Zhang, W. (2014). Anomalous Hysteresis in Perovskite Solar Cells. *The Journal of Physical Chemistry Letters,* 5 (9): 1511–1515.

[197] Unger, E. L., Hoke, E. T., Bailie, C. D., Nguyen, W. H., Bowring, A. R., Heumuller, T., Christoforo, M. G., McGehee, M. D. (2014). Hysteresis and transient behavior in current-voltage measurements of hybrid-perovskite absorber solar cells. *Energy & Environmental Science,* 7: 3690–3698.

[198] Noel, N. K., Stranks, S. D., Abate, A., Wehrenfennig, C., Guarnera, S., Haghighirad, A. A., Sadhanala, A., Eperon, G. E., Pathak, S. K., Johnston, M. B., Petrozza, A., Herz, L. M., Snaith, H. J. (2014). Lead-free organic–inorganic tin halide perovskites for photovoltaic applications. *Energy & Environmental Science,* 7 (9): 3061.

[199] Sha, W. E. I., Ren, X., Chen, L. and Choy, W. C. H. (2015). The Efficiency Limit of $CH_3NH_3PbI_3$ Perovskite Solar Cells. *Applied Physics Letters,* 106 (22): 221104.

[200] Ren, X., Wang, Z., Sha, W. E. I., Choy, W. C. H. (2017). Exploring the Way to Approach the Efficiency Limit of Perovskite Solar Cells by Drift-Diffusion Model. *ACS Photonics,* 4 (4): 934–942.

[201] Mosconi, E., Amat, A., Nazeeruddin, M. K., Grätzel, M., Angelis, F. D. (2013). First-Principles Modeling of Mixed Halide Organometal Perovskites for Photovoltaic Applications. *The Journal of Physical Chemistry C,* 117 (27): 13902–13913.

[202] Lang, L., Yang, J. H., Liu, H. R., Xiang, H. J., Gong, X. G. (2014). First-principles study on the electronic and optical properties of cubic ABX_3 halide perovskites. *Physics Letters A,* 378 (3): 290–293.

[203] Gonzalez-Pedro, V., Juarez-Perez, E. J., Arsyad, W.-S., Barea, E. M., Fabregat-Santiago, F., Mora-Sero, I., Bisquert, J. (2014). General Working Principles of $CH_3NH_3PbX_3$ Perovskite Solar Cells. *Nano Letters,* 14 (2): 888–893.

[204] Umari, P., Mosconi, E., Angelis, F. D. (2014). Relativistic GW calculations on $CH_3NH_3PbI_3$ and $CH_3NH_3SnI_3$ Perovskites for Solar Cell Applications. *Scientific Reports,* 4: 4467.

[205] Agarwal, S., Nair, P. R. (2015). Device engineering of perovskite solar cells to achieve near ideal efficiency. *Applied Physics Letters,* 107 (12): 123901.

[206] Minemoto, T., Murata, M. (2014). Device modeling of perovskite solar cells based on structural similarity with thin film inorganic semiconductor solar cells. *Journal of Applied Physics,* 116 (5): 054505.

[207] Sun, X., Asadpour, R., Nie, W., Mohite, A. D., Alam, M.A. (2015). A Physics-Based Analytical Model for Perovskite Solar Cells. *IEEE Journal of Photovoltaics,* 5 (5): 1389–1394.

[208] Sivaram, V., Stranks, S. D., Snaith, H. J. (2015). Outshining Silicon. *Scientific American,* 44–46.

[209] Chen, W., Wu, Y., Yue, Y., Liu, J., Zhang, W., Yang, X., Chen, H., Bi E., Ashraful, I., Gratzel, M. and Han, L. (2015). Efficient and stable large-area perovskite solar cells with inorganic charge extraction layers. *Science,* 350: 944–948.

[210] Saliba, M., Matsui, T., Seo, J. Y., Domanski, K., Correa-Baena, J. P., Nazeeruddin, M. K., Zakeeruddin, S. M., Tress, W., Abate, A., Hagfeldt, A. and Grätzel, M. (2016). Cesium-containing triple cation perovskite solar cells: improved stability, reproducibility and high efficiency. *Energy Environ. Sci,* 9 (6): 1989–1997.

[211] Zhang, F., Shi, W., Luo, J., Pellet, N., Yi, C., Li, X., Zhao, X., Dennis, T. J. S., Li, X., Wang, S., Xiao, Y., Zakeeruddin, S. M., Bi, D. and Grätzel, M. (2017). Isomer-Pure Bis-PCBM-Assisted Crystal Engineering of Perovskite Solar Cells Showing Excellent Efficiency and Stability. *Advanced Materials,* 1606806.

[212] Grancini, G., Roldán-Carmona, C., Zimmermann, I., Mosconi, E., Lee, X., Martineau, D., Narbey, S., Oswald, F., Angelis, F. D., Graetzel, M. & Nazeeruddin, M. K. (2017). One-Year stable perovskite solar cells by 2D/3D interface engineering. *Nature Communications,* 8 (15684): 15684.

[213] Cruz, A. M., Perreira, M. D. (2018). The New Generation of Photovoltaic Cells Entering the Market. *Leitat, Barcelona.*

[214] Snaith, H. J., Abate, A., Ball, J. M., Eperon, G. E., Leijtens, T., Noel, N. K., Wang, J. T. W., Wojciechowski, K., Zhang, W., Zhang, W. (2014). Anomalous Hysteresis in Perovskite Solar Cells. *The Journal of Physical Chemistry Letters,* 5 (9): 1511–1515.

[215] Zimmermann, E., Wong, K. K., Mueller, M., Hu, H., Ehrenreich, P., Kohlstaedt, M., Würfel, U., Mastroianni, S., Mathiazhagan, G., Hinsch, A., Gujar, T. P., Thelakkat, M., Pfadler, T., Schmidt-Mende, L. (2016). Characterization of perovskite solar cells: Towards a reliable measurement protocol. *APL Materials,* 4 (9): 091901.

[216] O'Regan, B. C., Barnes, P. R. F., Li, X., Law, C., Palomares, E. and Marin-Beloqui, J. M. (2015). Optoelectronic Studies of Methylammonium Lead Iodide Perovskite Solar Cells with Mesoporous TiO_2: Separation of Electronic and Chemical Charge Storage, Understanding Two Recombination Lifetimes, and the Evolution of Band Offsets during J – V Hysteresis. *Journal of the American Chemical Society,* 137: 5087–5099.

[217] Kim, H. S., Jang, I. H., Ahn, N., Choi, M., Guerrero, A., Bisquert, J. and Park, N. G. (2015). Control of I – V Hysteresis in $CH_3NH_3PbI_3$ Perovskite Solar Cell. *The Journal of Physical Chemistry Letters,* 6: 4633–4639.

[218] Noel, N. K., Abate, A., Stranks, S. D., Parrot, E. S., Burlakov, V. M., Goriely, A., Snaith, H. J. (2014). Enhanced Photoluminescence and Solar Cell Performance via Lewis Base Passivation of Organic–Inorganic Lead Halide Perovskites. *ACS Nano,* 8 (10): 9815–9821.

[219] Abate, A., Saliba, M., Hollman, D. J., Stranks, S. D., Wojciechowski, K., Avolio, R., Grancini, G., Petrozza, A., Snaith, H. J., (2014). Supramolecular Halogen Bond Passivation of Organic–Inorganic Halide Perovskite Solar Cells. *Nano Letters,* 14 (6): 3247–3254.

[220] Xiao, Z., Bi, C., Shao, Y., Dong, Q., Wang, Q., Yuan, Y., Wang, C., Gao, Y., Huang, J. (2014). Efficient, High Yield Perovskite Photovoltaic Devices Grown by Interdiffusion of Solution-Processed Precursor Stacking Layers. *Energy & Environmental Science,* 7 (8): 2619.

[221] Kadro, J. M., Pellet, N., Giordano, F., Ulianov, A., Müntener, O., Maier, J., Grätzel, M. and Hagfeldt, A. (2016). Proof-of-concept for facile perovskite solar cell recycling. *Energy Environ. Sci.,* 9 (10): 3172–3179.

[222] Zhang, J., Gao, X., Deng, Y., Li, B. and Yuan, C. (2015). Life Cycle Assessment of Titania Perovskite Solar Cell Technology for Sustainable Design and Manufacturing. *Chem Sus Chem,* 8: 3882–3891.

[223] Babayigit, A., Thanh, D. D., Ethirajan, A., Manca, J., Muller, M., Boyen, H.-G. and Conings, B. (2016). Assessing the toxicity of Pb- and Sn-based perovskite solar cells in model organism Danio rerio. *Scientific Reports,* 6: 18721.

[224] Fabini, D. (2015). Quantifying the Potential for Lead Pollution from Halide Perovskite Photovoltaics. *The Journal of Physical Chemistry Letters,* 6: 3546–3548.

[225] Hailegnaw, B., Kirmayer, S., Edri, E., Hodes, G. and Cahen, D. (2015) Rain on Methylammonium Lead Iodide Based Perovskites: Possible Environmental Effects of Perovskite Solar Cells. *The Journal of Physical Chemistry Letters,* 6: 1543–1547.

In: Perovskite Solar Cells
Editor: Murali Banavoth
ISBN: 978-1-53615-858-8
© 2019 Nova Science Publishers, Inc.

Chapter 2

MECHANISMS OF RADIATION-INDUCED DEGRADATION OF HYBRID PEROVSKITES BASED SOLAR CELLS AND WAYS TO INCREASE THEIR RADIATION TOLERANCE

Boris L. Oksengendler[1],*, Nigora N. Turaeva[2], Marlen I. Akhmedov[3] and Olga V. Karpova[4]

[1]Institute of Ion-Plasma and Laser Technologies,
Tashkent, Uzbekistan
[2]Department of Biological Sciences,
Webster University, St. Louis, MO, US
[3]National University of Uzbekistan,
Tashkent, Uzbekistan
[4]Turin Polytechnic University in Tashkent,
Tashkent, Uzbekistan

---

* Corresponding Author's Email: oksengendlerbl@yandex.ru.

And now remains
That we find out the cause of this effect;
Or rather say, the cause of this defect,
For this effect defective comes by cause:
Thus it remains, and the remainder thus.

(W. Shakespeare, Hamlet)

ABSTRACT

The basic processes of perovskite radiation resistance are discussed for photo- and high-energy electron irradiation. It is shown that ionization of iodine ions and a staged mechanism of elastic scattering (upon intermediate scattering on light ions of an organic molecule) lead to the formation of a recombination center Ii. The features of ionization degradation of interfaces with both planar and fractal structures are considered. A special type of fractality is identified, and its minimum possible level of photodegradation is predicted. By using the methodology of classical radiation physics (including chaos ideas), the Hoke effect was studied, as well as the synergetics of cooperative phenomena in tandem systems. The principal channels for counteracting the radiation degradation of solar cells based on hybrid perovskites have been revealed.

Keywords: hybrid perovskite, radiation and degradation of solar cells, tandems, fractal interfaces, Hoke effect, synergetics, scale-free network

1. INTRODUCTION

In the long history of the creation of solar cells (SCs) based on various semiconductor materials, it has become absolutely clear that the issue of radiation stability of these semiconductor structures has always remained one of the most challenging [1-4]. At present, the established views are that the issue of SC radiation degradation has three aspects:

- -degradation of an SC under the action of solar light in a wavelength range considered as the operating one;

- -degradation of a SC under the action of high-energy radiation;
- -degradation of a SC under the conditions of radiation technology;

The first aspect is important for common operating conditions when photochemical processes are in the foreground. The second aspect is vital for operation in space and under special terrestrial conditions connected with intense radiation (atomic power plants, conditions of a nuclear explosion, etc.). The third aspect is the degradation that always takes place immediately in the moment of constructive radiation technology, and it is necessary to establish when radiation technological processes begin to be leveled by radiation degradation processes.

The mechanisms implementing all these three aspects of radiation physics are utterly different, since, different ratios between the four factors of radiation action (ionization, shock wave, heat release, and elastic scattering) occur in each of them [5–7].

The so-called organic-inorganic perovskites, which combine atoms of metals, haloids, and organic molecules in their regular structure [8], were an absolutely particular new type of semiconductor and appeared to be promising at the present stage of the conversion of solar energy into electrical energy. This result refers to both: one transitional and tandem structures [9–12]. After its synthesis, the importance of the problem of SC radiation degradation, which remains no less important (as compared to silicon, gallium arsenide, and cadmium–mercury–tellurium alloys), looks much more complicated already from a priori notions. But what is this connected with?

First, it is connected with the multicomponent composition of organic-inorganic perovskite material, which is such that strong asymmetry of the masses of components takes place.

Second, in this type of substance, completely different types of chemical bonds are implemented.

Third, states with quite a high band-gap energy are implemented.

Fourth, the properties of both inner and outer interfaces begin to play a special role.

Fifth, both ordered and disordered atomic structures are often implemented in these materials.

Such a combination of properties, which resemble systems with high-temperature conductivity to some extent [13], becomes the basis for diversity of radiation physics and chemistry [14]. The latter, in turn, requires detailed knowledge of the micro mechanisms of the radiation response of this class of perovskites and the devices based on them in order to use the approaches of radiation technology, as well as a search for ways to reliably control the high radiation stability. The goal of the present work is to describe the approach to the mechanisms of radiation degradation of organic-inorganic perovskites and some device structures based on them in order to formulate the main trends of the future construction of the radiation physics of this class of materials.

2. General Methodology of Radiation Physics

Radiation physics of solid states [5, 7] arose in connection with the problem of the atomic bomb in the middle of the twentieth century. Over the past years, this science has become a deep and rich field of physics, chemistry and biology, which has major achievements and introduced new concepts to all physics. The methodology of modern radiation physics, in addition to classical quantum and statistical physics of condensed matter, is fundamentally connected with such concepts as: "nano," "fractal," "synergetics," and "chirality." At the same time, in studying radiation effects in condensed matter, the use of classical concepts of various channels of radiation energy transfer to the irradiated medium, as well as concepts of elementary radiation-stimulated atomic rearrangements (radiation defect formation, RDF; radiation-stimulated diffusion, RSD; radiation-stimulated quasi-chemical reactions, RSQCR; radiation-stimulated movement of boundaries, RSMB; radiation-stimulated movement of dislocations, RSDM; the radiation stimulated disordering, RSDis). All these elementary atomic rearrangements can be classified as thermal and athermal radiation effects due to their thermal and electron-stimulated origin, and low-scaled or large-scaled radiation effects.

Table 1. Dominant channels of energy transfer from radiation to materials

Radiation→ Excitements ↓	Light	X-ray	γ-ray	Electron beam	Energetic ions	Fission fragment	Neutron
Elastic displacement			x	x	x		x
Ionization	x	x	x	x	x	x	
Heat releasing				x	x	x	x
Elastic waves			x		x	x	x
Shock waves					x	x	x

Table 2. Basic elementary processes of atomic reconstructions in condensed matter under irradiation

Energy Transfer channels → Basic atomic processes ↓	Elastic displacements	Ionisation	Heat release	Elastic waves	Shock waves
RDF	x	x	x	x	x
RSD	x	x	x	x	x
RSQCR		x	x	x	x
RSMB		x	x		x
RSDis	x		x		x

RDF (Radiation defect formation), RSD (Radiation simulated diffusion), RSQCR (radiation simulated quantum chemical reactions), RSMB (Radiation stimulated motion of boundaries), RSDis (Radiation stimulated disordering).

These basic concepts are summarized in two Tables 1 and 2. Further, it must be noted that the formation of radiation defects is the primary and most important step of the overall radiation effect. The final resolution of the radiation-physical problem is considered to be the establishment of the

3. MECHANISMS OF DEGRADATION UNDER THE EFFECT OF SOLAR RADIATION

A vital feature of organic-inorganic perovskites is that this material features a large fraction of complexity in chemical bonds. In fact, its structural formula looks as $CH_3 - NH_3^+ Pb^{2+} I_3^-$. It is thus evident that the binding energy (potential well) results from the electrostatic interaction manifesting itself in a high Madelung energy value [9, 12]. In particular, the binding energy of an I^- ion in its regular position turns out to be 7.75 eV.

Since the basic effect of photogeneration in perovskite $CH_3 - NH_3^+ Pb^{2+} I_3^-$ is electron transfer from the valence band to the conduction band of the perovskite and these bands are formed from the states of I^- and Pb^{2+}, correspondingly, then, from the point of view of crystal chemistry [15], such a photogeneration act converts the negative iodine ion into a neutral atom, which eliminates the Madelung potential well for regular iodine, and its slightest fluctuation leads to displacement into an interstitial position I_i. Therefore, the schematic representation of successive photoelectron- ion processes looks like:

$$I_S^- + h\nu \rightarrow I_S^0 + e^- \rightarrow I_i^0 + e^-. \qquad (1)$$

However, implementation of the latter part of the chain (which would always take place in a gas phase) is impeded by the process determined by the quantum nature of a formed hole [16]. In fact, a neutral iodine atom (in the scheme of crystal chemistry) is a hole localized on incidental iodine. Note that such a duality (crystal chemistry and band approach) is a typical dichotomy popular in solid-state physics [17].

Next, the initially localized hole is delocalized along the iodine chain; therefore, the characteristic delocalization time can be estimated from the Bohr–Heisenberg uncertainty relation [18]

$$\tau_h \approx \frac{\hbar}{\Delta E_V}, \quad (2)$$

where ΔE_V is the valence-band width.

Consequently, an alternative of either translation of a hole and back conversion of I_S^0 into I_S^- or displacement of I_S^0 into an interstice due to the quantum nature of a hole will be determined by an expression [5]

$$\eta = \exp\left(-\frac{\Delta E_V}{\omega_D \hbar}\right) = \int_{\tau_\dagger = 1/\omega_D}^{\infty} W(t) dt. \quad (3)$$

The described mechanism of defect formation, which constitutes a fundamental act of photodegradation, is essentially a variant of the so-called Dexter–Varley mechanism [21] in radiation physics of ionic crystals [22] in its quantum variant developed 40 years ago by a research team in Tashkent as applied to subthreshold defect production [19, 23, 24]. Let us add that ω_D is the Debye frequency and η is the probability of formation of a defect (in this case, an interstitial I_i, which is the main recombination center [11]).

In relation to perovskites, we pointed out the special importance of this degradation mechanism earlier in [25], wherein it was shown that, in the bulk of a crystal, η is on the order of 10^{-4}. A number of new features of this mechanism are expected to be observed at higher quantum energy than that corresponding to the hump of the solar spectrum, when the initial ionization can proceed in a valence shell of perovskite (Figure 1(a)).

The main channel of decomposition of the formed "subvalent" hole is a band-to-band Auger process. As a result of an Auger process, a hole "floats" up to the valence shell yet "proliferating" to yield two holes.

The latter leads to still greater Coulomb instability upon the transition from the initial I_S^- into I_S^+, but how will it affect the probability of displacement η?

Studies show [25, 26] that the double charge on I_S^+ locally decreases the valence-band width ΔE_V. In addition, the time τ_+ needed for the knock-out of I_S^+ ion into an interstice also diminishes (Figure 1(b)). As a result, the probability of displacement $I_S \rightarrow I_i$ rises considerably; i.e., the fundamental degradation of a SC upon the addition of ultraviolet radiation increases.

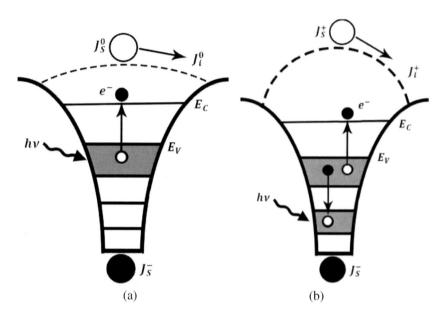

Figure 1. (a). Two-stage schematic representation of displacement of ion I_S^- in a lattice position into an interstice ($I_S^- \rightarrow I_S^0 \rightarrow I_i^0$) upon the absorption of a photon in the valence band. (b). Three-stage schematic representation of displacement of ion I in a lattice position into an interstice ($I_S^- \rightarrow I_S^0 \rightarrow I_S^+ \rightarrow I_i^+$) upon the absorption of a photon in the "subvalence" shell.

The summary of the above is also evident: in all places in a crystal where structural imperfections (locally) decrease ΔE_V, the probability of defect production is increased. It is necessary to add that the existence of the process described by the value $\eta \ll 1$ is evidence that the scheme of photodestruction, according to Schoonman [27], is apparently overestimated in the role of photodegradation. In fact, one may note that, in an experimental work [28], a serious delay in the decomposition of the state immediately following photogeneration of an electron–hole pair is pointed out. However, it is more correct here to speak not of the slowing of this stage but simply of the small fraction of the primary unstable state that is transformed into defects, which manifests as a "delay" of the defect-formation act.

4. MECHANISM OF DEGRADATION UNDER THE ACTION OF HIGH ENERGY PARTICLES (A SPECIAL ROLE OF AN ORGANIC MOLECULE)

Let us discuss the most popular mechanism of defect formation, viz., the displacement of a regular atom upon elastic scattering of a fast extraneous particle on it [5, 6], taking into account the features of the structure of organic-inorganic perovskites. The efficiency of the displacement of a regular atom upon elastic impact in a simplified form is characterized by the defect-formation cross-section [29]:

$$\sigma = \int_{E_d}^{E_{MAX}} d\sigma_T = \pi \sigma_0 \left(\frac{E_{MAX}}{E_d} - 1 \right), \qquad (4)$$

where σ_0 is a particular characteristic of Rutherford scattering of a charged extraneous particle (e.g., elec-tron) on an atom, E_d is the characteristic displace-ment energy (effective potential-well depth) of a regular atom in a crystal, and E_{MAX} is the maximum possible energy transfer from a fast extraneous particle to a regular atom in a crystal at head-on collision. It is of significance:

$$E_{MAX} = 4 \frac{m_e M_a}{(m_e + M_a)^2} E_e, \qquad (5)$$

where m_e is the mass of a fast electron, M_a is the mass of a regular atom, and E_e is the fast electron energy. Combination of equations (4) and (5) indicates the principal channel of increase in the efficiency of defect formation, which is a large E_{MAX} value.

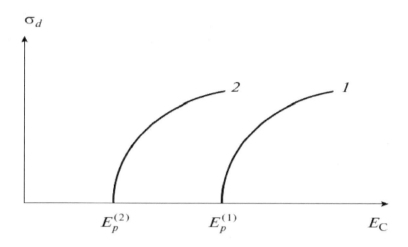

Figure 2. Energy dependence of the cross section (probability) of defect formation upon a (*1*) direct or (*2*) staged process of elastic displacement of a heavy atom (I) into an interstice.

Turning back to organic-inorganic perovskites, it is worth noting that, in such a material composed of atoms with a very great difference in masses of atoms (H and I, Pb), an extraordinarily efficient channel of defect formation arises: upon elastic scattering, fast electrons displace hydrogen atoms (M_H), which are next scattered on an iodine atom (with a mass of M_I) and displace it into an interstice. That is how the basic recombination center I_i is formed in perovskites [30]. It can be shown readily that the ratio between the maximum energy transfer to a regular iodine atom for the staged and direct processes has the form [5]

$$\frac{E_{MAX}^{(2)}}{E_{MAX}^{(1)}} = \frac{4M_I^2(m_e+M_H)^2}{[(M_I+M_H)^2(m_e+M_I)^2]} \approx 4\left(\frac{M_H}{M_I}\right)^2 \ll 1. \tag{6}$$

Note that the possibility of two-stage defect formation was qualitatively discussed as applied to hydrogen-doped germanium [31]. However, these are the organic-inorganic perovskites for which staged degradation is the basic fundamental degradation process, since, in their case, the concentration of the light component is the same as that of the others (and that is what makes the degradation in perovskites utterly different from this case [31]).

On the other hand, concern about the pronounced ionicity of the crystal components is of importance for the elastic mechanism of defect formation in perovskites. In fact, microscopic theory [5, 32], which makes it possible to express the displacement energy E_d (Eq. (3)) in terms of the depth of a true potential well of a regular atom and its capture radius R_0 (Vineyard–Koshkin "instability zone" [5, 29]), indicated a strong increase in E_d for ionic solids as compared to the other types of chemical bonds; therefore, $dE_d/d\alpha_i > 0$ (where α_i is the degree of ionicity in chemical bonds [17]). Therefore, defect formation in perovskites is expected to be problematic $\left(d\sigma_d/d\alpha_i < 0\right)$, which is indicative of the factor of their high native radiation resistance and, to some extent, should compensate for the efficiency of the staged channel of defect formation, turning the action of any energy radiation into internal proton irradiation. It should be mentioned that, in this case, the energy threshold of damage of a perovskite crystal is decreased drastically (Figure 2).

5. FEATURES OF RADIATION DEGRADATION OF SCS WITH FRACTAL INTERFACES

An unusual property of SCs based on organic-inorganic perovskites was recently discovered [33]: it turned out to be that, all other things being equal, the transition from a planar interface architecture (perovskite/HTM) to a rough (fractal) one increases the short-circuit photocurrent by 13%. Several reasons for such an extraordinary effect have been proposed [34] (an increase in the total path length of a solar beam in the vicinity of the interface, the facilitated passage of carriers in an extended junction, and the conversion of recombination levels in the interface into the class of attachment levels due to nanofractal fluctuations of the interface potential). Let us dwell on the third affecting channel in greater detail, taking into account both its greatest (probably) significance and rather intriguing demonstration in degradation processes, which has never been discussed before (see below). Since perovskites are materials with a high degree of ionicity, a Madelung model [35] is quite applicable in light of their electronic states at the interfaces. According to the model, the occurrence of the surface levels (Tamm levels) in the energy gap can be associated with a natural decrease in the Madelung energy at the surface. This idea was extended to the boundary with curvature [36]: it was found that the Tamm surface levels are shifted toward the center of the band gap in convex areas, while the shift occurs closer to the edges of the permitted energy bands in concave areas.

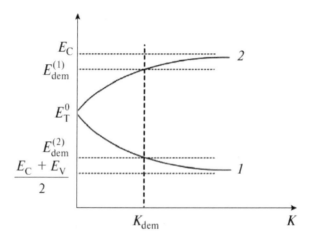

Figure 3. Schematic representation of arrangement of the Tamm surface level with respect to the surface (interface) curvature (E_T^0 is the Tamm level for a flat surface; curves 1 and 2 correspond to convex and concave areas of the surface; $E_{dem}^{(1)}$ and $E_{dem}^{(2)}$ the two possible variants of arrangement of the demarcation level). Upon intersection of the critical degree of curvature K_{dem}, the Tamm level passes from the recombination type to the case corresponding to attachment (curve 1) and vice versa (curve 2).

Generalizing this idea, one can say that, by controlling the fractality of the interface, it is possible in principle to transfer the local electronic levels of the boundary from one region to the other, in particular, the recombination levels into the region of attachment levels, for which it is sufficient just to provide the degree of curvature of the concave area such that it intersects the demarcation levels (Figure 3). We expect that this should decrease the recombination losses of the interface, which forms the basis of explanation of the results in a study [33]. From the viewpoint of radiation destruction of a SC through the modification of interfaces, one more important issue should be noted. In fact, let us consider the Tamm electronic orbitals at the fractal interface (Figure. 4).

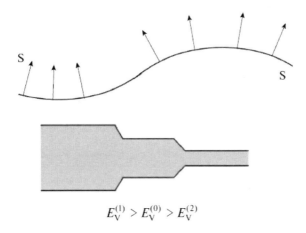

Figure 4. Dependence of the width of the surface band of Tamm states ($E_V^{(i)}$) on the curvature type: (0) intermediate region, (1) concave area, and (2) convex area.

As is well known [35], an overlap of neighboring Tamm orbitals creates bands of Tamm states. Evidently, in convex and concave areas of the interface, these overlaps are significantly different: they are smaller in the former and larger in the latter (as compared to a flat interface); the schemes of the valence bands of the corresponding areas agree with this (Figure 3). This result is fundamentally important for photodegradation (see Section 2).

According to its scheme, the probability of defect formation strongly depends on the valence-band width (Eq. (3)):

$$\eta = exp(-const \cdot \Delta E_V). \tag{7}$$

Therefore, in convex areas, the formation of interstitial I_i occurs much more intensely than in the concave areas which create diffusion atomic flows from the apices of the relief to the areas of concavities at a fractal surface, i.e., leads to smoothing of the relief and, accordingly, the decrease in the excess photocurrent (the above-mentioned 13% [33]). Finally, on the basis of the above reasoning, one can use interfaces with specially handpicked fractality (Figure 4), which serves the two noted functions at once: first, it shifts the levels of the interface toward the region of attachment; second, it makes it possible to obtain a low value, that is, radiation resistance (Figure 5). This new idea requires experimental verification. It is worth noting that, generally speaking, smoothing of a fractal surface under the action of radiation has already been observed experimentally in BaF_2, but under strong ionic irradiation [37]. It is thus undoubted that radiation engineering of both exterior and inner surfaces is a promising challenge of the near future [38–40].

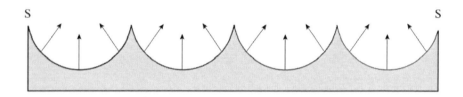

Figure 5. Special form of a rough (fractal) surface (interface) providing minimal photodegradation.

6. DEGRADATION MACROEFFECTS IN PEROVSKITES

In the second and third Sections, only those mechanisms of radiation defect formation that deal with the displacement of regular atoms of a lattice into interstitial positions were discussed. It is necessary to note that the total radiation macroeffect, in this case "radiation degradation," includes acts of defect formation only as the first stage. The second stage is radiation-stimulated diffusion (RSD) of atoms and defects in the sample, followed by the third stage, quasichemical reactions (QCRs) between atoms and defects, which result in modification of the electronic spectrum (especially in the band gap). If active recombination levels arise in this process, then radiation degradation appears [1 - 7, 29].

An efficient method to describe radiation degradation is the analysis of the kinetics of accumulation of active recombination centers. Such analysis is carried out based on a system of kinetic equations solved under certain initial and boundary conditions. A question arises: what features are imparted to the proceeding of the second and third stages by organic-inorganic perovskite itself and in what way do these features differ from the corresponding stages of the common radiation physics of semiconductors [1–7, 29]? It is significant that this issue is completely unsolved and has even not been posed yet [9–12]. Meanwhile, a range of features of organic-inorganic perovskites, which should appreciably affect the proceeding of the total radiation effect or degradation, is clear already. Let us discuss them schematically.

6.1. Radiation-Stimulated Diffusion

An essential feature of organic-inorganic perovskites is a wide band gap, which allows the electron-stimulated channels of RSD to be implemented upon recharging of defect levels [5–7]. The physics of these acts of RSD is based on the efficient transfer of energy of electron transitions to either deformation of the diffusion potential barrier, a sharp increase in the amplitude of local oscillations of defects, or a burst of temperature in the vicinity of a defect. In the most general case, six modes of intensification of diffusion of defects can be implemented in this process and can be written in the form of the expression for the RSD coefficient [5]:

$$D_{RSD} = D_0 exp\left(-\frac{Q^* - n\hbar\omega_{ph}}{k(T+\Delta T)}\right), \tag{8}$$

where Q^*, $\Delta E^* = n\hbar\omega_{ph}$, $T^* = T + \Delta T$ are the diffusion barrier modified with electron–hole junctions, the energy of electron transition, and the burst of temperature, respectively. Of vital importance is the case when $Q^* - \Delta E^* \leq 0$; this is the so-called mechanism of diffusion through inverse recharging [5, 26], a feature of which is the completely a thermal character of migration of defects. The experience of radiation physics of common semiconductors [1–7, 29] is evidence that such an effect can be observed with the so called negative-U centers. There is a high probability that such a negative-U defect is the iodine atom displaced into an interstice; therefore, consecutive iodine recharging is likely to be able to "drive" this defect a thermally (the idea that iodine is a negative-U center can be inferred from the features of change in stability of the configuration upon consecutive variation in charge [30]).

6.2. Quasi-Chemical Reactions

The high efficiency of defect interaction in perovskites, which leads to the creation of complexes, can be due to two reasons. First, since this material enjoys a considerable degree of ionicity, the opposite charges of defects entail their strong Coulomb interaction (which sharply increases the efficiency of attraction of defects to each other). Second, the great strength of electron–phonon interaction should be taken into account; it determines electron-stimulated atomic rearrangements in a local region [9–12]. In these electron-stimulated processes, a special role may be played by dipole molecules $CH_3NH_3^+$. In fact, any excitation of the regular component of the $CH_3NH_3^+$ lattice will immediately affect the value of the local band-gap energy, which is modulated when oscillations of separate modes of a $CH_3NH_3^+$ molecule occur (because the Madelung energy in an ionic crystal is an essential component of the band-gap energy) [35, 36]. Oscillations of a $CH_3NH_3^+$ molecule will immediately modulate the Madelung energy and, accordingly, the band-gap energy. It is also of importance that disturbances of the shape of the inner and outer interfaces (in particular, their transition from planar to rough geometry) dramatically change the efficiency of defect sinks. All of the mentioned circumstances will be reflected in change of the coefficients in the corresponding terms of the kinetic equations and, under strongly nonequilibrium conditions, may yield even a synergistic behavior of a crystal [38].

6.3. Radiation Annealing

An important feature of radiation action on materials is radiation annealing, the essence of which lies in elimination of the radiation effect and, accordingly, radiation degradation. On the basis of the pool of experience of the radiation physics of semiconductors [5], two classes of the processes of this kind should be differentiated: the effect of recombination of electrons and holes on a Frenkel pair $V_I + I_i$ with the subsequent reaction $V_I + I_i \to I_S$, and a special type of quasichemical reactions during irradiation, which is "a reaction of displacement" of an atom Cl_s positioned regularly in the lattice site by interstitial I_i (a case in point is mixed perovskite $CH_3NH_3Pb_{3-x}Cl_x$). This reaction can be written as $Cl_s + I_i \to I_s + Cl_i$ with $Cl_i \uparrow$ denoting rapid diffusion of the displaced atom towards various sinks (interfaces and surfaces). Note that the reaction of a displacement type plays a fundamental role in the radiation physics of all familiar semiconductor materials and is due to vibron effects [5]. Especially notable is the choice of the method of theoretical analysis of the potential relief for both types of reactions. The most adequate approach may be quantum chemistry on the basis of the DFT method [41]. In the Hohenberg–Kohn–Sham formulation, the Kohn–Sham equations reduce a many-particle problem of interacting electrons to a single-particle problem with an effective potential.

At present, this method is one of the most universal (ab initio) methods for calculation of the electronic structure and various characteristics of many-particle systems, which is applied in quantum chemistry and solid-state physics. In this method, the many-electron system is described not by a wave function, which would determine a very high dimensionality of the problem equal to at least 3N (the number of coordinates of N particles), but an electron-density function (the function of only three spatial coordinates), which leads to considerable simplification of the problem. In this case it turns out that the major properties of the system of interacting particles can be expressed with an electron-density functional; in particular, according to the Hohenberg–Kohn theorems, which provide a theoretical basis for the DFT method, such a functional is the ground-state energy of the system. As a result, a full picture of electron-stimulated quasi-chemical reactions (Figures. 6(a) and 6(b) can be constructed. Comprehensive knowledge of the potential relief of the reactions, together with the accompanying change in the spectrum of electronic levels in the band gap, allows one to choose the method of striving for an increase in the service life of a device. For example, the regular injection of an electron–hole plasma to the perovskite area can provide systematic purification from close Frenkel pairs (their recombination), according to Figure. 6(a) (arrow 2), similarly to that established in GaAs structures long ago [5].

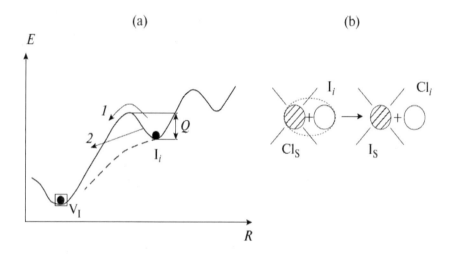

Figure 6. Schematic representation of rearrangement of defects compensating degradation processes: (a) channels of annealing of a close Frenkel pair in sublattice I: (1) thermoactivation channel and (2) recombination-stimulated channel; and (b) "dumbbell" reaction of displacement of regular atom Cl_s by interstitial I_i defect.

7. THE HOKE EFFECT IN HYBRID PEROVSKITES

Perovskite solar cells are a champion among other solar cells in the rate at which they increased their power conversion efficiency from 3% to 21% during 7-8 years [42-47]. Ongoing efforts to improve the efficiency beyond the Shockley-Queisser limit are focused on tuning their current bandgap around 1.5-1.6 eV to the range of 1.7-1.8 eV, which is ideal for a perovskite-

Si tandem solar cell. This can be achieved by replacing iodide components of $MAIPbI_3$ perovskite with bromide [48-51]. However, this increase does not result in a corresponding increase in open-circuit voltage [52, 53], which would be expected due to the increase of the band gap. Hoke et al. [53] showed that mixed I/Br perovskite fail to perform because of halide phase segregation induced by light, which is known as the Hoke effect. Solar cells made with mixed halide perovskites containing more than 20% bromide ($MAIPb(Br_xI_{1-x})_3$, $x > 0.2$) showed a decrease in open-circuit voltage with increasing bromide content [52]. It was shown [53] that for materials with $x > 0.2$, the initial PL intensity decreased and a new lower energy peak (as the initial PL peak energy of the $x = 0.2$) is appeared. From these data they assumed that upon illumination two separate phases corresponding to pure bromide and iodide perovskite crystals were developed. Due to the long lifetime and diffusion length of electrons and holes [47, 55], they cover multiple grains before the recombination. The thermolization of photo-generated carriers and their further trapping into I-rich low bandgap regions takes place on the picosecond time scale. With illumination, X-Ray diffraction showed the formation of phases with a larger and a smaller lattice constant [53] while without single-phase state. One of the important experimental results is the observation of long-range migration of halides, which could be linked to J-V hysteresis in solar cells and was observed in other halide perovskites [55]. PL and absorption spectra also demonstrated the reversibility of the segregation before and after light irradiation. Bischak et al. confirmed that segregation of I ions took place at grain boundaries and estimated the size of the iodide-rich clusters to be 8-10 nm in diameter [55]. Hoke et al. found out that the segregation occurred even at low temperatures [53], but for a longer time period. Bischak et al. also showed that mixed I/Br perovskites with well-mixed structure at room temperature undergo demixing transitions as a function of temperature with a critical temperature of 190K [55]. The phase segregation taking in the wide range of temperature follows Arrhenius law with the activation energy similar to halide conductivity activation energies for other halide perovskites [53, 55].

To explain this effect a few theoretical models have been developed [54-58]. Brivio et al. [54] used density functional theory (DFT) to show the thermodynamic characteristics of the solid $MAIPb(Br_xI_{1-x})_3$. Their theory showed that the phase segregation into thermodynamically stable I-rich and Br-rich phases occurred at room temperature upon the provision of necessary energy by illumination. However, the reversibility observed in experiments could not be explained in the frame of the theory. In the comprehensive Bischak model based on molecular dynamic simulations, polarons occur when the free electron and hole generated by light deform the surrounding lattice through electron-phonon coupling. Those polarons funnel into the reduced-band-gap I-rich domains [56]. Their study shows that the unique combination of mobile halides, substantial electron-phonon coupling, and long-lived charge carriers is required for photo-induced phase separation. The authors suggest that the concentrated polaron density switches the shape of free energy from one with one minimum to one with two minima, corresponding to the dark and light states. In this work, we present a mathematical description of the effect based on the Ising model, which was developed for phase transitions in the framework of statistical mechanics approaches.

7.1. The Ising Model

The Ising model enabled a mathematical description of phase transformations in ferromagnetics and antiferromagnetics, the order-disorder transitions in binary alloys and lattice gases [58]. In binary alloys AB there are two species A and B, each lattice site can be occupied by either species. Mapping the alloy onto the Ising model can be made by assigning A species to "up" spin ($\sigma_1 = +1$) and B species to the "down" spin($\sigma_1 = -1$). The Hamiltonian of the Ising model for binary alloys can be described by the general formula

$$E = -\sum_{ll'} J_{ll'}\sigma_l\sigma_{l'} - \mu H \sum_l \sigma_l. \tag{9}$$

For the ferromagnetic, µ is the magnetic moment, J is the exchange integral; H is the external magnetic field. By calculating the number of species A and B and the bonds between them, the total energy takes the following expression [58]:

$$E = -\frac{1}{2}zNJ + 2N_{AB}J - (2N_A - N)\mu H. \quad (10)$$

For binary alloys, we have

$$J = \frac{1}{2}\left(\varphi_{AB} - \frac{1}{2}\varphi_{AA} - \frac{1}{2}\varphi_{BB}\right), \quad (11)$$

$$\mu H = \frac{1}{4}z(\varphi_{BB} - \varphi_{AA}). \quad (12)$$

Here interaction energies between species such as AA, AB и BB are denoted by $\varphi_{AA}, \varphi_{AB}, \varphi_{BB}$. The exchange integral J of the Ising model plays the role of mixing energy, which gives the tendency for the system to keep different species apart ($J > 0$) or together ($J < 0$) giving a uniform mixture. In such binary alloys there is a critical temperature, above of which the lattice is disordered. Below the critical temperature the alloy tends to be ordered either by separation into the domains rich by A or B species($J > 0$), or with sites of one of sublattice mostly occupied by a given species in a regular arrangement($J < 0$). Note that at low temperatures the ferromagnetic Ising model ($J > 0$) gives phase separation into +1 – rich and -1 – rich domains, while in the antiferromagnetic Ising model ($J < 0$) corresponds to antiferromagnetic ordering. By minimization of free energy the relation between the long-range order parameter and the temperature is estimated as

$$\left(\frac{N_A - N_B}{N}\right) = th\left\{\frac{1}{kT}\left(\mu H + \left(\frac{N_A - N_B}{N}\right)zJ\right)\right\}. \quad (13)$$

When the order parameter is equal to ±1, A and B species form separate domains in the lattice. When it is zero, A and B species are mixed in random and there is no long-range order.

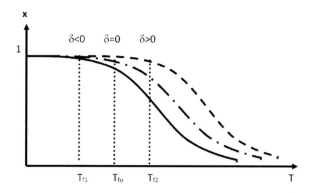

Figure 7. The dependence of ordering at different signs of δ.

We can apply the Ising model developed for binary alloys to the Hoke effect in perovskites described above. The lattice of the Ising model is then related to the lattice, the sites of which are occupied by halide ions. As species A and B introduced in the Ising model we imply two halide anions, iodide and bromide, respectively. In dark conditions, at room temperatures above the critical temperature (190K), the hybrid perovskites can be presented as a lattice with compositional disorder (sites occupied with A or B more and less at random). Under the light, the lattice transits into more ordered state with the rich in A and B phase separate domains, shifting the critical temperature up to higher temperatures. We can account it by adding the term δ responsible for the interaction of electron-hole pair generated by light with the iodide ions fluctuations discussed above into the formula (13):

$$X = th\left(\frac{1}{kT}(\delta + \mu H + zJX)\right). \quad (14)$$

Here $X = \frac{N_A - N_B}{N}$, $\delta = z(\varphi_{AB} - (\varphi_{AA} - \Delta\varphi_{AA}))$, $\Delta\varphi_{AA}$ is the interaction of iodide ions with an electron-hole pair. Since interaction between iodide ions fluctuations and electron (hole) polarons is attractive, and they interact with each other stronger compared to with bromide ions, we have $\delta > 0$, the curve $X(T)$ shifts to the right of room temperatures, to make segregation occurred at room temperatures upon illumination as observed in the experiments [12] (Figure 7).

7.2. The Size of Ordered Domains

It was suggested [55] that the extension of segregated halide atoms rich domains depends upon the extension of the polarons and is in the range of 8-10 nm. Using those data we can evaluate the correlation length ξ of iodide-rich (or bromide rich) domains. The correlation function at large distances R is defined by the formula

$$\Gamma(R) \propto R^{-1}e^{-R/\xi}. \tag{15}$$

The size of the domains L is defined from the expression

$$L^2 = \frac{\int R^2 \Gamma(R) d^3R}{\int \Gamma(R) d^3R}. \tag{16}$$

Taking into account the estimated value of L, we have that ξ is about several nm.

7.3. Synergetics of Perovskite Photosegregation: Intermittency Mode

Among the results on photo-segregation, it is very interesting to note some quasi-transitivity in the change in photoluminescence of mixed perovskites $MAlPbI_xBr_y$ [59]. This poses the problem, at least in principle, to modelly understand this quasi-periodicity within the framework of the phenomenology used. Looking at some kind of synergetics in this phenomenon, we will represent the evolution of the phenomenon of photo-aggregation in the form of a one-dimensional mapping for the long-range order parameter

$$X_{n+1} = \varphi(\mu, X_n), \tag{17}$$

where the index n divides the entire time scale into equal intervals, μ is the set of parameters of the problem. Using a certain renormalization, expression (9) can be rewritten in the form

$$X_{n+1} = \tilde{\varepsilon} + X_n + uX_n^2 - gX_n^3, \qquad (18)$$

which easily can be changed to the differential equation:

$$\frac{dX}{d\tau} = \tilde{\varepsilon} + uX^2 - gX^3, \qquad (19)$$

where τ is the time in a certain normalization, u, g are parameters that take into account both the internal arrangement of the system (u) and its response to the external photoeffect (g). This formulation of the question allows us to successfully apply the laws of nonequilibrium statistical mechanics, in particular to the entrance to the range of variation of the quantity $\rightarrow X_{in}$. This allows us to immediately implement the ideas of synergetics, for example using the Lamerey diagram (Figure 8).

This diagram shows that after entering the ordering process ($X \rightarrow X_{in}$ is the left point on the abscissa axis), the order parameter X slowly changes, which then goes into a quasi-oscillatory turbulent regime with large X drops on each jump. This Lamerey diagram corresponds to the kinetics of intermittency, where laminar and turbulent stages coexist (Figure 9).

Equations (18) and (19) easily allow us to estimate the average time of the laminar regime of an increase in the order parameter $\{X\}$ [60, 61]:

$$\langle \tau \rangle = \frac{1}{\sqrt{\tilde{\varepsilon}u}} arctg \left[\frac{c}{\sqrt{\tilde{\varepsilon}u}} \right]. \qquad (20)$$

Here $c \approx X_{max}$ (Figure 8). Note that for $\frac{c}{\sqrt{\tilde{\varepsilon}u}} \rightarrow 1$ we have:

$$\langle \tau \rangle \approx \frac{\pi}{2\sqrt{\tilde{\varepsilon}u}}. \qquad (21)$$

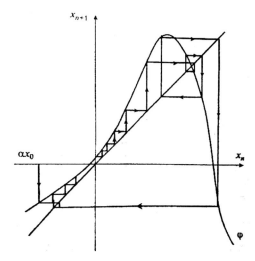

Figure 8. The Lamereya diagram, describing the mode of intermittency of the variation of the long-range order parameter in the photo segregation of mixed perovskites.

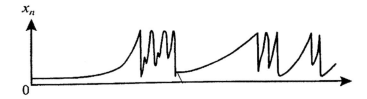

Figure 9. Kinetics of the order parameter for photosegregation of mixed perovskite (conjugation of laminar and turbulent phases of photoluminescence takes place).

Let us now discuss the "physical content" of the mathematics used. Note that the quadratic term X_n^2 in the formula (11) corresponds to such a picture: if there is a spherical region with an intermediate X_n, then on the surface of the sphere (that is, on the boundary between ordered and unordered) the value of the contacts is proportional to X_n^2 (this is reasonable for $uX_n^2 > gX_n^3$). On the other hand, when $X_n > u/g$, the role of photogenerated e-h pairs play the forefront, the interaction of which with ordered regions (which is modeled by g) destabilizes the ordered regions; And this effect is stronger the more X_n.

Thus, the developed synergetic model, corresponding to a certain extent and experiment [59], can point to dynamic chaos in the form of an intermittency regime, which is extremely trivial for a given perovskite system and to some extent supplements the number of examples obtained in other areas (for example, the Belousov-Zhabotinsky reaction [61]).

In conclusion, the Ising model of segregation of halide ions upon illumination in hybrid perovskites allows establishing the relationship between the long-range order and temperature describing the transition between demixed and mixed states with taking into account the interaction of iodide ions fluctuations with an electron-hole pair generated by light. The synergetic theory of segregation is also discussed.

8. Problems of Radiation Degradation of Tandens

Along with the development of the problem of radiation degradation of single junction solar cells, on the basis of hybrid perovskites, the complex of questions of tandems (tandem of type 2T with CZTSSe / PVSK structure, 4% conversion efficiency) was also studied, beginning from 2012. To date, both 2T and 4T tandems have been studied on very diverse systems: PVSK / PVSK; organics / PVSK; organics / PVSK (bilayer system); Silicon / PVSK. The structures of the last type are the most promising given the efficiency of 27%. In principle, tandems were designed to solve several problems. Firstly, spectral losses can be minimized by using two consecutive components with different electronic band gaps ($E_g^{(1)}$; $E_g^{(2)}$). Secondly, there is a possibility of selecting the composition of components to create so-called "Contour maps." Thirdly, it is also possible to choose the electrical connection between the components of the tandem (2T or 4T). As a result of these possibilities, according to De Vos [64] the theoretical limit of efficiency can be sharply increased up to 47%, in contrast, to 33% by Shockley–Queisser [65]. Against this background, the problems of stability and radiation degradation for tandems are still no less acute. The problem of stability and degradation of tandems has required broader research than a single-junction structure. Indeed, one should point out the most important aspects:

- (radiation) degradation of each component of the tandem separately;
- (radiation) degradation of the interfaces between the components;
- the problem of choosing the optimal symbiosis of a heterogeneous structure topology and electrical connection between the most important nodes of this structure; the latter should provide special stability to external influences of the common device;
- development of special methods of radiation annealing of defects at the moment of their inception (in situ).

Some new aspects of the problem of stability of solar cells based on hybrid perovskites are discussed below.

8.1. Features Associated with the Intermediate Layer

Types of intermediate layers can be different upon specific tasks. An important type of degradation occurs when the role of the intermediate layer is reduced to the object where carrier recombination occurs to provide the general neutrality.

As described above, the most effective neutralization (through carrier recombination) can be accomplished at the interfaces by selecting their fractal properties. This is accomplished by translating the sticking levels to the recombination zone. However, radiation-stimulated diffusion turns the fractal interface into the flat interface, which makes neutralization difficult and takes the device out of the desired mode of operation.

8.2. The Processes of Degradation of the Devices as a System of Complexity

The capabilities of various architectures and combinations of the types 2T and 4T allow identifying the parameters, characterizing the device as a whole. This leads to a consideration of some analogies of a complex ecosystem with external influences. As it is well-known, the most common analysis of the current type of ecosystem sustainability is based on network theory. In this theoretical construction, several characteristics appear that allow using the topology to describe the entire system. The most important concepts were the so-called clustering, "close world" and scale-free system. Focusing on our specific problem of degradation (in particular, radiation) of the complex device, we highlight the fact that the scale-free character of all elements and connections between them has a unique property of preserving the functionality of the entire system as long as possible, despite the fact that a large ratio of elements is out of operation. Moreover, the description of the damage threshold with the help of percolation ideas turned out to be adequate, and the percolation threshold of scale-free systems turned out to be very high. Applying to the object under study the following technology strategy can be offered - to build the architecture of electronic communication and the individual parts of the device in such a way to get a scale-free network.

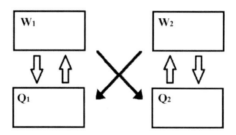

Figure 10. Simplified model of state variables of communicating hierarchical systems for two components of a tandem based on perovskite (1and 2). Dynamics at lower levels Q gives information about its collective properties to upper hierarchical levels; arrows show their direct interaction and cross correlation. Our scheme is an essential reduction of more general standard schemes applied in game theory [70-72].

It is interesting to note that the problem of resistance to external influences for the type of perovskite instruments under study can also be described in the framework of a simpler and generally accepted theory of ecosystems. This approach uses the analogy of ecosystem participants with electrons and holes of two parts of a common tandem. It was possible to trace the analogy for all three types of relationships between ecosystem participants (the predator – prey model, the symbiosis model, the restricted food model) with the electron – ionic processes of the tandem. A characteristic feature of this approach has become a large range of already known models of ecosystems [69], which gives hope for the usefulness of this direction.

It is especially interesting to point out the possibility of using the results of game theory [70]. When a tandem is modeled by the interaction of two hierarchical structures (Figure 10), each of the components of the tandem is a hierarchical (two-step) system (w_1 and Q_1; w_2 and Q_2). Events at each level of both subsystems are described by stochastic differential equations according to the methods of (Itoh, Stratonovich) [71].

From the Figure 10 it is clear that there are several types of interaction (arrows): the upper levels with the lower ones, and with "their own" and aliens; the latter type of interaction is called cross-correlation. The whole analysis is based on the fact that a certain characteristic (Φ) of the whole tandem is selected, which is constructed from the partial characteristics of the components (Φ_1 and Φ_2). There are two variants of these constructions: multiplicative ($\Phi = \Phi_1 * \Phi_2$) and additive ($\Phi = \Phi_1 + \Phi_2$). Further optimization occurs which either maximizes Φ (useful property) or minimizes Φ (when it is a harmful property).

It is interesting to note that as the possible variables for the {Q} and {W} levels, it would be possible to choose the populations of the electronic levels in the forbidden and allowed zones, respectively; at the same time, the cross-correlation interaction would describe the controlling role of the Fermi quasi-levels in both parts of the tandem. This approach allows us to study the role of various noises and has very rich results from the game theory (see [72]).

9. SUMMARY OF RESULTS ON METHODS TO PREVENT RADIATION DEGRADATION OF PEROVSKITES

We can distinguish two aspects of control radiation degradation:

1) controlling of the local electronic spectrum in the forbidden zone;

2) management of geometry and electronic states on the interfaces;

In general, these two aspects are implemented using the following methods:

1) injection annealing (recharging the level of defects);
2) radiation shaking (the liquidation of metastable states by an elastic wave);
3) the fractality of interfaces (the modulation of the width of the allowed zone of valence states);
4) the fractality of the crystal surface (modulation of the width of the allowed zones of Tamm levels);
5) acceleration of quasi-chemical reactions (the realization of reactions between defects through the use of electronic excitations);
6) the possibility of a wide choice of the type of halogen; the effect of changing –increasing- the width of the allowed valence band affects the settled time of the formed holes which results in less probability of defect production via ionization mechanism.
7) the usage of maximum heavy isovalent ionic components of perovskite will be the obstacle for defect formation due to elastic scattering mechanism
8) the use of information laws of game theory allows to obtain the optimal mode of tandems (the interaction of the hierarchical structures of the two components of the tandem allows using the methods of stochastic differential equations to choose the optimal mode for both positive and negative properties of the structure);
9) the organization of the complex structure of the device in the form of no large-scale networks (the essence of the effect is that without large-scale networks they are particularly resistant to destructive effects, as described by percolation theory);

CONCLUSION

Radiation degradation of organic-inorganic perovskites and device structures based on them, like any other physical radiation macro effect, is a combination of three consecutive stages: radiation defect formation, radiation-stimulated diffusion, and quasi-chemical reactions between defects. The implementation of these consecutive stages results in two effects: the electronic spectrum in the band gap and geometry of interfaces change. Both of these effects can dramatically affect the photoelectric parameters of perovskite materials and devices.

Among the processes of defect formation, the most essential is light ionization of the valence and subvalence bands constructed of the orbitals of the iodine components and elastic displacement of iodine anions through the two-stage channel of energy transfer from a fast particle to lattice atoms. For radiation-stimulated diffusion, the most important channels of stimulated displacement of defects are various kinds of recharging of deep-level centers. The effect of light induced smoothing of rough (fractal) interfaces can be due to modulation of the valence-band width by the change in an overlap of Tamm orbitals; this smoothing eliminates the increment in the short-circuit photocurrent discovered for rough interfaces.

It is necessary to point out the results of the research of tandems. The degradation properties depend both on the layer between the components and on the properties of the components themselves. Several original methods for analyzing the radiation degradation properties of tandems based on the game theory approach between two hierarchical structures - components of a tandem (Itoh, Stratonovich stochastic differential equations in the regime of general optimization of tandem properties) and the analysis within the model of scale-free networks have been proposed On the basis of the latter method, it was possible to identify the type of networks modeling tandem structures that are particularly resistive to their degradation.

We note that, in principle, the unique complexity of perovskite instruments and various types of radiation makes it possible to build a "complexity matrix." Horizontally, in such matrix of complexity, properties of perovskites are located, and vertically, various factors of radiation are positioned. The number of elements in such a complexity matrix will indicate the actual possibility of numerous radiation modifications of the device, but at the same time, unfortunately, too many channels of their degradation. With such a unique object, radiation physics encounters for the first time.

We also have to state that the presented research devoted to radiation effects in perovskites is a foreseeing one. Most of the present researchers focus on photo-chemical degradation and they really gained a lot of significant information about electron-ionic processes in different hybrid perovskites [73-76]. Undoubtedly, most of these results will be of particular importance when radiation physics with more energetic irradiation typical of space and earth conditions will be applied to the investigated material. At any rate, the intersection of these areas of physics and chemistry in implementation to hybrid perovskites is bound to bring a lot of unexpected and extraordinary results. Researchers in radiation physics have not worked with such an intriguing object so far.

Let's take off the hat, gentlemen!

ACKNOWLEDGMENTS

In the course of the research in the field of radiation degradation, the authors had the opportunity to discuss many issues with a number of prominent scientists in the field of condensed matter physics. Among them, we would especially like to thank Academician A. F. Andreyev (Russia), professors: A. A. Zakhidov (USA), N. R. Ashurov (Uzbekistan), Kh. B. Ashurov (Uzbekistan), D. U. Matrasulov (Uzbekistan), Dr. S. E. Maximov (Uzbekistan), Dr. D. R. Aristov (Russia), Dr. N. N. Nikiforova for support and useful discussion of key points. In addition, we would like to express our great gratitude to the students of the National University of Uzbekistan Akramova R. B. and Kobiljonov M. O. for assistance in the design of the chapter.

We would like to express our particular gratitude to the editorial board of Applied Solar Energy (Geliotechnika) for the permission to use illustrations from our articles published in this journal.

REFERENCES

[1] Vavilov, V. S., Ukhin, N. A. (1995). Radiation Methods in in Semiconductors and Semiconductor Devices. Springer US. DOI 10.1007/978-1-4684-9069-5.

[2] Special Issue on Radiation Effects in Semiconductor Devices, Editors, Hugh Barnaby, Simone Gerardin, Véronique Ferlet-Cavrois. (2017). IOP Publishing Ltd, Semiconductor Science and Technology, Volume 32, Number 8.

[3] Bourgoin, J. C. and Khirouni, K. (2013). Evaluation of the degradation of solar cells in space. *Journal of Space Exploration,* 2 (3): 165–169.

[4] Claeys, C, Simoen, E. (2002). Radiation Effects in Advanced Semiconductor Materials and Devices. Springer-Verlag Berlin Heidelberg.

[5] Oksengendler, B. L. and Turaeva, N. N. (2006). Radiation Physics for Condensed Media. V. 1. Tashkent: Fan (in Russian).

[6] Djurabekova F., Ashurov Kh. B., Maksimov S. E., Turaeva, N. N., Oksengendler B. L., (2013), Fundamental processes of radiation modification of semiconductor nanostructures, *Physica Status Solidi C*, 10: 685-688, DOI 10.1002/pssc. 201200751.

[7] Itoh, N. and Stoneham, A. M. (2001). Materials Modification by Electronic Excitation. Cambridge Univ. Press.

[8] Mitzi, D. B. (2004). Solution-processed inorganic semiconductors. *Journal of Material Chemistry,* 14: 2355.

[9] Frost, J. M., Butler, K. T., Brivio, F., et al. (2014). Atomistic orogins of high-performance in hybrid halide perovskite solar cells. *Nano Letters*, 14: 2584.

[10] Yin, W. J., Shi, T., and Yan, Y., (2015). Halide perovskite materials for solar cells: a theoretical review. *Journal of Material Chemistry*, 3: 8926–8942.
[11] Giorgi, G. and Yamashita, K. J. (2015). Organic-inorganic halide perovskites: an ambipolar class of materials with enhanced photovoltaic performances. *Material Chemistry A*, 3: 8981.
[12] Ashurov, N. R., Oksengendler, B. L., Rashidova, S. Sh., and Zakhidov, A. A. (2016). State and prospects of solar cells based on perovskites. *Applied Solar Energy*, 52: 5.
[13] Ginzburg, D. M. Ed. (1990). Physical Properties of High Temperature Semiconductors. World Sci.
[14] Karimov, Z. I. and Oksengendler, B. L. (1995). The Theory of Radiation-Physical Processes in High-Temperature Semiconductors. Tashkent: Ukituvchi (in Russian).
[15] Bube, R. H. (1960). Photoconductivity of Solids. New York - London: J. Willy & Sons Inc.
[16] Ansel'm, A. I. (1978). Introduction to Semiconductors Theory. Moscow: Nauka (in Russian).
[17] Phillips, J C. and Lucovsky, G. (2009). Bonds and Bands in Semiconductors. 2nd ed. Momentum Press.
[18] Bohm, D. (1989). Quantum Theory. Dover.
[19] Zaikovskaya, M. A., Oksengendler, B. L., Tokhirov, K. R., and Yunusov, M. S. (1977). Subthreshold defect production in silicon, in *Proc. Radiation Effect in Semiconductors*. Dubrovnik, Yugoslavia. 6-9 September, 1976. Ed. by Urli, N. B. and Corbett, J. W. *Institute of Physics Conference Series*, Bristol-London, 31: 279–283.
[20] Varley, J. (1962). Discussion of some mechanisms of F-centre formation in alkali halides. *Journal of Physics and Chemistry Solids*, 23: 985–1005.
[21] Dexter, D. (1960). Varley mechanism for defect formation in alkali halides. *Physical Review*, 118 (4): 934–935.
[22] Vinetskii, V. L. and Kholodar,' V. L. (1969). Statistical Interaction of Electrons and Defects in Semiconductors. Kiev: Naukova dumka (in Russian).

[23] Oksengendler, B. L. and Yunusov, M. S. (1975). Charged state longevity and subthreshold defect formation in solids. *Doklady Akademii Nauk Uzb SSR,* 34(6): 25–27.

[24] Yunusov, M. S., Zaykovskaya, M. A., Oksengendler, B. L., et al. (1976). Subthreshold defect production in silicon. *Physica Status Solidi A,* 35: 145–149.

[25] Oksengendler, B. L., Maksimov, S. E., and Marasulov, M. B. (2015). Degradation of perovskites and Dexter-Varley paradox. *Nanosystems: Physics, Chemistry, Mathematics,* 6 (6): 825–832.

[26] Oksengendler, B. L. (1990). Mechanisms and topology of radiation-stimulated atomic processes in solids, Extended Abstract of Doctoral Sci. (Phys.-Math.) Dissertation, Tashkent: Institute of Nuclear Physics AS RUz.

[27] Schoonman, R. (2015). Organic-inorganic lead halide perovskite solar cells materials: a possible stability problem. *Chemical Physics Letters,* 619: 193–195.

[28] Jemli, K., Diab, H., Ledee, F., et al. (2016). Using low temperature photoluminescence spectroscopy to investigate $CH_3NH_3PbI_3$ hybrid perovskite degradation. *Molecules,* 21: 885–897.

[29] Vinetskii, V. L. and Kholodar,' G. A., (1979). Radiatsionnaya fizika poluprovodnikov (Radiation Physics for Semiconductors), Kiev: Naukova dumka.

[30] Du, M. H. (2014). Efficient carrier transport in halide perovskites: theoretical perspectives. *Journal of Material Chemistry A,* 2: 9091–9093.

[31] Cheng, Y. and Mac Key, J. W. (1968). Subthreshold Electron damage in n-type germanium. *Journal of Physical Review,* 167: 745.

[32] Oksengendler, B. L., Turaeva, N. N., Maksimov, S. E., and Dzhurabekova, F. G. (2010). Peculiarities of radiation-induced defect formation in nanocrystals embedded in a solid matrix. *Journal of Experimental and Theoretical Physics,* 111 (3): 415.

[33] Zheng, L., Ma, Y., Chu, S., et al. (2014). Improved light absorption and charge transport for perovskite solar cells with rough interfaces by sequential deposition. *Nanoscale,* 6: 8171–8176.

[34] Oksengendler, B. L., Ashurov, N. R., Maksimov, S. E., et al. (2016). Role of Fractals in Perovskite Solar Cells. *Eurasian Chemical-Tecnological Journal,* 18 (4): 55–60.

[35] Davison, S. G. and Levine, J. D. (1970). Surface states, in Solid State Physics, Ehrenreich, F. and Turnbull, S. D., Eds., New York, London: Acad. Press 25.

[36] Oksengendler, B. L. and Turaeva, N. N. (2010). Surface Tamm states at curved surfaces of ionic crystals. *Doklady Physics,* 55: 477–479.

[37] Yadav, R. P., Manvendra K., Mittal, A. K., and Pandey, A. C. (2015). Fractal and multifractal characteristics of swift heavy ion induced self-affine nanostructured BaF_2 thin film surfaces. *Chaos,* 25: 083115(1-9).

[38] Maksimov, S. E., Ashurov, N. R., and Oksengendler, B. L. (2016). in Mater. 14-i Mezhdunar. nauch. tekhn. konf. "Bystrozakalennye materialy i pokrytiya" November 29-30 2016 MAI (Proc. 14th Int. Sci. Techn. Conf. "Fast-Quenched Materials and Coatings", Moscow Aviation Institute, Nov. 29–30, 2016), Moscow: Pro- bel-2000: 221–225.

[39] Bhattacharjee, B. S., Goswami, D. K., et al. (2003). Nanoscale self-affine smoothing by ion bombardment and the morphology of nanostructures grown on ion-bombarded surfaces. *Nuclear Instruments and Methods Physics Research B,* 230: 524– 532.

[40] Goswami, D. K. and Dev, B. N. (2003). Nanoscale self-affine surface smoothing by ion bombardment. *Physical Review B,* 60: 033401.

[41] Hohenberg, P. and Kohn, W. (1964). Inhomogeneous electron gas. *Physical Review B,* 136 (3): 864–871.

[42] Kojima, A., Teshima, K., Shirai, Y., and Miyasaka, T., (2009). Organometal halide perovskite as visible-light sensitizers for photovoltaic cells. *Journal of American Chemical Society,* 31: 6050.

[43] Polman, A., Knight, M., Garnett, E. C., et al. (2016). *Photovoltaic Materials: Present Efficiencies and Future Challenges. Science,* 352: 4424.

[44] Green, M. A., Emery, K., Hishikawa, Y. et al. (2016). Solar cell efficiency tables (version 48). *Progress in Photovoltaics,* 24: 905−913.

[45] Yin, W. J., Shi, T., Yan, Y. (2014). Unusual Defect Physics in $CH_3NH_3PbI_3$ Perovskite Solar Cell Absorber. *Applied Physics Letters,* 104: 063903.

[46] Steirer, K. X., Schulz, P., Teeter, G., et al. (2016). Defect Tolerance in Methylammonium Lead Triiodide Perovskite. ACS Energy Letters 1: 360−366.

[47] Stranks, S. D., Eperon, G. E., Grancini, G., et al. (2013). Electron-Hole Diffusion Lengths Exceeding 1 Micrometer in an Organometal Trihalide Perovskite Absorber. *Science* (Washington, DC, U. S.), 342: 341−344.

[48] Saliba, M., Matsui, T., Seo, J. Y., et al. (2016). Cesium-Containing Triple Cation Perovskite Solar Cells: Improved Stability, Reproducibility and High Efficiency. *Energy and Environmental Science,* 9: 1989−1997.

[49] Ahn, N., Son, D. Y., Jang, I. H., et al. (2015). Highly Reproducible Perovskite Solar Cells with Average Efficiency of 18.3% and Best Efficiency of 19.7% Fabricated via Lewis Base Adduct of Lead(II) Iodide. *Journal of American Chemical Society,* 137: 8696−8699.

[50] Zhou, Y., Yang, M., Pang, S., et al. (2016). Exceptional Morphology-Preserving Evolution of Formamidinium Lead Triiodide Perovskite Thin Films via Organic-Cation Displacement. *Journal of American Chemical Society,* 138: 5535−5538

[51] Yi, C., Luo, J., Melon, S., et al. (2016). Entropic Stabilization of Mixed A-Cation ABX 3 Metal Halide Perovskites for High Performance Perovskite Solar Cells. *Energy and Environmental Science,* 9: 656−662.

[52] Noh, J. H., Im, S. H., Heo, J. H., et al. (2013). Chemical Management for Colorful, Efficient, and Stable Inorganic- Organic Hybrid Nanostructured Solar Cells. *Nano Letters,* 13: 1764− 1769.

[53] Hoke, E. T., Slotcavage, D. J., Dohner, E. R., et al. (2015). Reversible Photo-Induced Trap Formation in Mixed-Halide Hybrid Perovskites for Photovoltaics. *Chemical Science,* 6: 613−617.

[54] Bischak, C. G., Hetherington, C. L., Wu, H., et al. (2016). Origin of Reversible Photo-Induced Phase Separation in Hybrid Perovskites. arXiv 2016, 1–6.

[55] Slotcavage, D. J., Karunadasa, H. I., McGehee, M. D. (2016). Light-induce phase segregation in halide perovskite absorbers. *ACS Energy Letters*, 1: 1199-1205.

[56] Wright, A. D., Verdi, C., Milot, R. L., et al. (2016). Electron–phonon Coupling in Hybrid Lead Halide Perovskites. Nature Communication DOI: 10.1038/ ncomms11755.

[57] Neukirch, A. J., Nie, W., Blancon, J. C., et al. (2016). Polaron Stabilization by Cooperative Lattice Distortion and Cation Rotations in Hybrid Perovskite Materials. *Nano Letters*, 16: 3809– 3816.

[58] Ziman, J. M. (1979). Models of disorder. Cambridge University Press.

[59] Gets, D. S., Tiguntseva, E. Yu, Zakhidov, A. A., et al. (2018). Photoinduced ions migration in optically resonant perovskite nanoparticles. *Pisma JETF* (in press).

[60] Oksengendler, B. L., Maksimov, S. E., Turaev, N. Yu. (2016). Synergetics of Catastrophic Failure of Semiconductor Devices under High-Energy Ion Irradiation. *Surface X-ray, synchrotron and neutron studies*, 4: 1-5.

[61] Schuster, H. G. (1984). Deterministic Chaos, Physik-Verlang Weinheim.

[62] Lal, N. N., Dkhissi, Y., Li, W. et al. (2017). Perovskite Tandem Cells. *Advanced energy Materials*, 1602761: 1-18.

[63] Lee, J. W., Hsieh, Y. T., Marco, N. De et al. (2017). Halide Perovskites for Tandem Solar Cells. *The Journal of Physical Chemistry Letters*, (8) 1999-2011.

[64] De Vos. (1980). Detailed balance limit of the efficiency of tandem solar cells. The *Journal of Physical Chemistry Letters*, 13 (5): 839.

[65] Shockley, W., Queisser, H. J. (1961). Detailed balance limit of efficiency of p-n junction solar cells. *Journal of Applied Physics*, 32 (3), 510-519.

[66] Pennekamp, F., Pontarp, M., Tabi, A. et al. (2018). Biodiversity increases and decreases ecosystem stability. *Nature*: 17 October 2018.

[67] Albert, R., Barabasi, A. L. (2002). Statistical mechanics of complex networks. *Reviews of Modern Physics*, 74: 47-97.

[68] Broadbent, S. R., Hammerslay, J. M. (1957). Percolation processes. I. Crystal and mazes. *Proceedings of Cambridge Philosophical Society*, 53 (5): 629.

[69] May, R. M. Will a large complex system be stable? Nature, 239: 413-414.

[70] Rapoport, A. (1966). Two-Person Theory, The Essential Ideas. University of Michigan Press, Ann Arbor.

[71] Non-Linear Transformation of Stochastic Processes (1965). Eds. P. I. Kuznetsov, R. L. Stratonovich, V. I. Tikhonov, Pergamon, London.

[72] Nicolis, J. S. (1980). Dynamics of Hierarchical System. Springer-Verlag.

[73] Correa-Baena, J. P., Saliba, M., Buonassisi, T., Grätzel, M., Abate, A. Tress, W., Hagfeldt. A. (2017). Promises and challenges of perovskite solar cells. *Science*, 358:739-744.

[74] Seok, S. I., Grätzel, M., Park, N. G. (2018). Methodologies toward Highly Efficient Perovskite Solar Cells, Small. Nano. Micro doi.org/10.1002/smll.2017 04177.

[75] Zhang, F., Bi, D., Pellet, N., Xiao, Ch., Li, Zh. Berry, J. J., Zakeeruddin, Sh. M., Zhu, K. Grätzel, M. (2013) Suppressing Defects Through Synergistic Effect of Lewis Base and Lewis Acid for Highly Efficient and Stable Perovskite Solar Cells. Energy & Environmental Science. DOI: 10.1039/C8EE02252F.

[76] Gregori, G., Yang, T. Y., Senocrate A., Graetzel, M., Maier J. (2016) Ionic Conductivity of Organic–Inorganic Perovskites: Relevance for Long-Time and Low Frequency Behavior. In book: Organic-Inorganic Halide Perovskite Photovoltaics. DOI: 10.1007/978-3-319-35114-8-5.

In: Perovskite Solar Cells
Editor: Murali Banavoth
ISBN: 978-1-53615-858-8
© 2019 Nova Science Publishers, Inc.

Chapter 3

PEROVSKITE: MATERIAL AND DEVICE OPTIMIZATION FOR SOLAR CELL APPLICATIONS

Antonio Frontera[1], Yaroslav Martynov[2], Rashid Nazmitdinov[3], and Andreu Moìa-Pol[4]*

[1]Departament de Química, Universitat de les Illes Balears, Palma de Mallorca, Spain
[2]Research and Production Corporation "Istok," Fryazino, Russia
[3]BLTP JINR and Dubna University, Dubna, Russia
[4]Departament de Fisica, Universitat de les Illes Balears, Palma de Mallorca, Spain

ABSTRACT

The first goal of this chapter is to discuss various ways to improve the efficiency and stability of perovskite structure by means of chemical composition of its constitutents. In particular, it describes the impact of either changing the anion, the cation or both upon several physical

---

* Corresponding Author's Email: rashid@theor.jinr.ru.

properties with respect to the standard methylammonium lead triiodide. Moreover, it also analyzes the influence over the power conversion efficiency of the substitution of lead by other elements, either from the same group or from adjacent groups. The second goal of this chapter is introducing from the first principles of semiconductor physics a phenomenological model. It provides some recommendations for improving the efficiency of a perovskite based photovoltaics element.

Keywords: mixed organic-inorganic perovskites, inorganic perovskites, photovoltaics, anion/cation substitution, current-voltage characteristics, power conversion efficiency, perovskite solar cell

1. INTRODUCTION

Perovskite solar cells (PSCs) have reached an unprecedented rate of ~ 23% power conversion efficiency in less than one decade, and have generated an over-flow of high impact work in the literature studying the properties of these promising materials. It is well-known that the highest performance has been observed for the methylammonium lead triiodide perovskites. Unfortunately, their low tolerance to moisture and rapid degradation when exposed to UV light and high temperature are the two major obstacles to their commercialization. Recently, the development of mixed halide hybrid perovskites where I^- is substituted with Br^- and Cl^- has given a new impulse to PSC research that is highlighted in the first part of Section I. An additional drawback to the development of PSCs is the lead toxicity that also obstructs the commercialization of this technology. Consequently, recent efforts have been devoted to investigate lead-free halide perovskite nanocrystals (NCs), as described in the second part of Section I. Moreover, we discuss our perspectives on the most exciting achievements in the materials design and photophysical properties of lead-free perovskite NCs paying special attention to the features of halide perovskite NCs; the adequate elements to replace lead and recent demonstrations of utilizing lead-free perovskite NCs in light-emitting devices. Finally, in the third part of Section I, we provide a short overview on mixed cation hybrid perovskites, where methylammonium is substituted by other cations like cesium or formamidinium. We discuss the optical characteristics of mixed cation perovskites and approaches to lower nonradiative recombination with cation substitution.

In Section II of this chapter we focus our attention on the analysis of the efficiency of perovskites within the physical model. In the first part of Section II, we discuss the main physical principles of transport properties of the carriers in the perovskite thin film. We will present the approach that starts from basic semiconductor physics and builds up a transport model with kinetic coefficients that are functions of carrier energy densities. We will treat electron and holes in collecting materials explicitly. This will allow us to trace the influence of permittivities of the p-i-n junction solar cell on the power conversion efficiency (PCE). Further efforts will be concentrated on the optimization of the triple of materials – electron transport material (ETM), perovskite and hole transport material (HTM), and also their geometry. Finally, taking into account all these factors, we will provide our estimations on the efficiency increase of the perovskite SC PCE.

1.1. Prelimiaries

Perovskite material takes its name from the Russian mineralogist L. A. Perovski [1]. It has a particular crystal structure with the ABX_3 formula where A is an organic cation located at the corners of the unit cell, B is the metal cation situated at the center, and X denotes the halide anion in the six-face center (see Figure 1). In spite of the fact that the most investigated perovskite are oxides due to their ferroelectricity properties, methylammonium (MA) lead iodide perovskite has attracted vast attention worldwide for photovoltaic and optoelectronic applications [2]. The pioneering work of Miyasaka and coworkers demonstrated that the three-dimensional (3D) perovskite $CH_3NH_3PbX_3$ as an inorganic sensitizer in their dye-sensitized solar cells exhibited 3.1% efficiency for bromide (X = Br) and 3.8% for iodide (X = I) [3]. In only two years, the PCE was doubled simply controlling the perovskite deposition parameters [4]. One year later, the substitution of the liquid electrolyte by a solid-state HTM increased the PCE to 9.7% [5]. Moreover, the introduction of a nonconducting mesoporous Al_2O_3 scaffold in a mixed halide perovskite $CH_3NH_3PbI_{3-x}Cl_x$ increased the PCE over 10% [6]. An intensive continuation of this line of research by the scientific community led to a device optimization using mixing of organic cations [7, 8]. It allowed to obtain the record PCE of hybrid PSCs, which reached 22.7% in 2017 slightly improving the 22.1% achieved in 2015 [9] using a mixture of formamidinium lead iodide ($CH(NH_2)_2PbI_3$) with 5% of methylammonium lead bromide ($CH_3NH_3PbBr_3$) [10, 11]. Surpassing 22% PCE by leveraging the chemical flexibility of the perovskite structure brought hybrid PSCs to be equal in efficiency to most polycrystalline solar absorbers, exceeded only by single-crystal silicon and III–V technologies.

To really use hybrid perovskites not only for solar cells [12, 13], but also light-emitting diodes [14], lasers [15], and photo- detectors [16], robust functional thin films should be constructed with high reproducibility. In general, hybrid perovskites exhibit electronic and chemical instability due to anomalous electronic hysteresis [17], and rapid decomposition. The latter occurs readily under exposure to thermal heat [18], UV or visible light [19], and moisture [20] thus presenting a critical issue for the utilization of hybrid perovskites for optoelectronics [21-23].

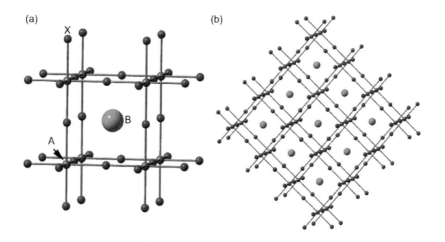

Figure 1. Structure of a ABX_3 cubic perovskite.

The first part of this chapter is devoted to analysis of the chemical flexibility of halide perovskites. In particular, we discuss the role of local variations in perovskite materials by increasing chemical complexity, progressing from effects seen in single halide to mixed halide materials and finally to a mixed A-site, mixed halide chemistries.

The recent investigations reveal a critical link between the local chemistry within perovskite thin films and charge collection. It must be controlled to develop robust, high-performance hybrid perovskite materials for optoelectronic devices, which are also described herein.

2. MATERIAL OPTIMIZATION

2.1. Halide Substitution

2.1.1. Chloride Perovskites

Several works have investigated the incorporation of chloride into the perovskite crystal structure [24-26]. It has been demonstrated a significant improvement in the conversion efficiency for organolead mixed halide $CH_3NH_3PbI_{3-x}Cl_x$ devices compared to that of a pure CH_3NHPbI_3 device [27]. The presence of chloride in the organolead mixed halide system gives more freedom to control the crystallization process, and, therefore, the quality and the morphology of perovskite film [2,28]. As a matter of fact, Stranks and coworkers [29] achieved a long-balanced electron-hole diffusion length exceeding 1 μm. In addition, Zhao and co-workers constructed the tri-iodide $CH_3NH_3PbI_3$ PSC by incorporating methylammonium chloride in a mixture of PbI_2/CH_3NH_3I [30]. Interestingly, the $CH_3NH_3PbI_3$ perovskite achieved an enhancement in the PCE from 2% to 12% in the presence of CH_3–NH_3Cl, thus suggesting that it greatly influences the kinetic of the crystallization process.

Two different techniques have been developed to integrate chloride ion into the perovskite film. One consists of dipping a PbI_2 film into a mixture of CH_3NH_3I and CH_3NH_3Cl solution [31]. The other one uses ammonium chloride (NH_4Cl) as a chloride source [32]. It has been shown experimentally that these approaches provide smooth morphology, improve the optical absorption and lifetime of the photo-excited excitons. Moreover, chloride ion enhances light harvesting, minimizes the series resistance of the device, and maximizes the shunt resistance of the solar cell.

It has been also demonstrated that the incorporation of chloride ions in the mixed iodide-bromide organolead perovskite [33] yields a more uniform and dense film, which is responsible for the tremendous kinetic alteration during the formation of trihalide intermediate including all the halide precursors. The explanation for this interesting behavior is still under discussion in the literature [34-38]. Moreover, it was shown that the long electron-hole diffusion length in the mixed chloride- iodide perovskite film is due to the significant improvement in the formation energy of the interstitial defects [39]. Therefore, the addition of an appropriate amount of chloride ions in the synthesis of $CH_3NH_3PbI_3$ perovskite improves the crystallization of perovskite film, its morphology and surface coverage.

2.1.2. Bromide Perovskites

One of the most investigated lead bromide perovskite is $CH_3NH_3PbBr_3$ ($MAPbBr_3$) [40-44], which is a good candidate as an absorber material for solar cells [45-53]. The electronic structure of this perovskite has been obtained experimentally by using photoemission measurements of isolated NCs in a synchrotron [54], and the optical band gap of 2.34 eV (530 nm) was obtained by the UV–vis absorption spectroscopy. By using the impulsive vibrational spectroscopy (IVS), Batignani et al., [55] have unveiled the phonon spectra of the ground and excited electronic states in $MAPbBr_3$ polycrystalline thin films. It should be mentioned that the understanding of the correlation between the phonon modes and electronic excitations is one of the major challenges in this field. The IVS [56-58] is a powerful technique without spectral limitations and, remarkably, without artefacts arising from elastic pump-scattering and photoluminescence. Using this technique, Batignani et al., revealed the key phonon modes, generated via the displacive excitation mechanism, and evidenced the polaronic nature of photo-excitation in methylammonium lead-bromide perovskite.

MAPbBr$_3$ is also the most studied perovskite for LED applications due to both its easy thin-film fabrication and its green emission. However, the low photoluminescence quantum yield of MAPbBr$_3$ limits its use in efficient light-emitting applications [59]. A convenient strategy to improve the emission in this material is to control the crystal growth by reducing the grain size to the nanoscale [60-62]. The physical mechanisms that control the optoelectronic performance provoked by the perovskite morphology are still under discussion. The free carriers [63], excitons [64], and their interplay are important players, and they have been studied at high carrier densities [65] and at low temperature [66]. In this context, the dependence of the photoluminescence on the excitation density needs to be investigated to understand the origin of the photoluminescence [67,68]. In this sense, Droseros et al., [59] have elucidated the photoluminescence properties of MAPbBr$_3$ perovskite thin films with different morphologies and crystal grain sizes by steady-state and time-resolved spectroscopy. They compared a polymer–MAPbBr$_3$ (50 ± 25 nm) composite [69], and (iii) a small molecule–MAPbBr$_3$ blend with crystal size in the weak quantum confinement regime (9.1 ± 1.6 nm). It was possible to demonstrate experimentally the coexistence of both free carriers and excitons in the polycrystalline perovskite, and monitor their interaction over a broad range of excitation densities. The increase of photoluminescence with a crystal size reduction was due to a bright excitonic population, even at low excitation density, and the reduction of the surface traps by passivation induced by the additives.

Organic-inorganic hybrid PSC has a well-known challenging problem related to crystal destruction upon persistent moisture and heating. Moreover, some solutions related to the use of state-of-the-art HTMs (layers) or noble metals are unacceptable solutions due to the increase in the solar cell cost for potential commercialization [70, 71]. One possible solution to this impasse is to replace organic methylammonium (MA) or formamidium (FA) species with inorganic Cs$^+$ ion, and substituting I$^-$ with Br$^-$. The resultant inorganic CsPbX$_3$ (X = I, Br) halides are featured with remarkable stability against heat and humidity, comparable light absorption ability and high carrier mobility [72-74]. Therefore, inorganic CsPbBr$_3$ solar cells with classical configuration of FTO/TiO$_2$/CsPbBr$_3$/carbon have attracted considerable attention due to their exceptional long-term stability [75]. However, the solar-to-electrical conversion ability is poorer compared to that of hybrid devices. Consequently, an urgent objective is to enhance the PCE of inorganic CsPbBr$_3$ to promote commercialization. The serious charge-carrier trap state within solar cells, including perovskite films and interfaces, limits the further improvement of photovoltaic performance. It has been proposed that the reduction of the interfacial energy-level differences and passivating interfacial defect would provide possibility to reduce a space charge accumulation and enhance a charge extraction [76-79].

The fully inorganic perovskite NCs (CsPbX$_3$) with luminous characteristics similar to those of hybrid organic-inorganic perovskites (MAPbX$_3$) have been extensively developed recently [80-85]. In spite of the high brightness of CsPbX$_3$ NCs, they have some drawbacks related to their instability and anion-exchange reaction [86-89] that complicates their practical application. In this sense, Song et al., reported the covering of green-emitting CsPbBr$_3$ NCs with ethyl cellulose, resulting in better stability [90-92]. However, it is complicated to control the thickness of the film and to stack the film layer by layer. A similar strategy was followed by Sun et al., [93] who synthesized silica-coated inorganic perovskite. It exhibited excellent photoluminescence properties and high quantum yield [93]; however, the deficient uniformity of the composite is also inconvenient for their fabrication. Recently, it has been reported [94] a simple and easy polymeric encapsulation method for green-emitting CsPbBr$_3$ NCs based on oxidized polyethylene wax, that keeps suitable barrier properties (i.e., low water vapor and oxygen transport rate) [95]. The CsPbBr$_3$ perovskite NCs have in a polyethylene microcapsule outstanding luminous characteristics, narrow emission width, high photoluminescence quantum yield, and thermal stability. In fact, compared with the photoluminescence quantum yield of uncovered CsPbBr$_3$ perovskite NCs, the polyethylene microcapsule enhances the yield to 40%. Therefore, this composite material is a promising candidate to build luminescence material for application in optoelectronic devices.

2.1.3. Mixed Bromide/Chloride Perovskites

Solid-state illumination from white-light-emitting diodes has significant advantages, such as higher efficiency, lower energy consumption, and safer application, relative to traditional light sources [96, 97]. Basically, two methods can be used to produce white light for emitting diodes: blue-light-emitting diodes, exciting a yellow phosphor [98-101]; and an ultraviolet light-emitting diode, exciting red, green, and blue phosphors [102, 103]. However, the first method has limitations related to the low color rendering index for the narrow emission band [104]. The second method causes efficiency losses and poor color purity related to the phosphor degradation [105]. Moreover, the phosphor, used in white-light-emitting diodes, are doped by rare-earth ions (Eu^{3+} or Ce^{3+}), and require very high temperature for the synthesis, increasing the production cost [106]. Accordingly, designing new white-light-emitting materials with broadband emission, excited by UV easy to synthesize is very important for their application [107-110]. It is worth mentioning, that obtaining a single-component broadband white-light-emitting is very challenging, and only a few successful examples are known [111-113].

Recently, different types of 2D layered perovskites (X = Cl and X = Br structures), were found to be white-light emitters [114-116], attributed to the self-trapped exciton resulting from the strong electron–phonon coupling in a deformable lattice. For instance, two families of 2D layered perovskites have been reported by Karunadasa et al., [117, 118]. They are based on organic cations: N-methyl-ethane-1,2-diammonium and 2,2'-(ethylenedioxy)bis(ethylammonium), with moderate photoluminescent quantum yields (0.5–1.5% for the first cation and 9% for the second one, respectively). Interestingly, a series of mixed-halide 2D layered perovskites $(C_6H_5C_2H_4NH_3)_2PbBr_xCl_{4-x}$ was prepared (using diffusion or precipitation methods), which exhibit white emission with high quantum yields up to 16.9% [119]. More interestingly, the emission can be adjusted from "warm" white light to "cold" white light through changing the loading amount of Br. This mixed-halide (Cl/Br) perovskite powders exhibit tunable white electroluminescent emission with very high color rendering index of 87–91. Steady-state fluorescent spectra at room and low temperature, as well as transient fluorescent characteristics, were studied to explore the possible white-emissive mechanism.

The phenylethylammonium cation has been also used for the preparation of a 2D mixed Cl/Br halide perovskite film of formula $(C_6H_5C_2H_4NH_3)_2$ $PbCl_2Br_2$ with equivalent quantities of Cl and Br [120]. The light-emitting diodes, based on 2D organic-inorganic mixed halide perovskite, exhibited bluish white electroluminescence at room temperature. Temperature and power density dependent photoluminescence measurements revealed that the white emission band consists of a free exciton and a self-trapped exciton transition. This perovskite light-emitting diode device was fabricated with a high turn-on voltage of 4.9V and a maximum luminance of ~70 cd/m^2 at 7V. This investigation certainly provides an approach to achieve enhanced white light emitting diodes, using 2D mixed halide perovskite materials.

As commented above (Section 2.2), inorganic cesium lead halide perovskite NCs have outstanding photoelectric properties and facile chemical tunability of the band gap [121-123]. In fact, CsPbX$_3$ based NCs are promising composites to substitute II–VI and III–V semiconductor quantum dots in various applications. For example, in the fabrication of low-threshold optically pumped lasers [124, 125], highly efficient light-emitting diodes and detectors [126, 127].

Many efforts have been also made to improve the stability and tune the optical properties of perovskite nanocomposites [128,129]. Recently, perovskite NCs or quantum dots, doped with transition metal ions, have been explored [130-133]. The transition metal dopants not only replace Pb^{2+} in perovskite NCs but also afford a way to introduce new optical, electronic, and magnetic properties [134]. In fact, the formation energies of CsPb$_{1-x}$X$_3$:xMn perovskite NCs are enhanced compared to pure ones (CsPb$_1$X$_3$) due to doping of Mn^{2+}. It results in improving their poor thermal stability [135]. However, compatible band alignment and efficient host exciton energy transfer to Mn^{2+} have not been achieved in CsPbBr$_3$. The possible reason is the interplay among several competing processes for a band edge emission and Mn^{2+} emission, including BE electron–hole recombination [136-138]. Remarkably, the introduction of chloride in the structure improves the properties of this Mn^{2+} doped material. In particular, perovskite

NCs of formula $CsPb_{1-x}(Cl_yBr_{1-y})_3:xMn^{2+}$ have been produced with strong Mn^{2+} emission via ion exchange engineering. This method takes advantage of the different rates between anion and cation exchange in lead halide perovskite NCs. The rate of anion exchange is controlled by the solubility of the halide precursors. Indeed, in highly soluble precursors the anion exchange is produced within a few seconds, whereas less soluble salts significantly decrease the exchange rate [139, 140]. These mixed Cl/Br perovskite NCs have been used successfully in the fabrication of a prototype white light-emitting diode device, based on a commercially available 365 nm LED chip [141]. This strategy to synthesize Br/Cl hybrid perovskites, based on the relative solubility of chloride/bromide salts, opens up new opportunities to fabricate doped halide perovskite NCs with diverse composition and tunable photoluminescence.

2.1.4. Mixed Bromide/Iodide Perovskites

All inorganic lead halide perovskite absorbers ($CsPbX_3$) were developed to solve the instability problems inherited from the cationic organic component [142-147]. It is well-known that the charge carrier mobilities of $CsPbX_3$ perovskites are similar to those reported for the hybrid perovskites, suggesting that the organic cation is not crucial conferring the fundamental advantage of perovskites [148]. Actually, the $CsPbX_3$ perovskites do not generate volatile decomposition products at elevated temperatures or under illumination. Therefore, they are intrinsically more stable than hybrid perovskites [149,150]. In particular, the most phase stable inorganic perovskites, based on $CsPbBr_3$ [89,151-153] and $CsPbBr_2I$ [154-156], are wide band gap semiconductors. Such perovskites are inadequate for harvesting light in the long wavelength spectrum (low solar cell efficiency). Moreover, the pure $CsPbI_3$ perovskite is unstable in the black perovskite phase and decomposes into a non-perovskite yellow phase even at room atmosphere [157, 158].

Decreasing size [159, 160] and doping [161] are useful methods for stabilizing its black phase, but the stability of $CsPbI_3$ is not totally satisfactory. The $CsPbBrI_2$ perovskite exhibits advantages in both its thermal stability and suitable band gap (1.91 eV). In particular, an all vacuum deposited CsPbBrI$_2$ device displays the PCE as high as 11.8% with acceptable device stability [162]. However, the utilization of easier solution deposition lowers the PCE to 10% [163,164]. Furthermore, the incorporation of strontium (Sr) cations (to generate a Sr-enriched surface on the CsPbBrI$_2$ film) increases again the PCE up to 11.3% [165]. Despite these progress, the PCE is still too low, leaving much room for further improvement.

In fact, by means of band alignment engineering, by gradually profiling the interface dimensionality, it is possible to improve the device performance. Using a 3D–2D–0D dimension profiled interface, one introduces an electric field in the rear side of the device that promotes carrier extraction [166]. This increases the current, and the interface allows for greater electron and hole quasi-Fermi level splitting at the maximum power point. It results in triggering the best performing device with the PCE record of 12.39% for the inorganic CsPbBrI$_2$.

Remarkably, all-inorganic perovskites-based solar cells are nowadays intensively investigated due to the PCE record of 13.43%. In addition, the application of all-inorganic perovskites for the construction of smart windows is also very attractive [167,168]. One of the promising methods to stabilize the α-phase of perovskite is increasing the surface/volume ratio. This can be achieved by synthesizing inorganic perovskites quantum dots to stabilize the desired phase [169]. Two additional methods can be employed. The first method introduces a hydroiodic acid into the perovskite precursor solution to improve the solubility of the reactants, and carries out a low-temperature phase-transition from a nonperovskite to a stable α-CsPbI$_3$ [160,170-174]. The second method deals with a substitution of lead by metals with a smaller radius (Bi^{3+}) [161], and strontium (Sr^{2+}, 1.18 Å) [165], and iodide by bromide to increase the stability of the black-phase CsPbI$_3$ [175]. Among them, CsPbI$_2$Br, with a 1.92 eV band gap is a promising candidate, combined with CuSCN as the inorganic HTL [176]. Thus, all-inorganic perovskites combined with inorganic transporter layers might provide a convenient way to solve the thermal instability problem [177,178].

In fact, the PCE of 13.21% was achieved with the structure of FTO/c-TiO$_2$/CsPb$_{0.96}$Bi$_{0.04}$I$_3$/CuI/Au. The thermal stability (up to 160°C) can be further improved by using this combination FTO/NiOx/CsPbIBr$_2$/MoOx/Au

[154], where a mixture of Br and I is used. However, the PCE decreases due to the wide band gap of the CsPbIBr$_2$ absorber layer (2.08 eV). Recently, the introduction of a bilayer formed using ZnO encapsulated into a C$_{60}$ fullerene (ZnO@C$_{60}$) as the ETM layer, has permitted the fabrication of a device with this composition: FTO/NiOx/CsPbI$_2$Br/ZnO@C$_{60}$/Ag, [178]. This all-inorganic perovskite shows the PCE of up to 13.3% with a remarkably stabilized power output at 12% within 1000 s. Importantly, this device without encapsulation exhibits high long-term thermal stability with only 20% PCE, quenched after being heated at 85°C for 360 h.

2.1.5. Mixed Fluoride/Bromide/Iodide Perovskites

It is quite clear that the ultimate way to avoid the disadvantages associated with organic cations is to completely replace them with inorganic cations. However, as is already commented above, the most appropriate Cs-based perovskite (CsPbI$_3$), that possesses a band gap of 1.73 eV, is not stable at room temperature. Unfortunately, the reverse phase transition from the yellow orthorhombic phase to the black cubic phase occurs at t > 328°C. Moreover, the α-CsPbI$_3$ quantum dots, which remain stable in ambient air, require complex and expensive synthetic procedures [159,179,180].

Taking advantage of compositional engineering, halogens such as chlorine, bromine and iodine can be incorporated to modulate the stability and light harvesting ability. Different ratios of iodine to bromine have been explored to balance stability and the PCE, which led to the CsPbBrI$_2$ composition as the most preferred one [155,163,156,181-185]. However, even though the phase stability of CsPbBrI$_2$ is favoured over CsPbI$_3$, it is of great importance to further improve the stability of the α-phase under an ambient atmosphere, and its performance. To this sense, it has been recently reported the utilization of the smallest halide (F), to investigate if the stability of the α-phase improves in the heterojunction [186]. In particular, a fluorine-modulated bulk-heterojunction of inorganic caesium lead halide with enhanced efficiency and stability has been constructed (see Figure 2).

Interestingly, when a built-in electric field is introduced, the bulk-phase heterojunction is better with respect to those without fluoride in terms of charge generation and transport. Also important is that this fluorine-modulated bulk-heterojunction presents a significant reduction of the recombination loss. The structural stability of the α-phase also improves, when I is partially substituted by F. As the ratio of F increases, the effective tolerance factor of $CsPbBrI_{2-x}F_x$ increases, significantly enhancing its structural-stability.

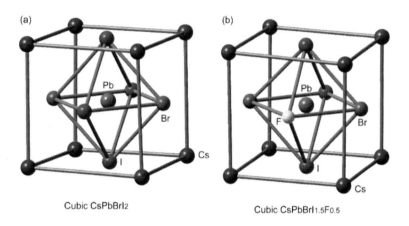

Figure 2. (a) Molecular structure of the cubic $CsPbBrI_2$ perovskite; (b) Molecular structure of the $CsPbBrI_{2-0.5}F_{0.5}$ perovskite.

The $CsPbBrI_{1.78}F_{0.22}$ composition achieved the enhanced PCE up to 10.26%. Without encapsulation, the phase heterojunction cell retained 69.81% of its initial PCE after 240 h of storage at room temperature, whereas the unsubstituted $CsPbBrI_2$ heterojunction in the α phase retained only 4.20% of its initial PCE.

3. LEAD SUBSTITUTION

Since the perovskites yielding the most efficient solar cells contain a significant percentage of lead (>30 wt%), manufacturing, deployment, and disposal of PSCs are harmful to the environment. To address this toxicity issue of the material, substitution at the lead-sites in $CH_3NH_3PbI_3$ has been proposed, and a number of potential metals have been considered in this direction.

3.1. Group-14 Elements

The obvious candidates for homovalent substitution of lead in the perovskite structure are tin and germanium. Both elements have a similar electronic configuration to lead, and also interesting optoelectronic properties. For both these alternatives, the stability of the corresponding halide perovskites is a critical concern as detailed below.

3.1.1. Tin

Due to the similar ionic radii, tin was the first metal considered as a replacement of lead in the MAPI structure. Tin-based halide perovskites have optical gaps that are close to the optimum requirement without compromising carrier mobility. Therefore, they are indeed very promising materials for the construction of solar cells and optoelectronic applications. In fact, pure tin-based perovskites $ASnI_3$, with A = Cs, methylammonium or formamidinium, have shown near infrared active direct optical gap of around 1.2–1.4 eV [187], which is lower than their Pb-analogues. However, such advantages are counterbalanced by their low air-stability. This is to some extent fixed by using mixtures of lead and tin at the M-sites of a perovskite [188].

In one of the pioneering works toward the utilization of lead-free perovskite solar cells was reported Noel et al., [189]. They fabricated a completely lead-free solution-processed $MASnI_3$-based perovskite solar cells, which was fundamental on a mesoporous TiO_2. The PCE of the lead-substituted solar cells was up to 6.4% under 1 sun illumination. Remarkably, this material exhibits a 1.23 eV optical gap and an open-circuit voltage over 0.88 V. Therefore, such a high voltage evidences that tin is a promising candidate for substituting lead as the metal cation.

Hao et al., [190] have also reported tin-based perovskites for the construction of lead-free solar cells. Similarly to Noel and co-workers work, they have demonstrated a reduction of the band gap in $MASnI_3$ (1.30 eV) compared to $MAPbI_3$ (1.55 eV). They have also obtained comparable data for the open-circuit voltage (0.68 V) and the PCE of 5.23% under 1 sun illumination. It has been proposed that tin halide perovskites provide lower open-circuit voltage due to the lowering of the conduction band edge upon tin substitution. Remarkably, the band gap can be further adjusted by using chemical engineering. That is, the chemical substitution of the iodide atom by bromide was used to vary the band gap. For instance, in the mixed halide MASnI$_{3-x}$Br$_x$ (x = 0–3) compounds, the compound varied chemically from black for MASnI$_3$ to dark brown for MASnI$_2$Br and to yellow for MASnBr$_3$. The highest efficiency of 5.73% was obtained with the MASnIBr$_2$ compound.

The poor photovoltaic performance of MASnI$_3$ raises serious concerns. Likely, the major one is the facile oxidation of Sn^{2+} to Sn^{4+} in contact with air destroying the charge neutrality and stability of the perovskite with a concomitant formation of oxides/hydroxides of the metal and MAI as sub-products [191, 192]. Furthermore, the formation of SnO$_2$, as a p-type dopant, makes the oxidized material metallic through a self-doping process [193]. The second concern is that one of the degraded products (SnI$_2$) is equally lethal compared to PbI$_2$ [194, 195]. Third, an additional problem of using MASnI$_3$ perovskite is its rapid crystallization [196], that makes very difficult to control the solution process [197]. This fact complicates the fabrication of homogeneous thin-films of these type of perovskites. These strong drawbacks indicate that tin-based perovskites are not feasible options towards the fabrication of lead-free perovskite solar cells.

Some of these serious concerns are solved by the partial substitution of lead by tin in hybrid perovskites, since a better performance in terms of the PCE and stability. The presence of tin in the mixed Sn/Pb-films reduces the band gap, and lead has an important role regarding the oxidation of tin. Following this strategy, MASn$_x$Pb$_{1-x}$I$_3$ mixed cation based solar cells were fabricated and characterized [198]. Depending on the lead to tin ratio, the perovskites exhibited optical band gaps varying from 1.51 eV (100% Pb) to 1.10 eV (100% Sn) in a device fabricated using mesoporous TiO$_2$ and poly(3-hexylthiophene-2,5-diyl) as a carrier. However, the efficiency reached its maxima of only 2.37% at a ratio of Pb : Sn = 0.5 : 0.5. A poor performance at the two extremities has been attributed to higher charge recombination in Sn-rich samples (low shunt resistance), and the energy level mismatch with poly(3-hexylthiophene-2,5-diyl) in the Pb-rich case (high series resistance).

A similar study, using hybrid tin and lead (ASn$_{1-x}$Pb$_x$I$_3$) perovskites and combining different cations (MA, FA, Cs), has demonstrated that substitution of lead by 50% of tin makes them more stable under ambient conditions. In addition, this investigation shows that the oxidation reaction depends upon the size of the cation. For example, the incorporation of a small cation like Cs$^+$ reduces the oxidation rate, increases the structural stability and improves the film morphology leading to a better device performance. Interestingly, Liu et al., [199] showed that a Pb : Sn binary PSCs of formula FAPb$_{0.75}$Sn$_{0.25}$I$_3$, with the PCE of 10.27% improves to 14.46% by substituting 20% of the FA cation by cesium. More importantly, these devices were stable at ambient conditions. This strategy has been used to construct perovskites with mixed organic cations (FA with MA) that give films with better morphology and increment the efficiency of the devices. Moreover, compositional engineering at cationic and anionic sites, combined with tin-perovskites, has been reported [200-202]. Compounds with quite complicated composition like MA$_{1-x}$(FA)$_x$Pb$_{1-y}$Sn$_y$I$_{3-z}$Br$_z$ may provide the desired combination of stability and efficiency.

3.1.2. Germanium

Germanium is another Group-14 element and, consequently, a possible candidate for the generation of lead-free PSCs. To anticipate the optical gaps of Ge-based perovskites with the standard AGeX$_3$ cubic structure, where A = Cs or MA, and X = Cl, Br, or I, theoretical calculations have been reported. The predicted trend for the band gap depending on the halide is AGeCl$_3$ > AGeBr$_3$ > AGeI$_3$ for A = Cs [203] and A = MA [204]. Moreover, the influence of the cation (A-site) on the band gap has been analyzed in Ge-based perovskites [205]. The band gap increases as the size of the cation increases. The substitution of Cs (1.6 eV) for MA (1.9 eV) or FA (2.2 eV) increases the band gap and does not changes the 3D structure of the perovskites. The utilization of even larger cations like iso-propylammonium (2.6 eV), guanidinium (2.7 eV), and trimethylammonium (2.8 eV) not only increases the band gap but also reduces the dimensionality to one dimensional, and changes the nature of the band gap from a direct to an indirect. Most of the Ge-perovskites exhibit a large band gap and, consequently, their role as an absorber is very limited. With MA and FA, the band gaps look more promising, however the inclusion of Ge^{2+} in perovskites leads to very poor conductivities. Therefore, unfortunately Ge-perovskites are not suitable photovoltaic absorber, in spite of the non-toxic nature and natural abundance of germanium. The PCE values are 0.11 and 0.20% for $CsGeI_3$ and $MAGeI_3$, respectively, incorporated in mesoscopic PSCs, which are limited mainly by the oxidation of germanium and defect chemistry in the material [206]. To date the maximum PCE of 3.2% has been published for a $CsGeI_3$ structure [207], with low stability of Ge^{2+}.

3.2. Alkaline Earth-Metals

The high abundance of these elements and their basically nontoxic nature make alkaline-earth metals suitable to replace lead in halide perovskites. Metals like strontium, calcium, etc have a stable +2 oxidation state, thus adequate to form conventional AMX_3 perovskites [208]. Indeed, Navas et al., [209], have suggested that strontium (Sr^{2+}) is a promising dopant, because it has the same ionic radius of lead, abundant availability and nontoxic nature. Moreover, strontium has been investigated using quantum mechanical and DFT calculations [210]. These results suggested that the perovskite crystal structure is basically unaltered by the Sr^{2+} substitution due to the identical charge, and almost identical radius of Sr^{2+} and Pb^{2+}. However, the predicted band gap is very large (3.6 eV), making CH_3-NH_3SrI_3 a bad candidate to replace MAPI. As revealed by the calculations and density of states (DOS) studies, the increase of the band gap after the inclusion of Sr^{2+} is attributed to the upward shift of the conduction band due to the different orbital contribution (5s in Sr versus 6s in Pb) [211]. The lower electronegativity of strontium (0.95) compared to lead (2.33) changes the magnitude of the local dipoles, which also affects the optical gap. It has been shown that the band gap increases with decreasing electronegativity of the M-cation [212]. The synthesis of $CH_3NH_3SrI_3$ is an additional drawback because the perovskite formation from the precursor solution has a larger energy barrier compared to lead.

Pazoki et al., have provided a theoretical insight into the effect of substituting Pb^{2+} by a series of alkaline earth metals Ca^{2+}, Ba^{2+} and Sr^{2+} on the electronic structure of perovskites [213]. Considering the standard $CH_3NH_3MI_3$ (M = Ca^{2+}, Ba^{2+}, or Sr^{2+}), DFT calculations were utilized to anticipate the formation of perovskite materials. It was predicted that all three metals form stable perovskite materials, similar to MAPI. The optical gap of these perovskites is influenced by the electronegativity difference between the halide and the alkaline-earth metal. The optical gap values reported by Pazoki et al., are 2.95 eV for $MACaI_3$, 3.3 eV for $MABaI_3$, and 3.6 eV for $MASrI_3$. The large optical gap in combination with a high conduction band energy and large difference between electron and hole effective masses envisage their application as a carrier selective contacts.

In spite of its very small ionic radius (72 pm), the possibility of utilization of magnesium (Mg^{2+}) as a suitable alternative to lead has been also explored by theoretical approaches [214]. Mg-perovskites may form stable 3D perovskites with a tunable direct band gap, according to the DFT calculations. For instance, Filip et al., [214] studied different A-cations in the cubic $AMgI_3$ structure. They found that the band gap decreases as the size of the A-cation increases: 1.7 eV for $CsMgI_3$, 1.5 eV for $MAMgI_3$ and 0.9 eV for $FAMgI_3$. These promising band gaps for solar cell is strongly handicapped by the strong sensitivity of Mg-perovskites to humidity.

3.3. Transition Metals

Klug et al., [215] have made a significant contribution in their investigation, replacing lead in the perovskite structure by transition metal atoms with the same oxidation state (2+). They have taken into consideration a series of divalent metals (Co, Cu, Fe, Mg, Mn, Ni, Sn, Sr, and Zn), and focusing their attention to the photovoltaic performance and optical properties. The replacement of lead by such variety of metals (including different ratios of replacement) showed a remarkable tolerance of the perovskite structure to substitution without altering the electronic properties excessively. This interesting study reveals that elements like copper, zinc, and cobalt are promising candidates in this context. Particularly Co showed an interesting performance (up to 10.2% depending on the amount of Co). The introduction of cobalt ions into the MAPI structure changed the Fermi level and the band edges without altering the pristine band gap.

A systematic study on the substitution of lead by Hg^{2+}, Cd^{2+}, Zn^{2+}, Fe^{2+}, Ni^{2+} and Co^{2+} dopants in the MAPI structure has been reported by Frolova et al., [216]. Unfortunately, most of them resulted in a deterioration of the perovskites. The bivalent elements acted as traps and corrupted the fundamental mechanisms of charge generation and transport in the perovskite The only element that improves the MAPI-device is mercury. Due to its high ability to undergo co-crystallization with lead, the mercury doped samples were stable without altering the structure until a high amount of doping (>30%). The presence of mercury influences the photovoltaic performance of planar solar cells. The PCE is maximum (13%) with a doping concentration of 10% Hg.

3.4. Lanthanides

The utilization of rare earth metals to replace lead in the perovskite structure results in the appearance of interesting optical properties. For instance, the introduction of Eu^{2+} at the metal sites (A) results in an intense blue photo-luminescence (Eu^{2+} emits at $\lambda = 460$ nm) at room temperature thus giving to Eu-based perovskites promising applications in hybrid optoelectronic devices, such as LEDs [217].

In addition to complete substitution of lead by Eu^{2+}, the role of rare earth metals as dopants in conventional perovskites has been analyzed. Lanthanides (Eu^{2+}, Tm^{2+} and Yb^{2+}) have been introduced as dopants in traditional AMX_3 perovskites, where A = Cs or K, M = Mg, Ca, or Sr and X = Cl, Br, or I). The introduction of Eu^{2+} dopant in $AMCl_3$ perovskites with A = Cs, or K, and M = Mg, Ca, or Sr, resulted in an intense photo-luminescence emission [218]. Suta et al., have reported similar results in

Eu^{2+}, activated iodide perovskite materials [219, 220]. Moreover, Yb^{2+}-doped perovskites CsMX$_3$ (M = Ca, or Sr, X = Cl, Br, or I) also present photo-luminescence properties. The doping of CsCaX$_3$ perovskites with Tm^{2+} ions has been also reported by Grimm et al., [221]. It was observed interesting light-emission and photophysical properties.

3.5. Antimony and Bismuth

Antimony and Bismuth are trivalent dopants that intrinsically have a concomitant challenge to keep the pristine crystal structure unaltered after doping. In the case of perovskites, the trivalent metal cations must also maintain the MX$_6$ octahedral symmetry. Abdelhady et al., [222] have studied such possibility. By the in situ chemical approach, they have recently incorporated a controlled amount of Bi^{3+}(among other trivalent ions like In^{3+} and Au^{3+}) in the conventional APbX$_3$ structure (A = MA, X = Br). The trivalent bismuth ion has the same (6s^{2}6p^0) electron configuration of lead, similar electronegativity (2.03) and ionic radius (Bi^{3+} = 103 pm), thus it is ideal as a dopant. Indeed, it is feasible to dope the pristine perovskite with a large amount of Bi^{3+}. Abdelhady et al. have shown that the band gap is reduced from 2.17 eV for the undoped perovskite to 1.89 eV upon incorporation of 10% bismuth. Also important, the incorporation of Bi^{3+} dopant increases the charge carrier concentration, significantly improves the conductivity and switches the sign of the majority carriers. Remarkably, there is a strong correlation between the doping concentration and the band gap value [223]. Moreover, the p-type semi-conductor MAPI switches to n-type after Bi^{3+} doping, as shown by band gap calculation through photoelectron spectroscopy and Hall effect measurement.

The first report of antimony substitution in the standard MAPI structure was reported by Zhang et al., [224]. In a sharp contrast to Bi^{3+} doping, that results in a band gap reduction, the Sb^{3+} doping increases the band gap of the perovskite. Moreover, there is not n→p semiconductor switching upon addition of Bi^{3+}, also in contrast to the behavior observed for bismuth. The increment of the band gap upon the addition of the Sb^{3+} ion to the structure can be visualized by the color change of the film. That is, the color varies from dark brown for pure MAPbI$_3$ (0% Sb, 1.55 eV) to brown under 50–75% MAPbI$_3$ (1.61–1.95 eV), and finally changes to orange for 90–100% Sb-doping (2.06 eV). This behavior has been explained by the orbital composition of the conduction band due to the electronic configuration of Sb^{3+}. Furthermore, the lower electronegativity of Sb^{3+} compared to Pb^{2+} results in an increment of the band gap upon doping. By means of the DFT calculations, it was demonstrated that the Fermi energy level moves towards the conduction band in the doped perovskites, unveiling the n-type doping. As a matter of fact, Sb^{3+} has more valence electrons than Pb^{2+}, the doping causes an excess of electrons in the perovskite system.

The effect of Sb^{3+} doping on the photovoltaic performance of MAPI perovskite (0–100% Sb) has been investigated. The Sb^{3+}-doped perovskites have been incorporated in a conventional regular device structure of perovskite solar cells between the layers of TiO$_2$ and spiro-OMeTAD. Quite remarkably, the perovskites showed a high efficiency of 15.6%, when doped with 1% Sb. Using this percentage of doping, the device reached almost 1 V of open circuit voltage. Since at this level of doping the change in the band gap is very small, this significant increment in the open circuit voltage is the upward movement of the quasi-Fermi level under illumination. The increment of antimony content in the perovskite structure beyond 1% provokes a reduction of the performance of the solar cells due to strong reduction of the transport properties of the perovskites with increasing antimony concentration. This is mainly due to a reduction of the effective diffusion coefficient and electron lifetime and, consequently, the diffusion length (higher probability of charge recombination).

4. DEVICE OPTIMIZATION

4.1. Primary Trends

Beside various chemical compositions of perovskites, one may improve the efficiency of perovskite based devices by analysing physical mechanisms responsible for their performance. In this section we will focus on the question of how to improve the efficiency of PSCs, considering the CH$_3$NH$_3$PbI$_3$ as a typical absorber. To this aim, we present a physical model based on the fundamental principles of semiconductor physics. The model enables us to evaluate the PCE for the PSC for the chosen perovskite. It also allows to evaluate energy losses in a photovoltaic element (PVE) due to well known mechanisms and reveal the corresponding energy dissipating channels that should be eliminated. With the aid of the considered model we will estimate the PCE improvement of the PSC due to various modifications.

4.2. The Model and the PCE

The most adequate simulation of basic physical mechanisms of a perovskite based solar cell can be done by numerical solution of the following set of equations for electron/hole density $\eta = n/p$:

$$\frac{\partial \eta}{\partial \tau} + \frac{1}{q}\vec{\nabla} \bullet \vec{j}_\eta = T - R, \tag{1}$$

$$\vec{j}_\eta = q_\eta \bullet \eta \bullet \mu_\eta \bullet \vec{E}_\eta - D_\eta(\vec{\nabla}\eta), \quad q_\eta = \begin{cases} e, & \eta = n \\ q = |e|, & \eta = p \end{cases} \tag{2}$$

$$\vec{\nabla} \bullet (\varepsilon\vec{E}) = -q(n - p + N_a - N_d), \tag{3}$$

$$\vec{j} = \vec{j}_n - \vec{j}_p, \tag{4}$$

$$\vec{E}_n = \vec{E} - (\vec{\nabla}E_C), \quad \vec{E}_p = \vec{E} - (\vec{\nabla}E_V) \tag{5}$$

Here, e- electron charge, $T(R)$ is the photoionization (the recombination) rate of carriers per unit volume, N_a/N_d- is the ionized acceptor/donor density, $E_C(E_V)$ is the edge energy of the conduction (valence) band, $\vec{j_n}(\vec{j_p})$ - electrons (holes) current densities, ε are semiconductors permittivities. Evidently, Eq.(3) transforms to the Poisson's equation, since the electric field strength $\vec{E} = -\vec{\nabla}\varphi$ is related to the scalar potential φ. It is common to use Eqs.(1-5) to calculate the current-voltage characteristics (CVC) and the PCE [225-227]. In order to trace the power density dissipated by electron and hole currents in the semiconductor lattice of the PSC, it is useful to add the continuity equations for the energy density $\Re_\eta = \eta \xi_\eta$ of electron-hole plasma. Note, in general, the mobility and the diffusion coefficient are variables that depend on the mean carrier energy ξ_η:

$$\frac{\partial \Re_\eta}{\partial \tau} + \vec{\nabla} \bullet \vec{S}_\eta = -P_\eta - \frac{\eta(\xi_\eta - \xi_0)}{\tau_{\xi_\eta}}. \tag{6}$$

Here τ_{ξ_η} is the energy relaxation time; $\xi_\eta = 3kT_0/2$, where T_0 is the lattice temperature, k is the Boltzmann constant; and $P_\eta = \vec{j_\eta} \bullet \vec{E_n}$. The energy flux density $\vec{S}_\eta$ has the following form

$$\vec{S}_\eta = \frac{1}{q} \bullet \gamma \bullet \xi_\eta \bullet \vec{j_\eta} - K_\eta(\vec{\nabla}\xi_\eta), \tag{7}$$

where γ is the differential thermo e.m.f. coefficient (see [228], p. 430), K_η is the thermal conductivity. The photoionization rate $T(x)$ at the depth x from the semiconductor surface in Eq.(1) is defined as

$$T(x) = \int_0^\infty \alpha(v)F(v)\exp[-\alpha(v)x]dv, \tag{8}$$

where ν and $F(\nu)$ is a photon frequency and a photon flux density in the unit spectral interval, existed at the depth x of the semiconductor. While the rate of non-radiative, Shockley-Read-Hall (SRH), recombination has a standard form

$$R = (np - n_i^2)/(\tau_p n + \tau_n p), \qquad (9)$$

where n_i is the carrier concentration in the intrinsic semiconductor. The main parameters that determine the PCE of the PSC are electron and hole recombination lifetimes (τ_η), band gaps (E_g) and electron affinities (χ) of perovskite, perovskite absorption coefficient spectra $\alpha(\nu)$, and properties of ETM and HTM. Most of these parameters are measured directly ($E_g, \chi, \alpha(\nu), \varepsilon$), or evaluated from some relations. In particular, τ_η can be estimated from the measured diffusion lengths $L_{dif} \Leftrightarrow L_\eta = \sqrt{\mu_\eta \dfrac{kT_0}{q} \tau_\eta}$.

Thus, once the above parameters are measured or calculated, by means of numerical solution of Eqs. (1-5) or (1-9) it is possible to evaluate the PCE of the PSC. For example, the thermal stability of $CH_3NH_3PbI_3$ can be improved noticeably, replacing the MA cation (CH_3NH_3+) by the formamidinium (FA) cation [$HC(NH_2)_2+$]. The band gap value 1.55 (1.61) eV for FA(MA) perovskite [229], their absorption coefficient spectra are shown in Figure 3.1(a), and the electron affinities = 4.2 (3.9) eV for FA (MA) [230]. Solving the system Eqs.(1-9) for the structure ZnO-Perovskite-CuI (see Figure 3.2), we determine the CVC for these perovskites [see Figure 3.1(b)]. These calculations yield the following values for FA(MA) perovskite: 1) the PCE is 17.5% (19.7%); 2) the short circuit current (Isc) is 18.7 (21) mA/cm^2; 3) the fill factor is 0.85(0.86); 4) the open circuit voltages (Uoc) are 1.1 V, i.e., it is approximately the same in the both perovskites. Evidently, the enhanced absorption leads to higher photocurrent in the PSC.

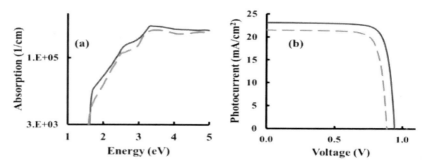

Figure 3.1. (a) Absorption coefficient spectra, (b) the current-voltage characteristics, of the illuminated PSC (the Air Mass 1.5 Sun spectrum). The solid (dashed) line is associated with the MA(FA) perovskite.

The built-in potential is 1.17 V. The width of ZnO and CuI layers is 0.2 μm, while the width of $CH_3NH_3PbI_3$ is 0.5 μm.

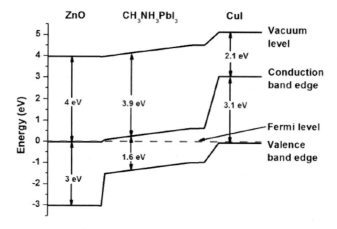

Figure 3.2. Band profile of ZnO-perovskite-CuI (p-i-n) solar cell.

4.3. Energy Losses

Energy losses can be analysed by means of a procedure based on the matching the measured CVC of the illuminated PSC and the simulated CVC with the aid of the circuit layout shown in Figure 4.1.

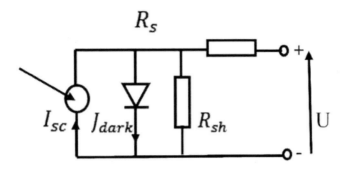

Figure 4.1. The circuit layout that includes: i) the PSC circuit; ii) R_s is a parasitic resistance of contacts and feeding conductors; and iv) R_{sh} is associated with a leakage resistance related to the local short circuiting of ETM/HTM layers.

The total CVC of the diode and the generator can be evaluated with the aid of a numerical solution of the transport equations (1-9) [231]. Such appoach enables us to establish the PCE dependence on the perovskite properties. In general, the absorbing properties are the overall result of several factors: i) the nonradiative recombination time of carries; ii) parasitic resistances of contacts and feeding conductors; iii) a current leakage, related to the local short circuiting of ETM/HTM layers; and iv) a degree of light reflection by the front surface.

Before to proceed, there are a few considerations to comment. First, the above four parameters can be determined only approximately. In particular, the teams, reported on highly efficient PSCs [27, 34, 37, 232-234], do not provide data on the absorption spectra of their perovskite absorbers. Therefore, in our calculations we use the perovskite ($CH_3NH_3PbI_3$) absorption coefficient spectra measured in Ref. [235]. Besides this, this perovskite is not stable under external conditions (see Section: Material optimization). We recall that in order to improve its stability, the researchers create a composition of $CH_3NH_3PbI_3$ and different additive compounds such as $CH_3(NH_2)_2PbI_3$, $CH_3NH_3PbCl_3$, or cations of Ce. As a result, these additives affect the band gap value. In our illustrative calculations, a single band gap value and absorption coefficient spectra [235] are only used. We believe that these approximations introduce only negligible errors in the definition of the four parameters employed in our example. Second, the physical quantity that may affect our results is the absorption material thickness W. In our calculations the absorption material thickness W is extracted from the publications that provide the corresponding cross section of the perovskite sample. Third, the impurity (acceptor or donor) concentration in the absorber material is another important physical quantity. Unfortunately, this parameter is not provided in the literature. However, as it will be shown later, this parameter strongly affects the diffusion length value extracted from the CVC.

Table 1. Parameter values for the SC materials in [27,34,37,232-234] used in the simulations. m_e is the mass of free electron

	n-TiO$_2$	CH$_3$NH$_3$PbI$_3$	p-SpiroOMeTAD	p-PTAA
Permittivity (ε)	60	60	3	13
Electron affinity (χ), eV	4	3.75	2.12	1.8
Band gap, eV	3.05	1.58	3.1	3.4
Electrons effective mass (relative to m_e)	1	1	1	1
Holes effective mass (relative to m_e)	1	1	1	1
Donor (N_d) or acceptor (N_a) concentration, cm^{-3}	10^{19}	0	3·10^{18}	3·10^{18}
Electrons mobility (μ_n), cm^2/(V s)	0.006	50	0.0001	0.01
Holes mobility (μ_p), cm^2/(V s)	0.006	50	0.0001	0.01
Electron (hole) life time ($\tau_n = \tau_p$), ns	1	varied	1	1

Fourth, the perovskite electron affinity is the characteristic that affects the PSC photocurrent. We assume that it is the same for all used perovskite compositions, and equal to 3.75 eV. Fifth, we assume that the mobility of electrons and holes in all perovskite compositions are 50cm^2/(V·s). The influence of the remaining physical quantities on the PSC photocurrent is much weaker, and introduce insignificant errors that are not crucial for our discussion. Material parameter values, used in the simulation, are displayed in Table 1. In spite of the denoted uncertainties, the following discussion will elucidate the basic mechanisms of the energy losses.

In order to illuminate the details of our approach let us consider as an example the PSC with the PCE of 11.4% manufactured in 2014 [233]. First, from the PSC cross section (see Figure 5) we extract the SpiroOMeTAD-$CH_3NH_3PbI_3$-TiO_2 layer thicknesses (400-500-250 nm, see Table 2). With the aid of these data, solving Eqs. (1-9) simultaneously, we evaluate the CVC of the PSC exposed to the sunlight illumination with AM 1.5 spectrum from the ETM side.

Further, we fit the calculated open circuit voltage (U_{oc}) by adjusting the electron-hole recombination time, i.e., the diffusion length of the carriers (see Figure 6), in order to reproduce the measured U_{oc}. Next, the calculated open circuit currents I_{sc} are fitted to the measured one by altering the overall reflection coefficient for the PSC (see Figure 7).

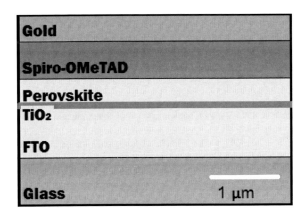

Figure 5. A sketch of the PSC cross section [233].

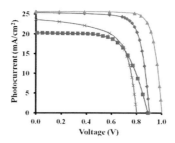

Figure 6. The comparison of the measured and the calculated short circuit current at different diffusion length of the carriers. Experimental results [233] are connected by filled squares. Simulated results with the different diffusion length are displayed as: triangles, $L_{dif} = 1\mu m$; rhombuses, $L_{dif} = 0.33\mu m$; crosses, $L_{dif} = 0.12\mu m$.

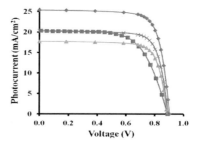

Figure 7. The comparison of the measured and the calculated short circuit current obtained by altering the reflection coefficient value. The experimental results [233] are displayed as filled squares. Simulated results with different reflection coefficients are displayed as: rhombuses, $R_{coeff} = 0.0$; crosses, $R_{coeff} = 0.2$; triangles, $R_{coeff} = 0.3$.

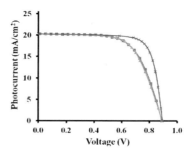

Figure 8. Fitting of the calculated and measured short circuit current by adjusting two parasitic resistances in accordance with the SC equivalent circuit (Figure 7.1). The filled square corresponds to the measured CVC [233]. Simulated CVC are displayed as: crosses, $R_{sh} = \infty$, $R_s = 0$, $R_{coeff} = 0.2$, $L_{dif} = 0.33\mu m$; triangles, $R_{sh} = \infty$, $R_s = 6.5$ Ohm· cm², $R_{coeff} = 0.2$, $L_{dif} = 0.33\mu m$.

Finally, our goal is to reach the overall good fitting of the measured and simulated CVC by adjusting two parasitic resistances, connected in parallel (R_{sh}) and in series(R_s) (see Figure 8); in accordance with the PSC equivalent circuit (see Figure 4).

4.4. Feasibility of the PCE Increase

The results of our analysis for the available PSCs [27, 34, 37, 232-234] are summarized in Table 2 The analysis is carried out under the assumption that all considered perovskites are undoped.

Our study implies a link between the PCE and the diffusion length of the corresponding PSCs, developed in recent years. Indeed, the increase of the diffusion length in the contemporary PSCs reflects the increase of the quality and the purity of the absorber. Note, however, that there is a slow progress in the diffusion length increase after reaching the PCE value 18%. It seems that a further increase of the PCE originates from the increase of the amount of a captured light. The reflection of this process is a noticeable increase of the short circuit current I_{sc}. The increase of I_{sc} can be related to: i)the increase of the absorber thickness W (from 300 [37] to 500 [232] nm); and also ii) the decrease of the effective reflection coefficient R_{coeff}. Evidently, it would be quite natural to decrease the energy losses in the PSC by minimizing the light reflections.

Our analysis indicates on the absorber thickness dependence of the CVC and the PCE of the considered PSCs, exhibited by the behaviour of the short circuit current (I_{sc}), the open circuit voltage (U_{oc}), the fill factor (FF) for different carriers diffusion lengths (see, e.g., [236]). The basic outcome implies that all these characteristics change drastically in the vicinity of the thickness value $W \approx 500$ nm. A further increase of the thickness decreases the efficiency of the p-i-n system [236]. We recall, however, that it was reported about 175 μm diffusion length observation in $CH_3NH_3PbI_3$ [237]. A charge diffusion length that greatly exceeds the absorption depth of photons with energy larger than the band gap of perovskites implies that Internal Quantum Efficency of essentially 100% may be achieved under the low internal electric fields at device working condition.

4.5. Impact of the PSC Absorber Doping

We recall that our investigations are based on the assumption of the undoped absorber. Here, we attempt to estimate the effect of doping. With the increase of a doping impurity concentration the p-i-n diode transforms gradually to the diode with one p-n junction. If we assume that the diode is always illuminated from the n⁺ contact side, the absorber doping by donors leads to the creation of the p-n junction, illuminated from the rear contact, i.e., from the side being far from the high-field region of the junction. In contrast, the absorber doping by acceptors leads to the creation of the p-n junction, illuminated from the front contact, i.e., from the side being near to the high-field region of the junction (see Figure 9).

We recall that the current J_{sc} depends on the absorber thickness and the band gap value in a SC with a p-n junction. The open circuit voltage U_{oc} of such a SC is obtained from the following equations [238]:

$$U_{oc} \propto \ln\left(\frac{I_{sc}}{I_0}+1\right), \quad I_0 = \left(\frac{\mu_p n_i^2}{L_p N_d} + \frac{\mu_n n_i^2}{L_n N_a}\right) k T_0, \qquad (10)$$

where I_0 is the dark current, $L_n(L_p)$ is the electrons (holes) diffusion length. In our case $\mu_n = \mu_p = \mu$, $\tau_n = \tau_p = \tau$, and $L_n = L_p = L_{dif} = \sqrt{\mu \tau k T_0 / q}$.

It can be seen from Eq. (10) that, in contrast to the U_{oc} of the SC with the p-i-n structure, the U_{oc} of the SC with p-n diode depends on the diffusion length and the absorber impurity concentrations.

Table 2. A summary table of the considered PSCs with their characteristics

PCE%	FF %	I_{sc} mA/cm^2	U_{oc} V	L_{dif} μm	R_s Ohm·cm^2	R_{sh} KOhm·cm^2	Rcoeff	W_{per} nm	W_{ETM} nm	W_{HTM} nm	Ref.
11.4	64	20.3	0.89	0.33	6.5	inf	0.19	500	400	400	[233]
15	73	20	0.993	2.05	5.4	inf	0.16	350	400	700	[34]
15.7	75	20.4	1.03	3.35	1.2	0.8	0.097	300	25	200	[234]
17.9	74	21.8	1.114	8	3	0.9	0.07	300	270	50	[37]
19.3	75	22.75	1.13	9.4	4.5	1.8	0.039	330	250	50	[27]
22.1	80	25	1.1	9.4	0.4	1.1	0	500	210	50	[232]

In fact, the U_{oc} increases with the increase of the both diffusion lengths and the impurity concentration (N_d or N_a). Thus, the measured U_{oc} (for example, in Ref.[37] U_{oc} = 1.114 V) can be obtained by means of the simulations with different interrelations of N_d, L_{dif}, or N_a, L_{dif}.

In the considered SC for the impurity concentration range 0-10^{15} cm^{-3} the value U_{oc} = 1.114 V is obtained at the diffusion length about 8 μm (see Figure 10). At these conditions the space charge region, and, hence, the region of a relatively high electric field occupies entire the absorber, and there is still a p-i-n diode in the SC (see Figure 9).

Once the acceptor concentration exceeds the value 10^{15} cm^{-3}, the size of the space charge region decreases and becomes less than the physical size of the absorber. The n-i-p diode transforms to the n-p diode irradiated from the face side, i.e., the high-field region becomes closer to the irradiated side. The region with the maximum electron-hole generation rate lies entirely within the high-field location, a large resistance region of the diode, depleted with carriers. The high electric field separates electrons and holes (generated by light), that can not recombine while travelling through the high-field region. As a result, the carriers are added to the photocurrent. The current in the region of the absorber with a low field is carried by a huge amount of carriers existing there. And this region serves as the electron/hole collecting layer. As the size of the high-field region decreases while N_a increases, it restricts the diffusion length needed for the creation of the same U_{oc}. If the absorber is doped by donors, the n-i-p diode transforms to the n-p diode, irradiated from the rear side (see Figure 9.). Hence, the greater diffusion length is needed in order to create the given U_{oc}. In this case, the generated holes travel almost the full absorber length to reach the high-field region situated near p+ contact (see Figures 9, 10).

For all pairs N_d, L_{dif}, or N_a, L_{dif}, that provide U_{oc} = 1.114 V (see Figure 10), we calculate also the short circuit current J_{sc} (see Figure 11). The short circuit current starts to decrease when the doping concentration exceeds some critical value. The increase of the doping concentration decreases the high-field region. Therefore, more electron-hole pairs are generated in the region with a low electric field. It results in the additional recombination of the electron-hole pairs. Moreover, the increase of the impurity concentration shrinks the diffusion length (see Figure 10), i.e., there is a rise of the recombination probability. Hence, less amount of holes succeeds to reach the HTM layer before the recombination. Besides that, in the case of the donor doped absorber, a significant portion of the electron-hole pairs are generated immediately in the low-field region. Hence, I_{sc} starts to decrease earlier than in the case of the acceptor doped absorber (see Figure 11).

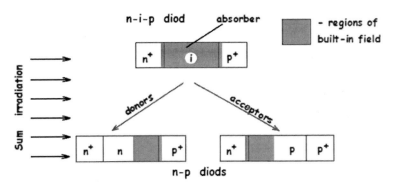

Figure 9. A sketch of the transformation of a n-i-p diode to a n-p diode at the perovskite doping.

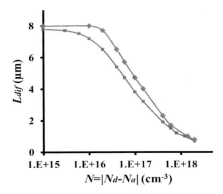

Figure 10. The diffusion length as a function of the impurity concentration: donor doped (rhombuses), acceptor doped (squares), - in the absorber that provides $U_{oc} = 1.114$ V in the SC [37].

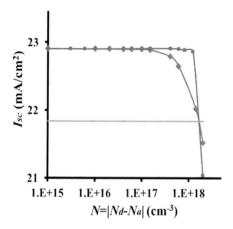

Figure 11. Short circuit current for all pairs of N_d, L_{dif} (filled square), or N_a, L_{dif} (rhombuses), provided by Figure 9.2. The measured I_{sc} = 21.84 mA/cm² [37] is marked by the horizontal line.

From our analysis it appears that the difference between simulated and measured short circuit current is due to some additional mechanisms. In particular, it could be related to some light reflectance from the PVE. Thus, our results demonstrate that the absorber impurity concentration affects the open circuit voltage of the PSC as well as the diffusion length.

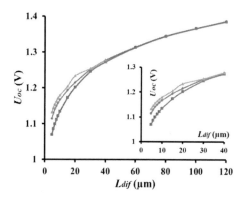

Figure 12. The dependence of the open circuit voltage on the absorber carriers diffusion length for the PSC with 300 nm thick absorber and different doping values: acceptor doped (triangles), donor doped (rhombuses), intrinsic absorber (squares).

Can we further increase U_{oc} with the absorber doping ? To answer this question we compute the dependence of U_{oc} on L_{dif}, at the absorber thickness 300 nm and the absorber donor concentration 1.5· 10^{17} (see Figure 12). This concentration is small enough and does not reduce noticeably the PSCs short circuit current. We obtain that the absorber doping, preferably with acceptors, can increase the value U_{oc} (~5.6% and, hence, the PCE) if the absorber diffusion length is less than 30 μm. The absorber doping has a little impact on materials with a high purity (absorber diffusion length is greater 30 μm) (see Figure 12).

CONCLUSION

The outstanding accomplishment of MAPI as a photovoltaic material anticipates that perovskite-based materials will have a prominent role as key players in solar cell industries. High-efficiency, durability, and low-cost are very important features. However, other features like semi-transparency, color, flexibility are also very desirable. Some major concerns are the stability and nontoxic nature, that prevents nowadays their commercialization. As commented in Section I of this chapter, some initial steps in this direction have been made, and some strategies are promising. The partial substitution of lead (doping) in the perovskite structure by chemical engineering allows to adjust its optical and electronic properties and, subsequently, the device performance. Among several metal dopants highlighted herein, the group-IV elements (Sn^{2+} and Ge^{2+}), which have been extensively studied by the scientific community, are not adequate due to the low efficiency. Other ions, like non-toxic antimony (Sb^{3+}) and cobalt (Co^{2+}), are attractive candidates and more research is needed to prove their usefulness. It should be kept in mind that a partial replacement cannot address the toxicity issue. It seems that the presence of some amount of lead is mandatory to achieve high-efficiency in solar cells with such partially substituted perovskites.

Halide substitution combined with cation substitution also gives good opportunities to obtain perovskites with higher stability and durability. The utilization of different ratios of halides from F to I in combination with different organic and inorganic cations provides infinite possibilities to chemical engineers to generate hybrid perovskites with enhanced properties.

The experimental efficiency of the PCE can be improved without complex refining of a standard cell structure, and could reach the theoretical limit. The increase of the sample uniformity and the decrease of the contact layer resistances can add ~2% to the PCE. The increase of the diffusion length up to 100 μm can add ~5 - 6% to the PCE. Yet, without the absorber purification, it is possible to improve the PCE by means of the absorber doping. We found that for the diffusion lengths >30 μm the influence of doping is small. However, for diffusion lengths <30 μm the increase of the doping value of the absorber leads to the increase of the open-circuit voltage without the decrease of the short circuit current. It results into ~5.6% PCE increase. Thus, only the uniformity of a sample, the decrease of the contact layer resistances, and the absorber doping could increase the PCE by an additional ~7%.

Application of perovskite materials in other optoelectronics devices is also a promising field. In particular, their use in photodetectors, light emitting diodes, lasers and transistors have been highlighted in this chapter. Organohalide lead perovskite materials offer an abundance of opportunities for low-cost photovoltaics technology and optoelectronic applications.

REFERENCES

[1] Graef, M. D. and McHenry, M. (2007) *Structure of Materials: An Introduction to Crystallography,* Diffraction and Symmetry (Cambridge University Press).

[2] Lee, M. M., et al., (2012). Efficient hybrid solar cells based on meso-superstructured organometal halide perovskites. *Science*, 338: 643.

[3] Kojima, A., et al., (2009). Organometal Halide Perovskites as Visible-Light Sensitizers for Photovoltaic Cells. *J. Am. Chem. Soc.*, 131: 6050.

[4] Im, J. H., et al., (2011). 6.5% efficient perovskite quantum-dot-sensitized solar cell. *Nanoscale,* 3: 4088.

[5] Kim, H. S., et al., (2012). Lead iodide perovskite sensitized all-solid-state submicron thin film mesoscopic solar cell with efficiency exceeding 9%. *Sci. Rep.* 2: 591.

[6] Ball, J. M., et al., (2013). Low-temperature processed meso-superstructured to thin-film perovskite solar cells. *Energy Environ. Sci.*, 6: 1739.

[7] Shin, S. S., et al., (2017). Colloidally prepared La-doped $BaSnO_3$ electrodes for efficient, photostable perovskite solar cells. *Science*, 356: 167.

[8] Malinauskas, T., et al., (2016). Branched methoxydiphenylamine-substituted fluorene derivatives as hole transporting materials for high-performance perovskite solar cells. *Energy Environ. Sci.*, 9: 1681.

[9] Liu, H., et al., (2016). Nano-structured electron transporting materials for perovskite solar cells. *Nanoscale* 8: 6209.

[10] Li, F. R., et al., (2017). Nanotube enhanced carbon grids as top electrodes for fully printable mesoscopic semitransparent perovskite solar cells. *J. Mater. Chem., A* 5: 10374.

[11] Jiang, Y., et al., (2016). Efficient Colorful Perovskite Solar Cells Using a Top Polymer Electrode Simultaneously as Spectrally Selective Antireflection Coating. *Nano Lett.*, 16: 7829.

[12] deQuilettes, D. W., et al., (2015). Impact of microstructure on local carrier lifetime in perovskite solar cells. *Science*, 348: 683.

[13] Wojciechowski, K., et al., (2014). Heterojunction Modification for Highly Efficient Organic–Inorganic Perovskite Solar Cells. *ACS Nano*, 8: 12701.

[14] Tan, Z. K., et al., (2014). Bright light-emitting diodes based on organometal halide perovskite. *Nat. Nanotechnol.*, 9: 687.

[15] Zhu, H., et al., (2015). Lead halide perovskite nanowire lasers with low lasing thresholds and high quality factors. *Nat. Mater.*, 14: 636.
[16] Wei, H., et al., (2016). Sensitive X-ray detectors made of methylammonium lead tribromide perovskite single crystals. *Nat. Photonics*, 10: 333.
[17] Snaith, H. J., et al., (2014). Anomalous Hysteresis in Perovskite Solar Cells. *J. Phys. Chem. Lett.*, 5: 1511.
[18] Dualeh, A., et al., (2014). Thermal Behavior of Methylammonium Lead-Trihalide Perovskite Photovoltaic Light Harvesters. *Chem. Mater.*, 26: 6160.
[19] Hoke, E. T., et al., (2015). Reversible photo-induced trap formation in mixed-halide hybrid perovskites for photovoltaics. *Chem. Sci.*, 6: 613.
[20] Yang, J., et al., (2015). Investigation of $CH_3NH_3PbI_3$ Degradation Rates and Mechanisms in Controlled Humidity Environments Using in Situ Techniques. *ACS Nano*, 9: 1955.
[21] Stranks, S. D. and Snaith, H. J., (2015). Metal-halide perovskites for photovoltaic and light-emitting devices. *Nat. Nanotechnol.*, 10: 391.
[22] Berry, J., et al., (2015). Hybrid Organic-Inorganic Perovskites (HOIPs): Opportunities and Challenges. *Adv. Mater.*, 27: 5102.
[23] Green, M. A., et al., (2014). The emergence of perovskite solar cells. *Nat. Photonics*, 8: 506.
[24] Xu, Q. et al., (2019). Low-temperature photoluminescence spectroscopy of $CH_3NH_3PbBr_xCl_{3-x}$ perovskite single crystals. *J. Alloys Compounds* 792: 185-190.
[25] Zhang, K. et al., (2019). Improve the crystallinity and morphology of perovskite films by suppressing the formation of intermediate phase of $CH_3NH_3PbCl_3$. *Org. Electronics*, 68: 96.
[26] Zhang, H., et al., (2019). Influence of Cl Incorporation in Perovskite Precursor on the Crystal Growth and Storage Stability of Perovskite Solar Cells. *ACS Appl. Mater. Interfaces*, 11: 6022.
[27] Zhou, H., et al., (2014). Photovoltaics. Interface engineering of highly efficient perovskite solar cells. *Science,* 345: 542.

[28] Li. S., et al., (2019) Boosting efficiency of planar heterojunction perovskite solar cells to 21.2% by a facile two-step deposition strategy. 484: 1191.

[29] Stranks, S. D., et al., (2013). Electron-hole diffusion lengths exceeding 1 micrometer in an organometal trihalide perovskite absorber. *Science*, 342: 341.

[30] Ke, W., et al., (2015). Efficient hole-blocking layer-free planar halide perovskite thin-film solar cells. *Nat. Commun.*, 6: 6700.

[31] Jeon, N. J., et al., (2014). Solvent engineering for high-performance inorganic-organic hybrid perovskite solar cells. *Nat. Mater.*, 13: 897.

[32] Grätzel, M., (2001). Photoelectrochemical cells. *Nature*, 414: 338.

[33] Bach, U., et al., (1998). Solid-state dye-sensitized mesoporous TiO_2 solar cells with high photon-to-electron conversion efficiencies. *Nature*, 395: 583.

[34] Burschka, J., et al., (2013). Sequential deposition as a route to high-performance perovskite-sensitized solar cells. *Nature*, 499: 316.

[35] Heo, J. H., et al., (2013). Efficient inorganic–organic hybrid heterojunction solar cells containing perovskite compound and polymeric hole conductors. *Nat. Photonics*, 7: 486.

[36] Ko, H. S., et al., (2015). 15.76% efficiency perovskite solar cells prepared under high relative humidity: importance of PbI_2 morphology in two-step deposition of $CH_3NH_3PbI_3$. *J. Mater. Chem. A*, 3: 8808.

[37] Jeon, N. J., et al., (2015). Compositional engineering of perovskite materials for high-performance solar cells. *Nature*, 517: 476.

[38] Liu, M., et al., (2013). Efficient planar heterojunction perovskite solar cells by vapour deposition. *Nature*, 501: 395.

[39] Seo, J., et al., (2014). Benefits of very thin PCBM and LiF layers for solution-processed p–i–n perovskite solar cells. *Energy Environ. Sci.*, 7: 2642.

[40] Endres, J., et al., (2016). Valence and Conduction Band Densities of States of Metal Halide Perovskites: A Combined Experimental–Theoretical Study. *J. Phys. Chem. Lett.*, 7: 2722.

[41] Philippe, B., et al., (2017). Valence Level Character in a Mixed Perovskite Material and Determination of the Valence Band Maximum from Photoelectron Spectroscopy: Variation with Photon Energy. *J. Phys. Chem. C*, 121: 26655.

[42] Lindblad, R., et al., (2015). Electronic Structure of $CH_3NH_3PbX_3$ Perovskites: Dependence on the Halide Moiety. *J. Phys. Chem. C*, 119: 1818.

[43] Zhang, Z., et al., (2017). Interplay between Localized and Free Charge Carriers Can Explain Hot Fluorescence in the $CH_3NH_3PbBr_3$ Perovskite: Time-Domain Ab Initio Analysis. *J. Am. Chem. Soc.*, 139: 17327.

[44] Wenger, B., et al., (2017). Consolidation of the optoelectronic properties of $CH_3NH_3PbBr_3$ perovskite single crystals. *Nat. Commun.*, 8: 590.

[45] Kovalenko, M. V., et al., (2017). Properties and potential optoelectronic applications of lead halide perovskite nanocrystals. *Science*, 358: 745.

[46] Akkerman, Q. A., et al., (2018). Genesis, challenges and opportunities for colloidal lead halide perovskite nanocrystals. *Nat. Mater.*, 17: 394.

[47] Hong, K., et al., (2018). Low-dimensional halide perovskites: review and issues. *J. Mater. Chem. C*, 6: 2189.

[48] Sichert, J. A., et al., (2015). Quantum Size Effect in Organometal Halide Perovskite Nanoplatelets. *Nano Lett.*, 15: 6521.

[49] Zhu, Q., et al., (2016). Correlating structure and electronic band-edge properties in organolead halide perovskites nanoparticles. *Phys. Chem. Chem. Phys.*, 18: 14933.

[50] Zheng, K., et al., (2016). Direct Experimental Evidence for Photoinduced Strong-Coupling Polarons in Organolead Halide Perovskite Nanoparticles. *Phys. Chem. Lett.*, 7: 4535.

[51] Telfah, H., et al., (2017). Ultrafast Exciton Dynamics in Shape-Controlled Methylammonium Lead Bromide Perovskite Nanostructures: Effect of Quantum Confinement on Charge Carrier Recombination. *Phys. Chem. C*, 121: 28556.

[52] González-Carrero, S., et al., (2018). Colloids of Naked $CH_3NH_3PbBr_3$ Perovskite Nanoparticles: Synthesis, Stability, and Thin Solid Film Deposition. *ACS Omega*, 3: 1298.

[53] Fu, X., et al., (2018). Controlled synthesis of brightly fluorescent $CH_3NH_3PbBr_3$ perovskite nanocrystals employing $Pb(C_{17}H_{33}COO)_2$ as the sole lead source. *RSC Adv.*, 8: 1132.

[54] Milosavljević, A. R., et al., (2018). Electronic Properties of Free-Standing Surfactant-Capped Lead Halide Perovskite Nanocrystals Isolated in Vacuo. *J. Phys. Chem. Lett.*, 9: 3604.

[55] Batignani, G., et al., (2018). Probing femtosecond lattice displacement upon photo-carrier generation in lead halide perovskite. *Nat. Commun.*, 9: 1971.

[56] Dhar, L., et al., (1994). Time-resolved vibrational spectroscopy in the impulsive limit. *Chem. Rev.*, 94: 157.

[57] Kahan, A., et al., (2007). Following Photoinduced Dynamics in Bacteriorhodopsin with 7-fs Impulsive Vibrational Spectroscopy. *J. Am. Chem. Soc.*, 129: 537.

[58] Schnedermann, C., et al., (2016). Vibronic Dynamics of the Ultrafast all-trans to 13-cis Photoisomerization of Retinal in Channelrhodopsin-1. *J. Am. Chem. Soc.*, 138: 4757.

[59] Droseros, N., et al., (2018). Origin of the Enhanced Photoluminescence Quantum Yield in MAPbBr3 Perovskite with Reduced Crystal Size. *ACS Energy Lett.*, 3: 1458-1466.

[60] Longo, G., et al., (2017). High Photoluminescence Quantum Yields in Organic Semiconductor–Perovskite Composite Thin Films. *Chem Sus Chem*, 10: 3788.

[61] La-Placa, M.-G., et al., (2017). White perovskite based lighting devices. *Chem. Commun.*, 53: 8707.

[62] Xiao, Z., et al., (2017). Efficient perovskite light-emitting diodes featuring nanometre-sized crystallites. *Nat. Photonics*, 11: 108.

[63] Richter, J. M., et al., (2016). Enhancing photoluminescence yields in lead halide perovskites by photon recycling and light out-coupling. *Nat. Commun.*, 7: 13941.

[64] He, H., et al., (2016). Exciton localization in solution-processed organolead trihalide perovskites. *Nat. Commun.*, 7: 10896.

[65] Zheng, K., et al., (2015). Exciton Binding Energy and the Nature of Emissive States in Organometal Halide Perovskites. *J. Phys. Chem. Lett.*, 6: 2969.

[66] Sarritzu, V., et al., (2018). Perovskite Excitonics: Primary Exciton Creation and Crossover from Free Carriers to a Secondary Exciton Phase. *Adv. Opt. Mater.*, 6: 1700839.

[67] Stranks, S. D., et al., (2014). Recombination Kinetics in Organic-Inorganic Perovskites: Excitons, Free Charge, and Subgap States. *Phys. Rev. Appl.*, 2: 034007.

[68] Draguta, S., et al., (2016). Spatially Non-uniform Trap State Densities in Solution-Processed Hybrid Perovskite Thin Films. *J. Phys. Chem. Lett.*, 7: 715.

[69] Guo, Y., et al., (2016). Polymer Stabilization of Lead(II). Perovskite Cubic Nanocrystals for Semitransparent Solar Cells. *Adv. Energy Mater.*, 6: 1502317.

[70] Duan, M., et al., (2017). Boron-Doped Graphite for High Work Function Carbon Electrode in Printable Hole-Conductor-Free Mesoscopic Perovskite Solar Cells. *ACS Appl. Mater. Interfaces*, 9: 31721.

[71] Liao, G., et al., (2018). Toward fast charge extraction in all-inorganic CsPbBr3 perovskite solar cells by setting intermediate energy levels. *Solar Energy*, 171: 279.

[72] Chen, H. Y., et al., (2017). All-Vacuum-Deposited Stoichiometrically Balanced Inorganic Cesium Lead Halide Perovskite Solar Cells with Stabilized Efficiency Exceeding 11%. *Adv. Mater.*, 29: 1605290.

[73] Rong, Y., et al., (2015). Beyond Efficiency: the Challenge of Stability in Mesoscopic Perovskite Solar Cells. *Adv. Energy Mater.*, 5: 1501066.

[74] Liang, J., et al., (2017). $CsPb_{0.9}Sn_{0.1}IBr_2$ Based All-Inorganic Perovskite Solar Cells with Exceptional Efficiency and Stability. *J. Am. Chem. Soc.*, 139: 14009.

[75] Liang, J., et al., (2016). All-Inorganic Perovskite Solar Cells. *J. Am. Chem. Soc.*, 138:15829.

[76] Chang, X., et al., (2016). Carbon-Based CsPbBr$_3$ Perovskite Solar Cells: All-Ambient Processes and High Thermal Stability. *ACS Appl. Mater. Interfaces*, 8: 33649.

[77] Li, H., et al., (2017). Carbon Quantum Dots/TiOx Electron Transport Layer Boosts Efficiency of Planar Heterojunction Perovskite Solar Cells to 19%. *Nano Lett.*, 17: 2328.

[78] Lau, C. F. J., et al., (2016). CsPbIBr$_2$ Perovskite Solar Cell by Spray-Assisted Deposition. *ACS Energy Lett.*, 1: 573.

[79] Zhou, S., et al., (2017). Slow-Photon-Effect-Induced Photoelectrical-Conversion Efficiency Enhancement for Carbon-Quantum-Dot-Sensitized Inorganic CsPbBr$_3$ Inverse Opal Perovskite Solar Cells. *Adv. Mater.*, 29: 1703682.

[80] Kim, Y. H., et al., (2016). Metal halide perovskite light emitters. *Proc. Natl. Acad. Sci.*, 113:11694.

[81] Veldhuis, S. A., et al., (2016). Perovskite Materials for Light-Emitting Diodes and Lasers. *Adv. Mater.*, 28. 6804.

[82] Zhang, J., et al., (2017). Low-Dimensional Halide Perovskites and Their Advanced Optoelectronic Applications. *Nano-Micro Lett.*, 9: 36.

[83] Sutherland, B. R. and Sargent, E. H., (2016). Perovskite photonic sources. *Nat. Photonics,* 10: 295.

[84] Zhao, Y. and Zhu, K., (2016). Organic–inorganic hybrid lead halide perovskites for optoelectronic and electronic applications. *Chem. Soc. Rev.*, 45: 655.

[85] Chen, Q., et al., (2015). Under the spotlight: The organic–inorganic hybrid halide perovskite for optoelectronic applications. *Nano Today*, 10: 355.

[86] Shi, Z., et al., (2017). High-Efficiency and Air-Stable Perovskite Quantum Dots Light-Emitting Diodes with an All-Inorganic Heterostructure. *Nano Lett.*, 17: 313.

[87] Yu, J. C., et al., (2016). Improving the Stability and Performance of Perovskite Light-Emitting Diodes by Thermal Annealing Treatment. *Adv. Mater.*, 28: 6906.
[88] Wu, C., et al., (2017). Improved Performance and Stability of All-Inorganic Perovskite Light-Emitting Diodes by Antisolvent Vapor Treatment. *Adv. Funct. Mater.*, 27: 1700338.
[89] Kulbak, M., et al., (2016). Cesium Enhances Long-Term Stability of Lead Bromide Perovskite-Based Solar Cells. *J. Phys. Chem. Lett.*, 7: 167.
[90] Song, Y. H., et al., (2017). Design of long-term stable red-emitting. $CsPb(Br_{0.4}, I_{0.6})_3$ perovskite quantum dot film for generation of warm white light. *Chem. Eng. J.*, 313: 461.
[91] Song, Y. H., et al., (2016). Long-term stable stacked $CsPbBr_3$ quantum dot films for highly efficient white light generation in LEDs. *Nanoscale*, 8:19523.
[92] Song, Y. H., et al., (2016). Design of water stable green-emitting $CH_3NH_3PbBr_3$ perovskite luminescence materials with encapsulation for applications in optoelectronic device. *Chem. Eng. J.*, 306: 791.
[93] Sun, C., et al., (2016). Efficient and Stable White LEDs with Silica-Coated Inorganic Perovskite Quantum Dots. *Adv. Mater.*, 28:10088.
[94] Song, Y. H., et al., (2018). Design of orange-emitting $CsPb_{0.77}Mn_{0.23}Cl_3$ perovskite film for application in optoelectronic device. *Chem Eng. J.*, 331: 803.
[95] Park, S. Y., et al., (2015). Long-term stability of CdSe/CdZnS quantum dot encapsulated in a multi-lamellar microcapsule. *Nanotechnology*, 26: 275602.
[96] Tonzani, S., (2009). Lighting technology: Time to change the bulb. *Nature*, 459: 312.
[97] Tsao, J. Y., et al., (2014). Toward Smart and Ultra-efficient Solid-State Lighting. *Adv. Opt. Mater.*, 2: 809.
[98] Im, W. B., et al., (2011). Efficient and Color-Tunable Oxyfluoride Solid Solution Phosphors for Solid-State White Lighting. *Adv. Mater.*, 23: 2300.

[99] Wu, J. L., et al, (2007). Structure–property correlations in Ce-doped garnet phosphors for use in solid state lighting. *Chem. Phys. Lett.*, 441: 250.

[100] Lee, S. P., et al., (2014). New Ce^{3+}-Activated Thiosilicate Phosphor for LED Lighting—Synthesis, Luminescence Studies, and Applications. *ACS Appl. Mater. Interfaces*, 6: 7260.

[101] Wu, Y. C., et al., (2011). A Novel Tunable Green- to Yellow-Emitting β-YFS:Ce^{3+} Phosphor for Solid-State Lighting. *ACS Appl. Mater. Interfaces*, 3: 3195.

[102] Roushan, M., et al., (2012). Solution-processable white-light-emitting hybrid semiconductor bulk materials with high photoluminescence quantum efficiency. *Angew. Chem., Int. Ed.*, 51: 436.

[103] Shen, C. C. and Tseng, W.-L., (2009). One-Step Synthesis of White-Light-Emitting Quantum Dots at Low Temperature. *Inorg. Chem.*, 48: 8689.

[104] Boonsin, R., et al., (2015). Development of rare-earth-free phosphors for eco-energy lighting based LEDs. *J. Mater. Chem. C*, 3: 9580.

[105] Ye, S., et al., (2010). Phosphors in phosphor-converted white light-emitting diodes: Recent advances in materials, techniques and properties. *Mater. Sci. Eng. R*, 71: 1.

[106] Silver, J. and Withnall, R., (2008). Color Conversion Phosphors for LEDS; John Wiley & Sons, Ltd: 75.

[107] Fang, X., et al., (2012). Tuning and Enhancing White Light Emission of II–VI Based Inorganic–Organic Hybrid Semiconductors as Single-Phased Phosphors. *Chem. Mater.*, 24: 1710.

[108] Liu, Y., et al., (2012). Dual-Emission from a Single-Phase Eu–Ag Metal–Organic Framework: An Alternative Way to Get White-Light Phosphor. *Chem. Mater.*, 24: 1954.

[109] Liu, Z. F., et al., (2013). Eu^{3+}-doped Tb^{3+} metal–organic frameworks emitting tunable three primary colors towards white light *J. Mater. Chem. C,* 1: 4634.

[110] Shang, M., et al., (2014). How to produce white light in a single-phase host?. *Chem. Soc. Rev.*, 43: 1372.

[111] Saidaminov, M. I., et al., (2016). Pure Cs_4PbBr_6: Highly Luminescent Zero-Dimensional Perovskite Solids. *ACS Energy Lett.*, 1: 840.

[112] Saidaminov, M. I., et al., (2017). Low-Dimensional-Networked Metal Halide Perovskites: The Next Big Thing. *ACS Energy Lett.*, 2: 889.

[113] Zhuang, Z., et al., (2017). Intrinsic Broadband White-Light Emission from Ultrastable, Cationic Lead Halide Layered Materials. *Angew. Chem., Int. Ed.*, 56: 14411.

[114] Hu, T., et al., (2016). Mechanism for Broadband White-Light Emission from Two-Dimensional (110). Hybrid Perovskites. *J. Phys. Chem. Lett.*, 7: 2258.

[115] Mao, L., et al., (2017). White-Light Emission and Structural Distortion in New Corrugated Two-Dimensional Lead Bromide Perovskites. *J. Am. Chem. Soc.*, 139: 5210-5215.

[116] Mao, L., et al., (2017). Tunable White-Light Emission in Single-Cation-Templated Three-Layered 2D Perovskites $(CH_3CH_2NH_3)_4Pb_3Br_{10-x}Cl_x$. *J. Am. Chem. Soc.*, 139: 11956.

[117] Dohner, E. R., et al., (2014). Self-Assembly of Broadband White-Light Emitters. *J. Am. Chem. Soc.*, 136: 1718.

[118] Dohner, E. R., et al., (2014). Intrinsic White-Light Emission from Layered Hybrid Perovskites. *J. Am. Chem. Soc.*, 136: 13154.

[119] Yang, S., et al., (2018). High Color Rendering Index White-Light Emission from UV-Driven LEDs Based on Single Luminescent Materials: Two-Dimensional Perovskites $(C_6H_5C_2H_4NH_3)_2PbBr_xCl_{4-x}$. *ACS Appl. Mater. Interfaces*, 10:15980.

[120] Cai, P., et al., (2018). Bluish-white-light-emitting diodes based on two-dimensional lead halide perovskite $(C_6H_5C_2H_4NH_3)_2PbCl_2Br_2$. *Appl. Phys. Lett.*, 112: 153901.

[121] van der Stam, W., et al., (2017). Highly Emissive Divalent-Ion-Doped Colloidal $CsPb_{1-x}M_xBr_3$ Perovskite Nanocrystals through Cation Exchange. *J. Am. Chem. Soc.*, 139: 4087.

[122] Wang, L., et al., (2017). Scalable Ligand-Mediated Transport Synthesis of Organic–Inorganic Hybrid Perovskite Nanocrystals with Resolved Electronic Structure and Ultrafast Dynamics. *ACS Nano*, 11: 2689.

[123] Zhang, F., et al., (2015). Brightly Luminescent and Color-Tunable Colloidal CH$_3$NH$_3$PbX$_3$ (X = Br, I, Cl) Quantum Dots: Potential Alternatives for Display Technology. *ACS Nano*, 9: 4533.

[124] Yakunin, S., et al., (2015). Low-threshold amplified spontaneous emission and lasing from colloidal nanocrystals of caesium lead halide perovskites. *Nat. Commun.*, 6: 8056.

[125] Su, R., et al., (2017). Room-Temperature Polariton Lasing in All-Inorganic Perovskite Nanoplatelets. *Nano Lett.*, 17: 3982.

[126] Zhang, X., et al., (2016). Enhancing the Brightness of Cesium Lead Halide Perovskite Nanocrystal Based Green Light-Emitting Devices through the Interface Engineering with Perfluorinated Ionomer. *Nano Lett.*, 16: 1415.

[127] Vashishtha, P., et al., (2017). Field-Driven Ion Migration and Color Instability in Red-Emitting Mixed Halide Perovskite Nanocrystal Light-Emitting Diodes. *Chem. Mater.*, 29: 5965.

[128] Lou, S., et al., (2017). Nanocomposites of CsPbBr$_3$ perovskite nanocrystals in an ammonium bromide framework with enhanced stability. *J. Mater. Chem. C*, 5: 7431.

[129] Xuan, T., et al., (2017). Highly stable CsPbBr$_3$ quantum dots coated with alkyl phosphate for white light-emitting diodes. *Nanoscale*, 9: 15286.

[130] Parobek, D., et al., (2016). Exciton-to-Dopant Energy Transfer in Mn-Doped Cesium Lead Halide Perovskite Nanocrystals. *Nano Lett.*, 16: 7376.

[131] Jahandar, M., et al., (2016). Highly efficient metal halide substituted CH$_3$NH$_3$I(PbI$_2$)$_{1-x}$(CuBr$_2$)$_X$ planar perovskite solar cells. *Nano Energy*, 27: 330.

[132] Abdi-Jalebi, M., et al., (2016). Impact of Monovalent Cation Halide Additives on the Structural and Optoelectronic Properties of CH$_3$NH$_3$PbI$_3$ Perovskite. *Adv. Energy Mater.*, 6: 1502472.

[133] Mir, W. J., et al., (2017). Colloidal Mn-Doped Cesium Lead Halide Perovskite Nanoplatelets. *ACS Energy Lett.*, 2: 537.

[134] Adhikari, S. D., et al., (2017). Chemically Tailoring the Dopant Emission in Manganese-Doped CsPbCl₃. Perovskite Nanocrystals *Angew. Chem., Int. Ed.*, 56: 8746.
[135] Zou, S., et al., (2017). Stabilizing Cesium Lead Halide Perovskite Lattice through Mn(II) Substitution for Air-Stable Light-Emitting Diodes. *J. Am. Chem. Soc.*, 139: 11443.
[136] Liu, W., et al., (2016). Mn^{2+}-Doped Lead Halide Perovskite Nanocrystals with Dual-Color Emission Controlled by Halide Content. *J. Am. Chem. Soc.*, 138: 14954.
[137] Rossi, D., et al., (2017). Dynamics of Exciton–Mn Energy Transfer in Mn-Doped CsPbCl₃ Perovskite Nanocrystals. *J. Phys. Chem. C*, 121: 17143.
[138] Li, F., et al., (2017). Optical properties of Mn^{2+} doped cesium lead halide perovskite nanocrystals via a cation–anion co-substitution exchange reaction. *J. Mater. Chem. C*, 5: 9281.
[139] Akkerman, Q. A., et al., (2015). Tuning the Optical Properties of Cesium Lead Halide Perovskite Nanocrystals by Anion Exchange Reactions. *J. Am. Chem. Soc.*, 137: 10276.
[140] Guhrenz, C., et al., (2016). Solid-State Anion Exchange Reactions for Color Tuning of CsPbX₃ Perovskite Nanocrystals. *Chem. Mater.*, 28: 9033.
[141] Li, F., et al., (2018). High Br⁻ Content $CsPb(Cl_yBr_{1-y})_3$ Perovskite Nanocrystals with Strong Mn^{2+} Emission through Diverse Cation/Anion Exchange Engineering. *ACS Appl. Mater. Interfaces*, 10:11739.
[142] Li, X., et al., (2017). All Inorganic Halide Perovskites Nanosystem: Synthesis, Structural Features, Optical Properties and Optoelectronic Applications. *Small*, 13: 1603996.
[143] Guo, Y., et al., (2017). Structural Stabilities and Electronic Properties of High-Angle Grain Boundaries in Perovskite Cesium Lead Halides. *J. Phys. Chem. C*, 121: 1715.
[144] Van Le, Q., et al., (2017). Investigation of Energy Levels and Crystal Structures of

Cesium Lead Halides and Their Application in Full-Color Light-Emitting Diodes. *Adv. Electron. Mater.*, 3: 1600448.

[145] Deng, W., et al., (2016). Organometal Halide Perovskite Quantum Dot Light-Emitting Diodes. *Adv. Funct. Mater.*, 26: 4797.

[146] Zhang, X., et al., (2016). All-Inorganic Perovskite Nanocrystals for High-Efficiency Light Emitting Diodes: Dual-Phase $CsPbBr_3$-$CsPb_2Br_5$ Composites. *Adv. Funct. Mater.*, 26: 4595.

[147] Le, Q. V., et al., (2017). Structural Investigation of Cesium Lead Halide Perovskites for High-Efficiency Quantum Dot Light-Emitting Diodes. *J. Phys. Chem. Lett.*, 8: 4140.

[148] Dastidar, S., et al., (2017). Slow Electron–Hole Recombination in Lead Iodide Perovskites Does Not Require a Molecular Dipole. *ACS Energy Lett.*, 2: 2239.

[149] Akbulatov, A. F., et al., (2017). Probing the Intrinsic Thermal and Photochemical Stability of Hybrid and Inorganic Lead Halide Perovskites. *J. Phys. Chem. Lett.*, 8: 1211.

[150] Xiao, C., et al., (2015). Mechanisms of Electron-Beam-Induced Damage in Perovskite Thin Films Revealed by Cathodoluminescence Spectroscopy. *J. Phys. Chem. C*, 119: 26904.

[151] Akkerman, Q. A., et al., (2016). Strongly emissive perovskite nanocrystal inks for high-voltage solar cells. *Nat. Energy*, 2: 16194.

[152] Zhang, X., et al., (2019) Photoresponse of nonvolatile resistive memory device based on all-inorganic perovskite $CsPbBr_3$ nanocrystals. *J. Phys. D: Appl. Phys.* 52: 125103.

[153] Jin, Y., et al., (2019) Morphology control of $CsPbBr_3$ films by a surface active Lewis base for bright all-inorganic perovskite light-emitting diodes. Appl. Phys. Lett. 114: 163302.

[154] Liu, C., et al., (2017). Ultra-thin MoO_x as cathode buffer layer for the improvement of all-inorganic $CsPbIBr_2$ perovskite solar cells. *Nano Energy*, 41: 75.

[155] Zhang, Q., et al., (2019). Light Processing Enables Efficient Carbon-Based, All-Inorganic Planar $CsPbIBr_2$ Solar Cells with High Photovoltages. *ACS Appl. Mater. Interfaces*, 11: 2997-3005.

[156] Ma, Q., et al., (2016). Hole Transport Layer Free Inorganic CsPbIBr$_2$ Perovskite Solar Cell by Dual Source Thermal Evaporation. *Adv. Energy Mater.*, 6: 1502202.

[157] Eperon, G. E., et al., (2015). Inorganic caesium lead iodide perovskite solar cells. *J. Mater. Chem. A*, 3: 19688.

[158] Frolova, L. A., et al., (2017). Highly Efficient All-Inorganic Planar Heterojunction Perovskite Solar Cells Produced by Thermal Coevaporation of CsI and PbI$_2$. *J. Phys. Chem. Lett.*, 8: 67.

[159] Swarnkar, A., et al., (2016). Quantum dot-induced phase stabilization of α-CsPbI$_3$ perovskite for high-efficiency photovoltaics. *Science*, 354: 92.

[160] Zhang, T., et al., (2017). Bication lead iodide 2D perovskite component to stabilize inorganic α-CsPbI$_3$ perovskite phase for high-efficiency solar cells. *Sci. Adv.*, 3: e1700841.

[161] Hu, Y., et al., (2017). Bismuth Incorporation Stabilized α-CsPbI$_3$ for Fully Inorganic Perovskite Solar Cells. *ACS Energy Lett.*, 2: 2219.

[162] Chen, C. Y., et al., (2017). All-Vacuum-Deposited Stoichiometrically Balanced Inorganic Cesium Lead Halide Perovskite Solar Cells with Stabilized Efficiency Exceeding 11%. *Adv. Mater.*, 29: 1605290.

[163] Niezgoda, J. S., et al., (2017). Improved Charge Collection in Highly Efficient CsPbBrI$_2$ Solar Cells with Light-Induced Dealloying. *ACS Energy Lett.*, 2: 1043.

[164] Nam, J. K., et al., (2017). Potassium Incorporation for Enhanced Performance and Stability of Fully Inorganic Cesium Lead Halide Perovskite Solar Cells. *Nano Lett.*, 17: 2028.

[165] Lau, C. F. J., et al., (2017). Strontium-Doped Low-Temperature-Processed CsPbI$_2$Br Perovskite Solar Cells. *ACS Energy Lett.*, 2: 2319.

[166] Zhang, J., et al., (2018). Solar Cells: 3D-2D-0D Interface Profiling for Record Efficiency All-Inorganic CsPbBrI$_2$ Perovskite Solar Cells with Superior Stability.*Adv. Energy Mat.*, 8: 1703246.

[167] Sanehira, E. M., et al., (2017). Enhanced mobility CsPbI$_3$ quantum dot arrays for record-efficiency, high-voltage photovoltaic cells. *Sci. Adv.*, 3: eaao4204.

[168] Lin, J., et al., (2018). Thermochromic halide perovskite solar cells *Nat. Mater.*, 1: 261.
[169] Wang, Q., et al., (2018). Solar Cells: Diffraction-Grated Perovskite Induced Highly Efficient Solar Cells through Nanophotonic Light Trapping. *Adv. Energy Mater.*, 8: 1800007.
[170] Xiang, S., et al., (2018). Highly Air-Stable Carbon-Based α-CsPbI$_3$ Perovskite Solar Cells with a Broadened Optical Spectrum. *ACS Energy Lett.*, 3: 1824-1831
[171] Zarick, H. F., et al., (2018). Mixed halide hybrid perovskites: a paradigm shift in photovoltaics. *J. Mater. Chem. A*, 6: 5507.
[172] Luo, P., et al., (2016). Solvent Engineering for Ambient-Air-Processed, Phase-Stable CsPbI$_3$ in Perovskite Solar Cells. *J. Phys. Chem. Lett.*, 7: 3603.
[173] Kumar, J., et al., (2019). Halide Perovskite Photovoltaics: Background, Status, and Future Prospects. Chem. Rev. 119: 3036-3103.
[174] Chen, G., et al., (2019) Stable α-CsPbI$_3$ Perovskite Nanowire Arrays with Preferential Crystallographic Orientation for Highly Sensitive Photodetectors, *Adv. Funct. Mat.* 29: 1808741
[175] Saliba, M., et al., (2016). Incorporation of rubidium cations into perovskite solar cells improves photovoltaic performance. *Science*, 354: 206.
[176] Arora, N., et al., (2017). Perovskite solar cells with CuSCN hole extraction layers yield stabilized efficiencies greater than 20%. *Science*, 358: 768.
[177] Liu, C., et al., (2019). Interfacial electronic structures of MoO$_x$/mixed perovskite photodetector. *Org. Electronics: Phys. Mat. App*: 75: 162-169
[178] Liu, C., et al., (2018). All-Inorganic CsPbI$_2$Br Perovskite Solar Cells with High Efficiency Exceeding 13%. *J. Am. Chem. Soc.*, 140: 3825.
[179] Chen, A., et al., (2018). Low-voltage all-inorganic perovskite quantum dot transistor memory *Appl. Phys. Lett.*, 112: 212101

[180] Li, W., et al., (2016). Addictive-assisted construction of all-inorganic CsSnIBr$_2$ mesoscopic perovskite solar cells with superior thermal stability up to 473 K. *J. Mater. Chem. A*, 4: 17104.

[181] Zhang, X., et al. (2019) Efficient and carbon-based hole transport layer-free CsPbI$_2$Br planar perovskite solar cells using PMMA modification *J. Mat. Chem. C: Mat. Optical Electronic Devices*, 7: 3852-3861.

[182] Dong, C., et al., (2019). Anti-solvent assisted multi-step deposition for efficient and stable carbon-based CsPbI$_2$Br all-inorganic perovskite solar cell *Nano Energy*, 59: 553-559.

[183] Guo, Y., et al., (2019). Efficient and hole-transporting-layer-free CsPbI$_2$Br planar heterojunction perovskite solar cells through rubidium passivation *ChemSusChem*, 12: 983-989.

[184] Tavakoli, M. M., et al., (2018) Multilayer evaporation of MAFAPbI$_{3-x}$Cl$_x$ for the fabrication of efficient and large-scale device perovskite solar cells, *J. Phys. D: Applied Phys.*, 52: 034005.

[185] Sutton, R. J., et al., (2016). Bandgap-Tunable Cesium Lead Halide Perovskites with High Thermal Stability for Efficient Solar Cells. *Adv. Energy Mater.*, 6: 1502458.

[186] Fu, L., et al., (2018). A fluorine-modulated bulk-phase heterojunction and tolerance factor for enhanced performance and structure stability of cesium lead halide perovskite solar cells. *J. Mater. Chem. A*, 6: 13263.

[187] Stoumpos, C. C., et al., (2013). Semiconducting Tin and Lead Iodide Perovskites with Organic Cations: Phase Transitions, High Mobilities, and Near-Infrared Photoluminescent Properties. *Inorg. Chem.*, 52: 9019.

[188] Eperon, G. E., et al., (2016). Perovskite-perovskite tandem photovoltaics with optimized band gaps. *Science*, 354: 861.

[189] Noel, N. K., et al., (2014). Lead-free organic-inorganic tin halide perovskites for photovoltaic applications. *Energy Environ. Sci.*, 7: 3061.

[190] Hao, F., et al., (2014). Lead-free solid-state organic-inorganic halide perovskite solar cells *Nat. Photonics*, 8: 489.

[191] Manser, J. S., et al., (2016). Intriguing Optoelectronic Properties of Metal Halide Perovskites. *Chem. Rev.*, 116: 12956.

[192] Leijtens, T., et al., (2017). Mechanism of Tin Oxidation and Stabilization by Lead Substitution in Tin Halide Perovskites. *ACS Energy Lett.*, 2: 2159.

[193] Takahashi, Y., et al., (2011). Charge-transport in tin-iodide perovskite $CH_3NH_3SnI_3$: origin of high conductivity. *Dalton Trans.*, 40: 5563.

[194] Babayigit, A., et al., (2016). Toxicity of organometal halide perovskite solar cells. *Nat. Mater.*, 15: 247.

[195] Babayigit, A., et al., (2016). Assessing the toxicity of Pb- and Sn-based perovskite solar cells in model organism Danio rerio. *Sci. Rep.*, 6: 18721.

[196] Yokoyama, T., et al., (2016). Overcoming Short-Circuit in Lead-Free $CH_3NH_3SnI_3$ Perovskite Solar Cells via Kinetically Controlled Gas-Solid Reaction Film Fabrication Process. *J. Phys. Chem. Lett.*, 7: 776.

[197] Konstantakou, M., et al., (2017). A critical review on tin halide perovskite solar cells. *J. Mater. Chem. A*, 5: 11518.

[198] Ogomi, Y., et al., (2014). $CH_3NH_3Sn_xPb_{(1-x)}I_3$ Perovskite Solar Cells Covering up to 1060 nm. *J. Phys. Chem. Lett.*, 5: 1004.

[199] Liu, X., et al., (2016). Improved efficiency and stability of Pb-Sn binary perovskite solar cells by Cs substitution. *J. Mater. Chem. A*, 4: 17939.

[200] Hsu, H. Y., et al., (2016). Optimization of Lead-free Organic-inorganic Tin(II) Halide Perovskite Semiconductors by Scanning Electro-chemical Microscopy. *Electrochim. Acta*, 220: 205.

[201] Yang, Z. B., et al., (2016). Mechanosensing Controlled Directly by Tyrosine Kinases. *Nano Lett.*, 16: 7739.

[202] Hu, M., et al., (2016). Stabilized Wide Bandgap $MAPbBr_xI_{3-x}$ Perovskite by Enhanced Grain Size and Improved Crystallinity. *Adv. Sci.*, 3 : 1500301.

[203] Tang, L. C., et al., (2009). First Principles Calculations of Linear and Second-Order Optical Responses in Rhombohedrally Distorted Perovskite Ternary Halides, $CsGeX_3$ (X = Cl, Br, and I). *Jpn. J. Appl. Phys.*, 48: 112402.

[204] Sun, P. P., et al., (2016). Theoretical insights into a potential lead-free hybrid perovskite: substituting Pb^{2+} with Ge^{2+}. *Nanoscale,* 8: 1503.
[205] Stoumpos, C. C., et al., (2015). Hybrid Germanium Iodide Perovskite Semiconductors: Active Lone Pairs, Structural Distortions, Direct and Indirect Energy Gaps, and Strong Nonlinear Optical Properties. *J. Am. Chem. Soc.*, 137: 6804.
[206] Krishnamoorthy, T., et al., (2015). Lead-free germanium iodide perovskite materials for photovoltaic applications. *J. Mater. Chem. A,* 3: 23829.
[207] Huang, C., et al., (2014). *Patent* CN201410173750.
[208] Shannon, R. D., (1976). Revised effective ionic radii and systematic studies of interatomic distances in halides and chalcogenides. *Acta Crystallogr., Sect. A: Cryst. Phys., Diffr., Theor. Gen. Crystallogr.,* 32: 751.
[209] Navas, J., et al., (2015). New insights into organic-inorganic hybrid perovskite $CHNH_3PbI_3$ nanoparticles. An experimental and theoretical study of doping in Pb^{2+} sites with Sn^{2+}, Sr^{2+}, Cd^{2+} and Ca^{2+}. *Nanoscale,* 7: 6216.
[210] Jacobsson, T.J., et al., (2015). Goldschmidt's Rules and Strontium Replacement in Lead Halogen Perovskite Solar Cells: Theory and Preliminary Experiments on $CH_3NH_3SrI_3$. *J. Phys. Chem. C,* 119: 25673.
[211] Giorgi, G., et al., (2014). Cation Role in Structural and Electronic Properties of 3D Organic-Inorganic Halide Perovskites: A DFT Analysis. *J. Phys. Chem. C,*118: 12176.
[212] Yusoff A. B. and Nazeeruddin M. K., (2016). Organohalide Lead Perovskites for Photovoltaic Applications. *J. Phys. Chem. Lett.*, 7: 851.
[213] Pazoki, M., et al., (2016). Effect of metal cation replacement on the electronic structure of metalorganic halide perovskites: Replacement of lead with alkaline-earth metals. *Phys. Rev. B,* 93: 144105.
[214] Filip M. R. and F. Giustino, F., (2016). Computational Screening of Homovalent Lead Substitution in Organic-Inorganic Halide Perovskites. *J. Phys. Chem. C,* 120: 166.

[215] Klug, M. T., et al., (2017). Tailoring metal halide perovskites through metal substitution: influence on photovoltaic and material properties. *Energy Environ. Sci.*, 10: 236.
[216] Frolova, L. A., et al., (2016). Exploring the Effects of the Pb^{2+} Substitution in $MAPbI_3$ on the Photovoltaic Performance of the Hybrid Perovskite Solar Cells. *J. Phys. Chem. Lett.*,7: 4353.
[217] Liang, K. and Mitzi, D. B., (1999). US pat. US5882548A.
[218] Gahane, D. H., et al., (2012). in Proceedings of the 16th International Conference on Luminescence and Optical Spectroscopy of Condensed Matter: Posters, ed. S. C. Rand, T. Carmon and M. Jarrahi, Elsevier Science Bv, Amsterdam: 42–45.
[219] Suta, M., et al., (2016). Photoluminescence properties of Yb^{2+} ions doped in the perovskites $CsCaX_3$ and $CsSrX_3$ (X = Cl, Br, and I) - a comparative study. *Phys. Chem. Chem. Phys.*,18: 13196.
[220] Suta M. and C. Wickleder, C., (2015). Photoluminescence of $CsMI_3$: Eu^{2+} (M = Mg, Ca, and Sr) - a spectroscopic probe on structural distortions. *J. Mater. Chem. C,* 3: 5233.
[221] Grimm, J., et al., (2006). Light-Emission and Excited-State Dynamics in Tm^{2+} Doped $CsCaCl_3$, $CsCaBr_3$, and $CsCaI_3$.*J. Phys. Chem. B,* 110: 2093.
[222] Abdelhady, A. L., et al., (2016). Heterovalent Dopant Incorporation for Bandgap and Type Engineering of Perovskite Crystals. *J. Phys. Chem. Lett.*, 7: 295.
[223] Palankovski, V., et al., (1999). Study of dopant-dependent band gap narrowing in compound semiconductor devices. *Mater. Sci. Eng.*, B, 66: 46.
[224] Zhang, J., et al., (2016). n-Type Doping and Energy States Tuning in $CH_3NH_3Pb_{1-x}Sb_{2x/3}I_3$ Perovskite Solar Cells. *ACS Energy Lett.*, 1: 535.
[225] Yadav, P., et al., (2015). Exploring the performance limiting parameters of perovskite solar cell through experimental analysis and device simulation. *Solar Energy*, 122: 773.
[226] Zhang, A., et al., (2016). Silicon-Based Integrated Microwave Photonics. *IEEE Journal of Quantum Electronics,* 52: 1.
[227] http://scaps.elis.ugent.be.

[228] Kalashnikov S. G. and Bonch-Bruevich V. L. (1990) *The Physics of Semiconductors* (Russian Edition) (Moscow: Main Editorial Board for Physical and Mathematical Literature)

[229] Kato, M., et al., (2017). Universal rules for visible-light absorption in hybrid perovskite materials. *J. Appl. Phys.*, 121: 115501.

[230] Gao, P., et al., (2014). Organohalide lead perovskites for photovoltaic applications. *Energy Environ. Sci.*, 7: 2448.

[231] Martynov, Y. B., (1999). *Comput. Math. Math. Phys.*, 39: 292.

[232] Yang, W. S., et al., (2017). Iodide management in formamidinium-lead-halide-based perovskite layers for efficient solar cells. *Science*, 356: 1376.

[233] Eperon, G. E., et al., (2014). Morphological Control for High Performance, Solution-Processed Planar Heterojunction Perovskite Solar Cells. *Adv. Funct. Mater.*, 24: 151.

[234] Liu, D. and Kelly, T. L., (2014). Perovskite solar cells with a planar heterojunction structure prepared using room-temperature solution processing techniques. *Nature Photonics*, 8: 133.

[235] Sun, S., et al., (2014). The origin of high efficiency in low-temperature solution-processable bilayer organometal halide hybrid solar cells. *Energy Environ. Sci.*, 7: 399.

[236] Martynov, Y. B., et al., (2017). On the efficiency limit of ZnO/CH$_3$NH$_3$PbI$_3$/CuI perovskite solar cells. *Phys. Chem. Chem. Phys.*, 19: 19916.

[237] Dong, Q., et al., (2015). Electron-hole diffusion lengths >175 μm in solution-grown CH$_3$NH$_3$PbI$_3$ single crystals. *Science*, 347: 967.

[238] Nelson, J. (2003) *The Physics of Solar Cells* (London: Imperial College Press).

In: Perovskite Solar Cells
Editor: Murali Banavoth

ISBN: 978-1-53615-858-8
© 2019 Nova Science Publishers, Inc.

Chapter 4

MOLECULAR ENGINEERING OF THE PEROVSKITES: DYNAMICS/KINETICS OF THE PHOTOVOLTAIC BEHAVIORS

Foroogh Arkan and Mohammad Izadyar[*]
Computational Chemistry Research Laboratory,
Department of Chemistry, Faculty of Science,
Ferdowsi University of Mashhad, Mashhad, Iran

ABSTRACT

Halide perovskites as a unique class of the solar cells are of interest, due to their especial physicochemical properties, low energy barrier of the crystal formation, low cost synthesis and high efficiency. Here, we investigate the optoelectronic characteristics through the quantum chemistry reactivity indices, structural properties and the analysis of the excitons behavior of the perovskites. The size and functionality of the cations/anions of the perovskite structure affect the dimensions, stability and photovoltaic behavior of the perovskite. Tolerance factor, frontier molecular orbital energies, absorption modes of the sensitizer/TiO_2, the

---

[*] Corresponding Author's Email: izadyar@um.ac.ir.

interaction of the photosensitizers and surrounding play a key role in the response of the photosensitizers. Also, the likeness of the perovskite, electrodes and hole/electron transporting materials can decrease the contact resistance and charge injection barrier in the PSCs. Moreover, current and voltage are affected by the structure of the solar cell, electron lifetime, regeneration and recombination processes. The investigation of the possible correlations between the photovoltaic properties and quantum chemistry parameters is helpful to have an insight into the perovskite solar cells efficiency from the molecular approach.

Keywords: halide perovskite, quantum reactivity indice, tolerance factor, exiton, energy gap

INTRODUCTION

Physical Chemistry and Performance of the Solar Cells

The physicochemical properties of the materials of the solar cells manage their performance in different ways. For example, structural modifications, sensitiser/TiO$_2$ absorption modes, the response of the sensitizers to the light, different interactions between the photosensitisers and other materials involved in the system affect the performance. Due to the dependency of the main photovoltaic parameters of the sensitisers to molecular orbital properties and intramolecular electronic transitions, ground and excited state properties play an important role in the efficiency of the solar cells. For example, structural modification of the photosensitizers changes the frontier molecular orbital energy levels and alters the distribution of the HOMO/LUMO, which in turn affects the electron-hole separation indices, electronic charge transfer and spectroscopic properties. Moreover, the exciton behaviours, such as the rate of exciton formation and dissociation change the final voltage of the solar cell.

Structural Properties of the Perovskites

Solid state perovskite solar cells (PSCs) are promising photovoltaic devices originated from the dye-sensitized solar cells (DSSCs) [1, 2]. There are three outstanding issues which are important in designing efficient PSC devices. One of the most interesting issues is related to the noteworthy growth of the efficiency of the PSCs 3.8% in 2009 and 38% in 2017 [3, 4]. The other issue returns to perovskites applications in different technologies such as transistors, light-emitting diodes and solar cells [5-7]. The last issue is related to their unique characteristics such as superior optoelectronic properties, long carrier lifetime, direct band gap, low cost, high carrier mobility [8-11]. Despite these properties, there are some serious disadvantages including the presence of a toxic lead element in the perovskite structure and poor stability upon long-term exposure to the ambient atmosphere [12].

Before discussing the molecular design and photo-physicochemical properties of the perovskites, an overview of the types of the perovskite structures and the factors affecting their efficiency is necessary.

Generally, there are different classes of the solid structures, in which non close packed class such as TiO_2 and perovskites are of great importance [13] The crystal structure of $CaTiO_3$ is the first perovskite discovered in 1839 [14]. The possible elements of the ABO_3 perovskite structure are shown in Figure 1 [15].

In the case of organic-inorganic hybrid perovskites, ABX_3, A cation is organic. Some of A cations incorporated in these perovskite structures consist of Ammonium $[NH_4]^+$, Hydroxylammonium $[H_3NOH]^+$, Thiazolium $[C_3H_4NS]^+$, Formamidinium $[NH_2(CH)NH_2]^+$, Dimethylammonium $[(CH_3)_2NH_2]^+$, Imidazolium $[C_3N_2H_5]^+$, Dabconium $[C_6H_{14}N_2]^{2+}$, Methylammonium $[CH_3NH_3]^+$, 3-Pyrollinium $[NC_4H_8]^+$, Ethylammonium $[C_2H_5NH_3]^+$, Azetidinium $[(CH2)_3NH_2]^+$, and Tropylium $[C_7H_7]^+$ (Figure 2) [16].

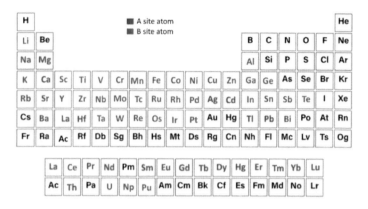

Figure 1. The elements incorporated at A/B cation site in the perovskite structures.

Halide-based perovskites, as a unique class emerged in the photovoltaic devices, have a low-cost synthesis and high efficiency. These perovskites are prepared at low temperature through the self-assembly mechanism from the solution phase into the semiconductor thin films. They have a low energetic barrier of the crystal formation within an electronic/optical tunability [17, 18]. Moreover, the extended range of absorption spectra of the mixed halide perovskites is useful for multi-junction PVSCs and fabrication of hybrid perovskite tandem cells with silicon [19]. However, these perovskites are not stable in a humid environment. Because of a large ionic radii, which precludes the incorporation of small metal ions in the octahedral coordination geometry, metal cations can be selected only from the elements including the alkaline earth, bivalent rare earth, and the heavier elements of group 14 (Ge^{2+}, Sn^{2+}, Pb^{2+}) [20].

Molecular engineering of the perovskite structures is interesting, because microscopic properties of the PSCs are tuned through a subatomic process. Some of these microscopic information consists of quantum coherence, intrinsic behaviour of the materials, molecular orbital energy, electronic charge transfer and photovoltaic properties [21-23]. Moreover, through this approach, it is possible to analyse various spatial orientations of the A/B cations and X anion relative to each other in some organic-inorganic halide perovskite molecules. In Figure 3 shows different configurations of the perovskite elements within the relative electronic energy [21].

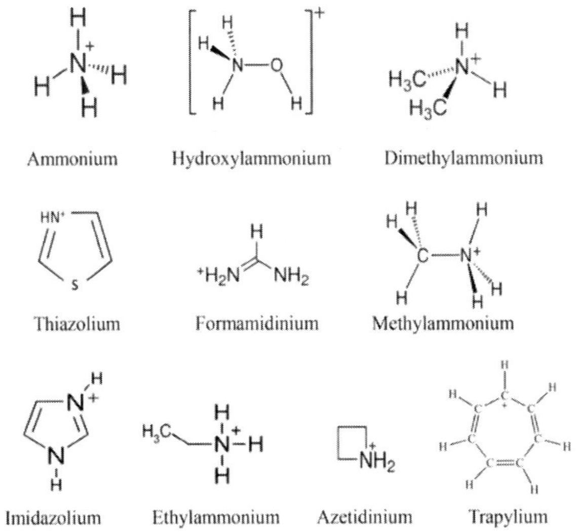

Molecular Engineering of the Perovskites 179

Figure 2. Chemical structures of some organic A cations applied in PSC.

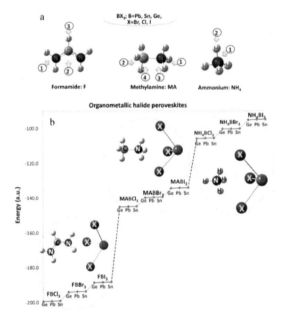

Figure 3. a) Different probable configurations of MAPbX$_3$, FBX$_3$ and NH$_4$BX$_3$ perovskites. b) Relative electronic energies [21].

According to this figure, a cation in contrast to B strongly affects the stability of these perovskites, and chloride-based perovskites are the most stable structures in comparison to the other halide-based ones. Different types of perovskite networks including layered, double perovskites and metal-deficient perovskites are constructed through the diversity of octahedral unit connections as shown in Figure 4 [24].

The unit cell of the ABX_3 perovskite structure is a BX_6 octahedron, where B is at the centre surrounded by six bonded Xs. The crystal network of the perovskite is created through the iteration of the unit cells by corner connection. The cations are located in 12 coordinated holes between the consecutive unit cells [25]. Geometrical structure of perovskite is shown in Figure 5.

The torsion of the angles and symmetry changes affect B-X-B angle in the perovskite structure. Also, the size of the cation, cation/metal interaction and octahedral unit linkage are other important factors in the perovskite stability [24].

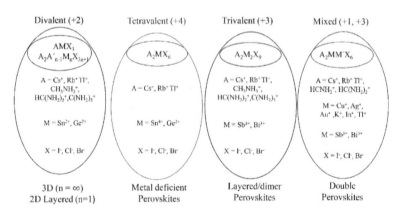

Figure 4. Different types of perovskite networks.

There are different crystal structures of the perovskites. For example, crystal structures of cubic, tetragonal (a=b≠c; α=β=γ=90) and orthorhombic (a≠b≠c; α=β=γ=90) have been indicated for $MAPbX_3$ by using X-ray diffraction method (XRD) [27]. The metal/anions ratio, solvent reactivity, temperature and pressure of the reaction and synthetic route are other effective factors on the structural and physicochemical characteristics [28, 29].

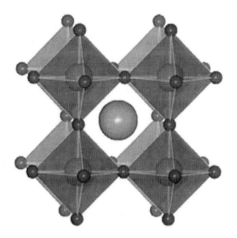

Figure 5. ABX₃ perovskite structure (A cations (blue), B cations (green), and X anions (red)) [26].

The structural perfection of the ABX_3 perovskite is determined by the tolerance factor, t (Eq. 1) [30]:

$$t = \frac{r_A + r_X}{\sqrt{2}(r_B + r_X)} \tag{1}$$

Where, r_A, r_B, and r_X are the ionic radii of A, B, and halide. An increase in the radius of B and a decrease in A cation decreases the tolerance factor. The perovskite will be cubic if $0.9 < t < 1.0$, orthorhombic if $0.75 < t < 0.9$, and hexagonal if the value of t drops below 0.75 [31]. For example, studies showed that the FAPbI₃ perovskite with a cubic structure contains a greater t factor in comparison to the MAPbBr₃ and MAPbI₃ with the tetragonal structure. A large deviation of t from 1 decreases the stability of the perovskite structure. For example, it was shown that ABO₃ perovskites exhibit enhanced electronic/magnetic properties, if their $t \approx 1$ [30]. Among the perovskites hybrid iodide type has a high tolerance factor ($1.06 < t < 1.07$) [32]. Pseudohalogen anions having a similar ionic radius and chemical properties with halogens affect the tolerance factor [32].

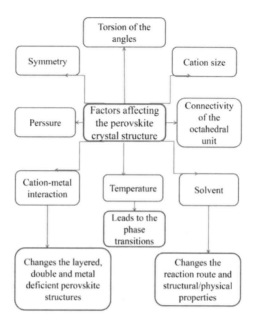

Scheme 1. The main factors affecting the perovskite structures.

Organic cation and *t* factor in ABX$_3$ affect the structural formation, stability and optoelectronic properties of the perovskites [33]. The size and functionality of A cation play a main role in dimensionality/stability of perovskite lattice. Moreover, an efficient cation substitution to make a more stable and appropriate position of the CB of the perovskite is selected based on the *t* factor on the BX6 octahedral. [34]. The tolerance factor for double perovskites, with the mixed *A*-site A′$_{2-x}$A″$_x$B′B″O$_6$ can be obtained from Eq. 2 [30]:

$$t = \frac{\left(1-\frac{x}{2}\right)r_{A'} + \frac{x}{2}r_{A''} + r_O}{\sqrt{2}\left(\frac{r_{B'}}{2} + \frac{r_{B''}}{2} + r_O\right)} \quad (2)$$

where, $r_{A'}$, $r_{A''}$, $r_{B'}$, and $r_{B''}$ are the ionic radii of the respective ions. For example, the calculated *t* factor of Sr$_2$FeMoO$_6$ is 0.990. Substitution at *A* site, to bring the *t* factor closer to 1, increases the low-field room-temperature magnetoresistance of Sr$_2$FeMoO$_6$, which is of interest in possible applications [30].

Finally, tolerance factor strongly influences the stability of the perovskite structure, crystalline phase (cubic, orthorhombic and hexagonal), the CB of the perovskites, optoelectronic properties, electronic/magnetic properties, performance of the perovskites, selection of efficient substitutions and lifetime of carriers, the interaction between nS orbital of the B cation and nP orbital of the X anion, band gap and in turn absorption spectra and light harvesting capability of the perovskites. For example, an increase in the band gap decreases the interaction of B and X ions in the perovskites. Scheme 1 shows different parameters that affect the perovskite structures.

The Dependence of the Optoelectronic Characteristics on the HOMO-LUMO Energies

Generally, main parameters in predicting the optoelectronic characteristics of the perovskites include the radius of A cation, cell volume and symmetry, electronic properties and bond length [35]. A fine tuning of the electronic properties of organic/inorganic molecular semiconductors requires adjustment of the HOMO-n/LUMO+n level and their bandgap energies. This is crucial for the perovskite solar cells, because the functionality of the materials is evaluated through the alignment of their energy levels. On the other hand, a mismatch between the hole/electron transporting materials (HTM/ETM) increases the contact resistance and injection barrier in the photovoltaic routes [36]. Moreover, investigation of the frontier energy levels alignment improves our knowledge about the probability of the photovoltaic processes. Such information can be obtained through the natural bond orbital (NBO) analysis, which provides a deep understanding of the charge transfer processes at the perovskite/TiO$_2$ and perovskite/solid electrolyte interfaces. Therefore, it is of great importance to optimize the positions of the energy levels of the materials involved in the PSC. This is possible through a change in an efficient substitution of the elements in the perovskite structures.

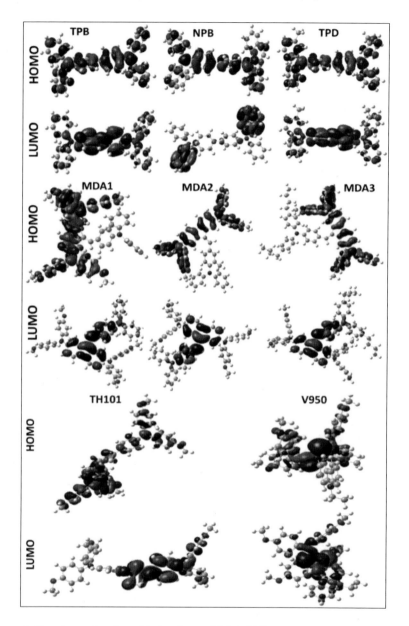

Figure 6. Frontier molecular orbitals of some HTMs [21].

In the most cases, a change in the electron donors leads to a tuning in the HOMO energy levels. In the case of the molecules having the strong electron-withdrawing groups, a change in this group affects the LUMO energy levels. Moreover, the extension of the π-spacer moiety increases the intramolecular charge transfer (ICT), decreases the electron-hole (e-h) overlap index and facilitates the e-h separation, which in turn improves the efficiency of the PSCs. However, such a route extension in the π-conjugation is considered as a barrier to the charge transfer in some cases.

HOMO/LUMO distribution on HTMs such as TPB, DPT, NPB, MDA1, MDA2, MDA3, TH101 and V950, obtained through DFT calculations, indicates the quality of their functionality (Figure 6) [21]. A greater interorbital overlap in the HOMO levels of the neighbouring HTM molecules improves material the tendency of hole transport [37, 38]. Therefore, the delocalisation of the HOMO electron densities of these HTMs on the whole molecule, homogeneously, makes them more appropriate for hole transfer.

Also, another factor for indicating the HTM performance in the PSC is the possibility of the perovskite regeneration calculated by Gibbs energy criterion. Moreover, global electrophilicity (ω) as one of the quantum chemistry indices can affect the regeneration processes. Figure 7 indicates an inverse linear relationship of Gibbs energy of regeneration as a function of ω. This means that perovskites having a lower electrophilicity can be better regenerated because of a greater exciton production and lower charge recombination process.

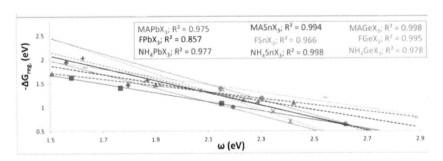

Figure 7. The correlation of $\Delta G_{reg.}$ and ω of the organic-inorganic halide perovskites [21].

Generally, in order to have a positive response to the charge injection in the PSCs, LUMO energy levels of the perovskites should be higher than the conduction band (CB) edge of TiO_2 to facilitate the photoexcited electron injection. On the other hand, the HOMO levels of the perovskites should be lower than HOMO levels of the HTMs to have an efficient hole extraction from the perovskites and regenerate the oxidised perovskites. A favourable alignment of the energy levels of the materials mentioned above is shown in Figure 8.

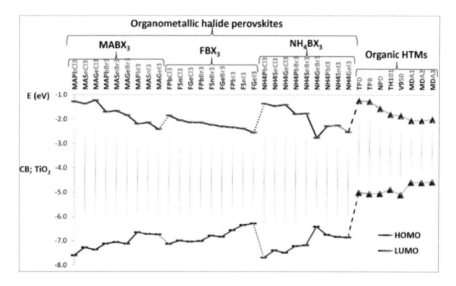

Figure 8. Energy diagram of the organic HTMs, organic-inorganic halide perovskites and TiO$_2$ [21].

A reduction in the dimensionality of the perovskite structure leads to the creation of layered 2H-type and other lower-dimensionality compounds, which in turn increase the band gap energy (E$_g$) of the structures [39, 40]. Electronic energy transition of a series of perovskite structures with the same metal halide is according to: E$_g$,3D < E$_g$,2D < E$_g$,1D < E$_g$,0D. Quantum confinement effects increase the disparity between the optoelectronic properties of 3D and layered structures [41].

As mentioned above, structural properties also affect the optoelectronic characteristics of the perovskites. A decrease in the radius of A cation and the cell volume decreases the band gap. However, there are some exceptions based on researches. For example, CsPbX$_3$ has a larger band gap in comparison with CH$_3$NH$_3$PbX$_3$ and HC-(NH$_2$)$_2$PbX$_3$ perovskites at room temperature despite a smaller Cs$^+$ cation. Also, a larger formamidinium cation has a smaller band gap relative to its methylammonium analogue [21, 22]. This contradiction is due to the size/nature effect of A-site cation on the reorientation of BX_6 octahedral units [35].

The analysis of the molecular structures of organic-inorganic halide perovskites of $CH_3NH_3PbX_3$, NH_4PbX_3 and $CH(NH_2)_2PbX_3$ also showed that the trend of their band gap is according to: $FPbX_3 < MAPbX_3 < NH_4PbX_3$ [21]. Also, the band gap is dependent on the X anion radius. DFT calculations of some organic-inorganic halide perovskites show that an increase in the halides radius decreases the band gap. But, due to the advantage of the Cl-based perovskites in regarding the other physico-chemical/photovoltaic parameters, they are superior candidates for these PSCs.

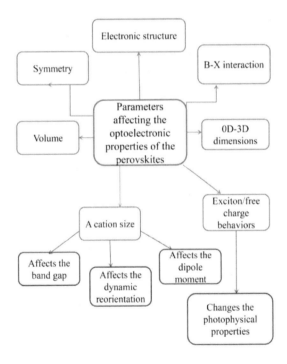

Scheme 2. Determinative parameters of the optoelectronic characteristics of the perovskites.

Research on A_2BI_6 (B=Sn, Pd, Pt, Te) perovskites by the hybrid HSE06 method demonstrated that band gaps and effective masses of the cubic structures decrease when the A site cation size decreases. Theoretical trends were evaluated for 81 combinations of A_2BX_6 halide (A=K, Rb, Cs; B=Si, Ge, Sn, Pb, Ni, Pd, Pt, Se, Te; and X=Cl, Br, I) through the semi-local GGA-PBE calculations. Based on the results, a decrease in the size of halides increases the band gaps of the cubic/tetragonal structures [42].

The effect of A-site cations is more complex. In the case of cubic structure, decrease in the size of A-site cation leads to a decrease in the band gap, while the phase transformation of the octahedral BX_6 to tetragonal structure increases the band gap.

Scheme 2 illustrates determinative parameters of the optoelectronic characteristics of the perovskites.

Analysis of the Excitons Behavior in the PSCs

Charge generation in the perovskites due to photon absorption leads to electron-hole pair (exciton) formation. Coulomb attraction of the electron and hole is calculated according to Eq. 3 [43]:

$$V = \frac{e^2}{4\pi\varepsilon\varepsilon_0 r} \qquad (3)$$

where e is the electron charge, ε is the dielectric constant, ε_0 is the vacuum permittivity, and r is the e-h distance.

In order to form the free charge carriers, the electron driving force of the system must be overcome to this attraction. Considering the lower value of ε of the organic materials (ε=2-4) in comparison inorganic materials (e.g., ε=11, 21 in GaAs and $CsPbCl_3$, respectively), free carrier generation in organic materials seems more difficult. On the other hand, r value is also an important parameter in competition of the e-h recombination and charge carrier separation processes. Moreover, an increase in the temperature and applied electric field accelerates efficiently free charge carrier formation.

Generally, there is an interesting circumstance, where excitons and free carriers are in equilibrium described by Eq. 4 [35]:

$$e + h \rightleftharpoons X \quad (4)$$

where e/h are free electrons/holes and X represents a Coulombically binded e-h pair (exciton).

The effects of different factors on the free charge carrier populations in a 3D-perovskite are shown through the well-known Saha-Langmuir equation (Eq. 5) [44]. These factors consist of the lattice and carrier temperature and the density of photogenerated species, which are dependent on the energy/intensity of absorbed photons by the perovskite.

$$\frac{\varphi_{fc}^2}{1-\varphi_{fc}} = \frac{1}{n}\left(\frac{2\pi m_r k_B T}{h^2}\right)^{3/2} \exp\left[-\frac{E_b}{k_B T}\right] \quad (5)$$

Where n is total excitation density (carriers/volume), φ_{fc} is the total excitation density in a 3D-perovskite, E_b is exciton binding energy, T is temperature, h is Planck's constant, m_r is the reduced effective mass, $k = (4\pi\varepsilon_0)^{-1} = 9 \times 10^9$ Nm^{2}C^{-2}. The number of free carriers increases with an enhancing in the temperature, due to an increase in phonon-induced ionization.

The excitonic behaviour of the perovskites strongly affects the PSC architecture and efficiency. Incident light to the active layer of the PSCs forms the excitons, and a Coulombically bound electron-hole pair, which diffuses in the interface, followed by dissociation and finally charge is injected towards the electrodes. The kinetics of these processes can be investigated by using their corresponding interaction operators, transition matrix element and Fermi's Golden rule as exciton formation, diffusion and dissociation rates, which would be discussed in the following [45].

The rate of photon absorption for a singlet excitation in the perovskite due to exciton-photon interaction, R^S_a, and triplet excitation due to exciton-spin-orbit-photon interaction, R^T_a, are calculated according to Eqs. 6 and 7, respectively [46]:

$$R^S_a = \frac{4ke^2(E_{LUMO}-E_{HOMO})^3 a_x^2}{3c^3\varepsilon^{1.5}\hbar^4} \quad (6)$$

$$R_a^T = \frac{32e^6 Z^2 k^2 \varepsilon (E_{LUMO} - E_{HOMO})}{c^7 \varepsilon_0 \mu_x^4 a_{xt}^4} \tag{7}$$

where S denotes singlet and T triplet, Z is the atomic number of a heavy metal atom, c is the light speed and a_x is the excitonic radius, which indicates e-h separation. The excitonic Bohr radius for the singlet (a_{xS}) and triplet excitons (a_{xT}) are according to Eqs. 8 and 9, respectively [47]:

$$a_{xs} = \frac{\alpha^2 \mu \varepsilon}{(\alpha-1)^2 \mu_x} a_0 \tag{8}$$

$$a_{xt} = \frac{\mu \varepsilon}{\mu_x} a_0 \tag{9}$$

Where a_0 is Bohr radius and μ is the reduced mass of a hydrogen atom and α is a material-dependent constant, which shows the ratio of Coulomb and exchange interactions between the excited electron and hole.

For singlet/triplet exciton dissociation rate, R_d^S / R_d^T, are expressed as [48]:

$$R_d^S = \frac{8\pi^2}{3\hbar^3 \varepsilon^2 E_B^S} [E_{LUMO}^D - E_{CB}^A - E_B^S]^2 (\hbar \omega_\vartheta) \mu_x a_{xs}^2 \tag{10}$$

$$R_d^T = \frac{8\pi^2}{3\hbar^3 \varepsilon^2 E_B^T} [E_{LUMO}^D - E_{CB}^A - E_B^T]^2 (\hbar \omega_\vartheta) \mu_x a_{xt}^2 \tag{11}$$

where ω_v (c/λ_{max}) is the phonon frequency and λ_{max} is the maximum wavelength of the perovskite.

A Coulombically bound electron-hole pair has an excitation binding energy, EBE. EBE can be considered as a criterion of the charge separation in the excitonic solar cells, which is calculated in different ways. One of the ways is related to the difference between the electronic and optical band gap energies of the photosensitizer (Eq. 12) [36].

$$EBE = (E_{LUMO} - E_{HOMO}) - E_{0-0} \tag{12}$$

where, E_{0-0} is the vertical excitation energy of the perovskite, extracted from TD-DFT output, directly.

Calculation of the material-dependent constant is required to obtain the excitonic radius. For this goal, the correlation of EBE and α parameters is important according to Eqs. 13 and 14 [23]:

$$E_B^S = \frac{(\alpha-1)^2 \mu_x e^4 k^2}{2\alpha^2 \hbar^2 \varepsilon^2} \tag{13}$$

$$E_B^T = \frac{\mu_x e^4 k^2}{\alpha^2 \hbar^2 \varepsilon^2} \tag{14}$$

Where ε is the dielectric constant of the donor, μ_x is the reduced mass of exciton, ℏ is the reduced Planck's constant, ε_o is the vacuum permittivity. In organic materials, the effective masses of electrons and holes are equal to the free electron mass, m_e. ($m_e^* = m_h^* = m_e \rightarrow \mu_x = 0.5 m_e$).

Another equation of the strength of the Coulombic e-h attraction, the hydrogen Rydberg analogous (Ry(H)=13.6 eV) is represented by Eq. 15 [35]:

$$E_b = \frac{m_r e_0^4}{2\hbar^2 (4\pi\varepsilon\varepsilon_0)^2} \frac{1}{n^2} = \frac{(m_r/m_0)}{\varepsilon^2} \frac{1}{n^2} R_y(H) \tag{15}$$

The associated exciton radius (r_B) is given by Eq. 16 [35]:

$$a = \frac{\hbar^2 (4\pi\varepsilon\varepsilon_0)}{m_r e_0^2} = \frac{\varepsilon}{(m_r/m_0)} a_B \tag{16}$$

The quantities of carrier lifetime (τ), carrier mobility and carrier diffusion length are important in the solar cell technologies. Particularly the carrier lifetime of the electrons and holes of the perovskite in the photovoltaic process play an important role. This parameter can be calculated by Eq. 17 [21]. A more stable excited state of the perovskite improves the power conversion efficiency of the PSC, while a shorter lifetime of the excited state leads to lower efficiency. However, the carrier lifetime of the perovskite materials depends on the nature, purity and dimension of the materials. The lifetime of the perovskite single crystal is much longer than the corresponding thin film polycrystalline, which represents superiority of the single crystal, due to a lower quantity of defects.

$$\tau = (\frac{4e^2 \Delta E_{k,k'}^2}{3\hbar^4 c^3} |r_{k,k'}|^2)^{-1} \tag{17}$$

where $\Delta E_{k,k'}$ and $r_{k,k'}$ are the transition energy and transition dipole moment from k to k′ states, respectively.

EBE values of some organic-inorganic halide perovskites are affected by A cation radius, according to: $F^+ < NH_4^+ < MA^+$ -based perovskites, due to a direct effect of the electronic band gap on this parameter. Accordingly, FBX_3 perovskites have the least values of the band gap in comparison with other compounds, which leads to a decrement in the exciton binding energy. Also an electron density distribution of the involved molecular orbitals controls the electron-hole separation/overlap of the system. In scheme 3, the main parameters that affect the competition of the exciton/free charge carrier generation in the PSCs are represented.

Quantum Chemistry Reactivity

The importance of the conceptual DFT reactivity indices of the photosensitizers in the dynamics/kinetics of charge transfers is indicated by considering the frontier energy levels and band gap effects of the photoactive layer on the photovoltaic processes in the PSCs. Quantum reactivity descriptors consist of electrophilicity index (ω), electronic chemical potential (μ_e), and chemical hardness (η_q), which are obtained within the conceptual framework of the DFT [22]. The electronic chemical potential represents the escaping tendency of an electron from the equilibrium and global chemical hardness can be considered as the resistance of the perovskite against to charge transfer. Electrophilicity index shows the stabilisation energy of the saturated systems by electrons of the surroundings, using the Koopmans theorem [22].

Dynamic/kinetic parameters, which affects the performance of the PSCs are consist of the electron driving force, e V_{oc}, Gibbs energy change of the electron injection, ΔG_{inj}, the electron affinity of the perovskites, EA, the coupling constant of the perovskite/TiO$_2$, $|V_{RP}|$, and charge transfer rate constant between the perovskite and TiO$_2$, k_{inj}, light harvesting efficiency, LHE, the quantum chemistry descriptors, the incident photon conversion to current efficiency (IPCE). The effectiveness of different parameters can be indicated through the evaluation of their correlations and PSC efficiency.

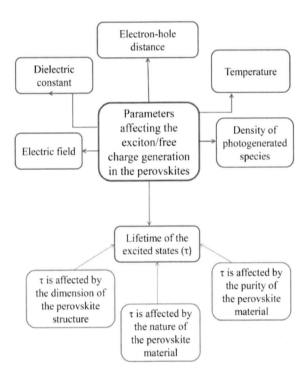

Scheme 3. Main parameters that affect the competition of the exciton/free charge carrier generation in the PSCs.

One of the intrinsic properties, which affect the dynamics of charge transfer in the PSCs is global electrophilicity. For example, in the MAPbX$_3$, FBX$_3$, EABX$_3$ and NH$_4$BX$_3$ perovskites, a decrease in the electrophilicity increases the driving force of the electron toward TiO$_2$, due to a weaker strength of the perovskite in electron capture. Also, the perovskite tendency of electron injection is dependent on X anions, according to this trend: chloride- > bromide- > iodide-based perovskites. This is originated from the dependence of eV_{OC} on the LUMO energy levels of the perovskites distributed on the X anions by different densities. Moreover, these perovskites, due to a stronger coupling with TiO_2, exhibit a greater negative character of Gibbs energy of charge injection in the PSCs.

Among these organic-inorganic halide perovskites, the chloride-based types are better candidates to develop photovoltaic processes at the perovskite/TiO_2 interface. Some characters, such as high electronegativity of the Cl⁻ and electronic cloud distribution on the halide ions leads to the superior position of the X anions.

In these series of perovskite, the rate constant of charge transfer at the perovskite/TiO_2 interface can be described by a polynomial function of the electron driving force [21]. According to the differential forms of these functions, perovskites having a higher eV_{oc} represent a stronger dependence of the $k_{inj.}$ to eV_{oc} in these type of perovskites. Theoretical trends of $eV_{oc}/k_{inj.}$ dependence are according to: EA^+ > F^+ > MA^+ > NH_4^+ -based perovskites and Cl- > Br- > I-based perovskites. It is clear that both of A cations and X anions affect the rate constant of charge transfer [21].

Physical Chemistry Management vs J_{SC}, V_{OC} the Efficiencies of the Solar Cells

The overall efficiency (η) of the solar cells is obtained by the current-voltage diagram, which is a standard technique, based on Eq. 18 [49]:

$$\eta = \frac{J_{sc}V_{OC}FF}{P_{in}} \tag{18}$$

Where J_{SC} is the short-circuit current density, V_{oc} is the open-circuit voltage, P_{in} is the power of the incident light and FF is the fill factor.

The functionality of the PSCs can be interpreted via several efficiency terms, which are discussed in the following statements. LHE parameter that is affected by the structural/excited state properties of the materials is calculated by Eq. 19 [50]:

$$LHE = 1 - 10^{-f} \tag{19}$$

Where, f is the oscillator strength, which represents the ability of the charge transfer between the frontier energy levels. Also, a theoretical value of IPCE can be evaluated by Eq. 20 [51]:

$$IPCE = LHE(\lambda) \cdot \Phi_{inject} \cdot \eta_{collect} \tag{20}$$

Where ϕ_{inj} is the net electron injection efficiency and η_{coll} is the electron collection efficiency, which is only determined by the structure of the applied material. Therefore, it can be considered to be constant in the case of a photoanode with the same material and different photosensitizers. Net electron injection efficiency, Φ_{inj}, is evaluated by Eq. 21 [52]:

$$\phi_{inj} = k_{inj}/(k_{inj} + k_{decay}) \tag{21}$$

where k_{decay} is the rate constant of the intramolecular recombination or decay and k_{inj} can be derived from the Marcus theory for electron transfer.

The absorbed photon-to-current conversion efficiency (APCE) as the internal quantum efficiency is related to a part of the incident light which is absorbed and leads to photoelectron generation in a PSC (Eq. 22). According to Eq. 22, if APCE value is low, it indicates the drawback either in a poor charge injection, or has some difficulties in electron delivery to the current collectors [23].

$$APCE = \Phi_{inj} \cdot \eta_{coll.} \tag{22}$$

External quantum efficiency, EQE, is expressed according to Eq. 23 [53]:

$$EQE = \eta_{abs} \cdot \eta_{diff} \cdot \eta_{diss} \cdot \eta_{tr} \cdot \eta_{cc} \tag{23}$$

where η_{diff}, η_{abs}, η_{diss}, η_{tr} and η_{cc} are the yields of exciton diffusion, absorption yield, exciton dissociation, charge transport and charge collection at the electrodes, respectively.

Different substitutions on the photosensitizers, molecular orbital energy levels of the materials involved in PVSCs, surrounding media, synthesis methods, the sensitizer surface coverage and the ratio of the photosensitizer molecules and hole transfer species affect the J_{sc}-V_{oc} characteristics and the light exposure performance mechanism [54]. Some factors, such as the architecture of the solar cell, solvent viscosity, electron lifetime, E_{CB} of TiO_2, photo-oxidation of the sensitizers and regeneration of the photosensitizers also play a key role in J-V plots [55, 54]. Moreover, charge injection/recombination processes and solar cell performance are affected by additives.

Also, the current of the solar cell can be enhanced by the improvement of the crystallinity of the perovskite active layer [56], the charge carrier mobility and the exciton dissociation efficiency or charge collection efficiency. If the carrier mobility of the ETM/HTM is low, the voltage can also be dependent on the carrier collection [57].

To describe the design effects on the V_{oc}, Br-I mixed perovskite of $CH_3NH_3PbI_2Br$ and TiO_2 is a good example, which shows a higher PCE than that of $CH_3NH_3PbI_3$ originated from a higher V_{OC} [58]. Also, TiO_2/$CH_3NH_3PbBr_3$/poly[N-9-hepta-decanyl-2,7-carbazole-alt-3,6-bis-(thiophen-5-yl)-2,5- dioctyl-2,5-dihydropyrrolo[3,4-]pyrrole-1,4-dione] (PCBTDPP) system shows a V_{oc} of 1.2 V [59], while TiO_2/$CH_3NH_3PbBr_3$/P3HT has a V_{oc} of 0.5 V. Such a difference in the voltage is attributed to the efficiency of the light filtering, degree of chemical interaction and electron lifetime [57].

On the basis of the reports, PCE can be improved by incorporation of C60 self-assembled monolayer to the ITO/TiO_2/$CH_3NH_3PbI_3$-xClx/P_3HT originated from a significant increase in both J_{sc} and V_{oc} [60, 57]. Also, according to Schokeley-Quisser limit arguments [61], it is indicated that the perovskites containing a lower band gap would lead to better efficiencies.

Generally, EQE and IPCE values are nearly equal. They are often expressed as the percentage of the incident photons that are absorbed by the photoactive material and converted into free electrons as a function of the wavelength.

In ABX$_3$ perovskites (A=CH$_3$NH$_3$, CH$_4$, CH(NH$_2$)$_2$, B= Pb, Ge, Sn, and X= Cl, Br, I), an improvement in LHE is dependent to the electronic chemical potential of the perovskites (Figure 9a). Since, differential plots are useful in understanding the rate of LHE growth as a function of μ, based on Figure 9b, a main dependence of the LHE to the electronic chemical potential is observed for the perovskites having a lower value of μ.

Transition dipole moment, $r_{k,k'}$, as a factor related to the excited states also controls the ability of the sun light absorption by the perovskite. Figure 10. is an example of the correlation of LHE/IPCE and $r_{k,k'}$ in the mentioned perovskite solar cells. It means that perovskites having a higher $r_{k,k'}$ have a greater ability for electron transfer among the involved molecular orbitals. A higher electric transition dipole moment is associated with the transitions between two states means that there is a greater strength of the interaction between the perovskites and light, due to charge distribution. When the perovskites absorb a photon and interact with the oscillating electric dipole moment, induced by the sunlight, their electrons are excited to a higher energy level. On the basis of quantum mechanics, a higher $r_{k,k'}$ means a greater intrinsic transition probability and oscillating strength in the perovskite molecules.

The correlation of the IPCE and other photovoltaic parameters is shown in Figure 10. Here, in addition to electron transition probability, some properties, such as the dynamic/kinetic of electron injection, oxidized perovskite regeneration and chemical hardness also affects this efficiency. For example, in spite of good regeneration of NH$_4^+$ and Cl- based perovskites, their electronic chemical hardness is greater, making them a resistant to electron transfer. Based on previous studies [21, 22], on molecular engineering of the organometallic halide perovskites, it is confirmed that this is an effective technique to tune the band gap. The electronic structure analysis represents that an increase in the cation size decreases the band gaps of the corresponding perovskites.

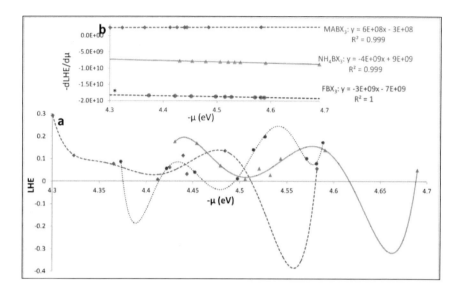

Figure 9a. LHE changes as a function of μ. b. The corresponding differential plot of organic-inorganic halide peroveskites. *The values of –dLHE/dμ for FBX_3 have been multiplied by 10 for better presentation of Figure 9b [21].

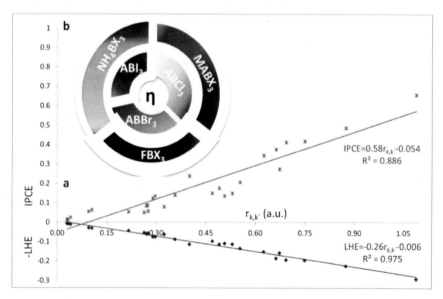

Figure 10a. The correlation of LHE and IPCE as the functions of η, b. Schematic view of the η changes in perovskites [21].

CONCLUSION

The structural and electronic properties of the perovskite as a photosensitizer affect the photovoltaic behaviour of the PSCs. The surrounding, structural geometry, type and interaction of the components in the perovskite structure are important to form a perovskite crystal network. These factors change the properties of the perovskite as a photosensitizer through the phase transition and the change in the physicochemical properties. Optoelectronic properties of the perovskites are dependent on the perovskite network dimension, exciton kinetic/dynamic behaviour and excited state properties of the perovskite. One of the main factors in determining the efficiency of the PSCs, the competition of electron recombination and transfer processes, managed by the excitons formation/dissociation. Such competition is a function of the excited state lifetime, the density of photogenerated species, electric field and physical properties of the perovskite structure. Finally, the performance of the PSC is determined by different photovoltaic processes and the quantum chemistry reactivity indices of the perovskite photosensitizers.

REFERENCES

[1] Noel, Nakita K, Samuel D Stranks, Antonio Abate, Christian Wehrenfennig, Simone Guarnera, AmiAbbas Haghighirad, Aditya Sadhanala, Giles E Eperon, Sandeep K Pathak, and Michael B Johnston. 2014. "Lead-Free Organic–Inorganic Tin Halide Perovskites for Photovoltaic Applications." *Energy & Environmental Science*, 7:3061-68.

[2] Tress, Wolfgang, Nevena Marinova, Thomas Moehl, Shaik Mohammad Zakeeruddin, Mohammad Khaja Nazeeruddin, and Michael Grätzel. 2015. "Understanding the Rate Dependent J–V Hysteresis, Slow Time Component, and Aging in $CH_3NH_3PbI_3$

Perovskite Solar Cells: The Role of a Compensated Electric Field." *Energy & Environmental Science*, 8:995-1004.

[3] Kojima, Akihiro, Kenjiro Teshima, Yasuo Shirai, and Tsutomu Miyasaka. 2009. "Organometal Halide Perovskites as Visible-Light Sensitizers for Photovoltaic Cells." *Journal of the American Chemical Society*, 131:6050-51.

[4] Futscher, Moritz H, and Bruno Ehrler. 2017. "Modeling the Performance Limitations and Prospects of Perovskite/Si Tandem Solar Cells under Realistic Operating Conditions." *ACS Energy Letters*, 2:2089-95.

[5] Saliba, Michael, Simon M Wood, Jay B Patel, Pabitra K Nayak, Jian Huang, Jack A Alexander-Webber, Bernard Wenger, Samuel D Stranks, Maximilian T Hörantner, and Jacob Tse-Wei Wang. 2016. "Structured Organic-Inorganic Perovskite toward a Distributed Feedback Laser." *Advanced materials*, 28:923-29.

[6] Tiazkis, Robertas, Sanghyun Paek, Maryte Daskeviciene, Tadas Malinauskas, Michael Saliba, Jonas Nekrasovas, Vygintas Jankauskas, Shahzada Ahmad, Vytautas Getautis, and Mohammad Khaja Nazeeruddin. 2017. "Methoxydiphenylamine-Substituted Fluorene Derivatives as Hole Transporting Materials: Role of Molecular Interaction on Device Photovoltaic Performance." *Scientific reports*, 7:150.

[7] Albrecht, Steve, Michael Saliba, Juan Pablo Correa Baena, Felix Lang, Lukas Kegelmann, Mathias Mews, Ludmilla Steier, Antonio Abate, Jörg Rappich, and Lars Korte. 2016. "Monolithic Perovskite/Silicon-Heterojunction Tandem Solar Cells Processed at Low Temperature." *Energy & Environmental Science*, 9:81-88.

[8] Xiao, Minyu, Suneel Joglekar, Xiaoxian Zhang, Joshua Jasensky, Jialiu Ma, Qingyu Cui, L Jay Guo, and Zhan Chen. 2017. "Effect of Interfacial Molecular Orientation on Power Conversion Efficiency of Perovskite Solar Cells." *Journal of the American Chemical Society*, 139:3378-86.

[9] Wang, Jiayu, Kuan Liu, Lanchao Ma, and Xiaowei Zhan. 2016. "Triarylamine: Versatile Platform for Organic, Dye-Sensitized, and Perovskite Solar Cells." *Chemical reviews*, 116:14675-725.

[10] Cha, Mingyang, Peimei Da, Jun Wang, Weiyi Wang, Zhanghai Chen, Faxian Xiu, Gengfeng Zheng, and Zhong-Sheng Wang. 2016. "Enhancing Perovskite Solar Cell Performance by Interface Engineering Using $CH_3NH_3PbBr_{0.9}I_2.1$ Quantum Dots." *Journal of the American Chemical Society*, 138:8581-87.

[11] Sun, Jian-Ke, Zdravko Kochovski, Wei-Yi Zhang, Holm Kirmse, Yan Lu, Markus Antonietti, and Jiayin Yuan. 2017. "General Synthetic Route toward Highly Dispersed Metal Clusters Enabled by Poly (Ionic Liquid) s." *Journal of the American Chemical Society*, 139:8971-76.

[12] Zhou, Xin, Joanna Jankowska, Hao Dong, and Oleg V Prezhdo. 2017. "Recent Theoretical Progress in the Development of Perovskite Photovoltaic Materials." *Journal of energy chemistry*. 27:637-649.

[13] Bisquert, Juan, Yabing Qi, Tingli Ma, and Yanfa Yan. 2017. "Advances and Obstacles on Perovskite Solar Cell Research from Material Properties to Photovoltaic Function." *ACS Energy Letters*, 2:520-23.

[14] Rose, Gustav. 1839. *De novis quibusdam fossilibus quae in montibus Uraliis inveniuntur* (typis AG Schadii).

[15] Elumalai, Naveen, Md Mahmud, Dian Wang, and Ashraf Uddin. 2016. "Perovskite Solar Cells: Progress and Advancements." *Energies*, 9:861.

[16] Saparov, Bayrammurad, and David B Mitzi. 2016. "Organic–Inorganic Perovskites: Structural Versatility for Functional Materials Design." *Chemical reviews*, 116:4558-96.

[17] Manser, Joseph S., Makhsud I. Saidaminov, Jeffrey A. Christians, Osman M. Bakr, and Prashant V. Kamat. 2016. "Making and Breaking of Lead Halide Perovskites." *Accounts of Chemical Research*, 49:330-338.

[18] Misra, Ravi K., Bat-El Cohen, Lior Iagher, and Lioz Etgar. 2017. "Low Dimensional Organic-Inorganic Halide Perovskite: Structure, Properties, and Applications." *Chem Sus Chem*, 10:3712-3721.

[19] Bailie, Colin D., M. Greyson Christoforo, Jonathan P. Mailoa, Andrea R. Bowring, Eva L. Unger, William H. Nguyen, Julian Burschka, Norman Pellet, Jungwoo Z. Lee, Michael Grätzel, Rommel Noufi, Tonio Buonassisi, Alberto Salleo, and Michael D. McGehee. 2015. "Semi-Transparent Perovskite Solar Cells for Tandems with Silicon and CIGS." *Energy Environmental Science*, 8:956-963.

[20] Stoumpos, Constantinos C., and Mercouri G. Kanatzidis, 2015. "The Renaissance of Halide Perovskites and their Evolution as Emerging Semiconductors." *Accounts of Chemical Research,* 42:791-2802.

[21] Arkan, Foroogh, and Mohammad Izadyar. 2018. "Molecular Engineering of the Organometallic Perovskites/HTMs in the PSCs: Photovoltaic Behavior and Energy Conversion." *Solar Energy Materials and Solar Cells*, 180:46-58.

[22] Arkan, Foroogh, and Mohammad Izadyar. 2018. "Computational Modeling of the Photovoltaic Activities in $EABX_3$ (EA= ethylammonium, B= Pb, Sn, Ge, X= Cl, Br, I) Perovskite Solar Cells." *Computational Materials Science*, 152:324-30.

[23] Arkan, Foroogh, Mohammad Izadyar, and Ali Nakhaeipour. 2017. "Improvement in Charge Transfer Dynamic of the Porphyrin-Based Solar Cells in Water: A Theoretical Study." *Journal of Renewable and Sustainable Energy*, 9:023502.

[24] Chakraborty, Sudip, Wei Xie, Nripan Mathews, Matthew Sherburne, Rajeev Ahuja, Mark Asta, and Subodh G Mhaisalkar. 2017. "Rational Design: A High-Throughput Computational Screening and Experimental Validation Methodology for Lead-Free and Emergent Hybrid Perovskites." *ACS Energy Letters*, 2:837-45.

[25] Stoumpos, Constantinos C., Christos D Malliakas, and Mercouri G Kanatzidis. 2013. "Semiconducting Tin and Lead Iodide Perovskites with Organic Cations: Phase Transitions, High mobilities, and Near-Infrared Photoluminescent Properties." *Inorganic chemistry*, 52:9019-38.

[26] Chen, Qi, Nicholas De Marco, Yang (Michael) Yang, Tze-Bin Song, Chun-Chao Chen, Hongxiang Zhao, Ziruo Hong, Huanping Zhou, Yang Yang. 2015. "Under the Spotlight: The Organic-Inorganic

Hybrid Halide Perovskite for Optoelectronic Applications." *Nano Today*, 10:355-396.

[27] Baikie, Tom, Yanan Fang, Jeannette M Kadro, Martin Schreyer, Fengxia Wei, Subodh G Mhaisalkar, Michael Graetzel, and Tim J White. 2013. "Synthesis and Crystal Chemistry of the Hybrid Perovskite (CH$_3$NH$_3$) PbI$_3$ for Solid-State Sensitised Solar Cell Applications." *Journal of Materials Chemistry A*, 1: 5628-41.

[28] Liu, Xixia, Nengduo Zhang, Baoshan Tang, Mengsha Li, Yong-Wei Zhang, Zhi Gen Yu, and Hao Gong. 2018. "Highly Stable New Organic-Inorganic Hybrid 3D Perovskite CH$_3$NH$_3$PdI$_3$ and 2D Perovskite (CH$_3$NH$_3$) 3Pd$_2$I$_7$: DFT Analysis, Synthesis, Structure, Transition Behaviour and Physical Properties." *The journal of physical chemistry letters*, 9:5862-5872.

[29] Oku, Takeo. 2015. "Crystal Structures of CH$_3$NH$_3$PbI$_3$ and Related Perovskite Compounds Used for Solar Cells." *Solar Cells-New Approaches and Reviews* (InTech).

[30] Popov, Guerman, and Martha Greenblatt. 2003. "Large Effects of A-site Average Cation Size on the Properties of the Double Perovskites Ba$_2$À$_x$Sr$_x$MnReO$_6$: A d^5-d^1 System." *Physical Review*, 67:024406.

[31] Meyer, Edson, Dorcas Mutukwa, Nyengerai Zingwe, and Raymond Taziwa. 2018. "Lead-Free Halide Double Perovskites: A Review of the Structural, Optical, and Stability Properties as well as Their Viability to Replace Lead Halide Perovskites." *Metals*, 8:667.

[32] Sani, Faruk, Suhaidi Shafie, Hong Ngee Lim, and Abubakar Ohinoyi Musa. 2018. "Advancement on Lead-Free Organic-Inorganic Halide Perovskite Solar Cells: A Review." *Materials*,11:1008.

[33] Boix, Pablo P., Shweta Agarwala, Teck Ming Koh, Nripan Mathews, and Subodh G. Mhaisalkar. 2015. "Mhaisalkar, S.G. Perovskite Solar Cells: Beyond Methyl-Ammonium Lead Iodide." *Journal of Physical Chemistry Letters*, 6:898-907.

[34] Wang, Ze, Zejiao Shi, Taotao Li, Yonghua Chen, and Wei Huang. 2016. "Stability of Perovskite Solar Cells: A Prospective on the Substitution of the A Cation and X Anion." *Solar Cells*, 55:2-25.

[35] Manser, Joseph S, Jeffrey A Christians, and Prashant V Kamat. 2016. "Intriguing Optoelectronic Properties of Metal Halide Perovskites." *Chemical reviews*, 116:12956-3008.

[36] Arkan, Foroogh, and Mohammad Izadyar. 2017. "The Role of Solvent and Structure in the Kinetics of the Excitons in Porphyrin-Based Hybrid Solar Cells." *Solar energy*, 146:368-78.

[37] Casanova, David, François P Rotzinger, and Michael Grätzel. 2010. "Computational Study of Promising Organic Dyes for High-Performance Sensitized Solar Cells." *Journal of chemical theory and computation*, 6:1219-27.

[38] Chi, Wei-Jie, Quan-Song Li, and Ze-Sheng Li. 2015. "Effects of Molecular Configuration on Charge Diffusion Kinetics within Hole-Transporting Materials for Perovskites Solar Cells." *The Journal of Physical Chemistry C*, 119:8584-90.

[39] Mosconi, Edoardo, Anna Amat, Md K Nazeeruddin, Michael Grätzel, and Filippo De Angelis. 2013. "First-Principles Modeling of Mixed Halide Organometal Perovskites for Photovoltaic Applications." *The Journal of Physical Chemistry C*, 117:13902-13.

[40] Christians, Jeffrey A, Pierre A Miranda Herrera, and Prashant V Kamat. 2015. "Transformation of the Excited State and Photovoltaic Efficiency of $CH_3NH_3PbI_3$ Perovskite upon Controlled Exposure to Humidified Air." *Journal of the American Chemical Society*, 137:1530-38.

[41] Sichert, Jasmina A, Yu Tong, Niklas Mutz, Mathias Vollmer, Stefan Fischer, Karolina Z Milowska, Ramon García Cortadella, Bert Nickel, Carlos Cardenas-Daw, and Jacek K Stolarczyk. 2015. "Quantum Size Effect in Organometal Halide Perovskite Nanoplatelets." *Nano letters*, 15:6521-27.

[42] Cai, Yao, Wei Xie, Hong Ding, Yan Chen, Krishnamoorthy Thirumal, Lydia H Wong, Nripan Mathews, Subodh G Mhaisalkar, Matthew Sherburne, and Mark Asta. 2017. "Computational Study of Halide Perovskite-Derived A_2BX_6 Inorganic Compounds: Chemical Trends in Electronic Structure and Structural Stability." *Chemistry of Materials*, 29:7740-49.

[43] Ostroverkhova, Oksana. 2016. "Organic Optoelectronic Materials: Mechanisms and Applications." *Chemical reviews*, 116:13279-412.
[44] Saha, Meghnad N. 1921. "On a Physical Theory of Stellar Spectra." *Proceedings of the Royal Society of London Series A*, 99:135-53.
[45] Narayan, Monishka Rita, and Jai Singh. 2013. "Effect of Exciton-Spin-Orbit-Photon Interaction in the Performance of Organic Solar Cells." *The European Physical Journal B*, 86:47.
[46] Arkan, Foroogh, and Mohammad Izadyar. 2018. "Recent Theoretical Progress in the Organic/Metal-Organic Sensitizers as the Free Dyes, Dye/TiO$_2$ and Dye/Electrolyte Systems; Structural Modifications and Solvent Effects on their Performance." *Renewable and Sustainable Energy Reviews*, 94:609-55.
[47] Narayan, Na, Monishka Rita, and Jai Singh. 2012. "Roles of Binding Energy and Diffusion Length of Singlet and Triplet Excitons in Organic Heterojunction Solar Cells." *physica status solidi (c)*, 9:2386-89.
[48] Narayan, Rita Monishka, and Jai Singh. 2013. "Study of the Mechanism and Rate of Exciton Dissociation at the Donor-Acceptor Interface in Bulk-Heterojunction Organic Solar Cells." *Journal of Applied Physics*, 114:073510.
[49] Ronca, Enrico, Mariachiara Pastore, Leonardo Belpassi, Francesco Tarantelli, and Filippo De Angelis. 2013. "Influence of the Dye Molecular Structure on the TiO$_2$ Conduction Band in Dye-Sensitized Solar Cells: Disentangling Charge Transfer and Electrostatic Effects." *Energy & Environmental Science*, 6:183-93.
[50] Arkan, Foroogh, Mohammad Izadyar, and Ali Nakhaeipour. 2016 "The Role of the Electronic Structure and Solvent in the Dye-Sensitized Solar Cells Based on Zn-Porphyrins: Theoretical Study." *Energy* 114:559-567.
[51] Wang, Jian, Fu-Quan Bai, Bao-Hui Xia, Lu Feng, Hong-Xing Zhang, and Qing-Jiang Pan. 2011. 'On the Viability of Cyclometalated Ru (II) Complexes as Dyes in DSSC Regulated by COOH Group, a DFT Study." *Physical Chemistry Chemical Physics*, 13:2206-13.

[52] Ardo, Shane, and Gerald J Meyer. 2009. "Photodriven Heterogeneous Charge Transfer with Transition-Metal Compounds Anchored to TiO_2 Semiconductor Surfaces." *Chemical Society Reviews*, 38:115-64.

[53] Yu, Yue, Dewei Zhao, Corey R. Grice, Weiwei Meng, Changlei Wang, Weiqiang Liao, Alexander Cimaroli, Kai Zhu, and Yanfa Yan. 2016. "Thermally Evaporated Methylammonium Tin Triiodide Thin Films for Lead-Free Perovskite Solar Cell Fabrication." *RSC Advances*, 6:90248-90254.

[54] He, Zhicai, Chengmei Zhong, Xun Huang, Wai-Yeung Wong, Hongbin Wu, Liwei Chen, Shijian Su, and Yong Cao. 2011. "Simultaneous Enhancement of Open-Circuit Voltage, Short-Circuit Current Density, and Fill Factor in Polymer Solar Cells." *Advanced Materials*, 23:4636-4643.

[55] Ahmed, Muhammad Imran, Amir Habib, and Syed Saad Javaid. 2015. "Perovskite Solar Cells: Potentials, Challenges, and Opportunities." *International Journal of Photoenergy*, 2015:1. DOI: 10.1155/2015/592308.

[56] Kim, Hui-Seon, Jin-Wook Lee, Natalia Yantara, Pablo P. Boix, Sneha A. Kulkarni, Subodh Mhaisalkar, Michael Grätzel, and Nam-Gyu Park, 2013. "High Efficiency Solidstate Sensitized Solar Cell-Based on Submicrometer Rutile TiO_2 Nanorod and $CH_3NH_3PbI_3$ Perovskite Sensitizer." *Nano Letters*, 13:2412–2417.

[57] Edri, Eran, Saar Kirmayer, David Cahen, and Gary Hodes, 2013. "High Opencircuit Voltage Solar Cells Based on Organic-Inorganic Lead Bromide Perovskite," *Journal of Physical Chemistry Letters*, 4:897-902.

[58] Jeng, Jun-Yuan, Yi-Fang Chiang, Mu-Huan Lee, Shin-Rung Peng, Tzung-Fang Guo, Peter Chen, and Ten-Chin Wen. 2013. "$CH_3NH_3PbI_3$ Perovskite/Fullerene Planar-Heterojunction Hybrid Solar Cells," *Advanced Materials*, 25:3727–3732.

[59] Shockley, William, and Hans J. Queisser. 1961. "Detailed Balance Limit of Efficiency of *p-n* Junction Solar Cells." *Journal of Applied Physics*, 32:510-519.

In: Perovskite Solar Cells
Editor: Murali Banavoth
ISBN: 978-1-53615-858-8
© 2019 Nova Science Publishers, Inc.

Chapter 5

TRANSITION FROM SMALL-AREA DEVICES TO LARGE-AREA MODULES FOR PEROVSKITE PHOTOVOLTAICS

Soonil Hong[1], Hongkyu Kang[1,2],*, Jinho Lee[1], Hyungcheol Back[1,2], Sooncheol Kwon[1,2], Heejoo Kim[1,2,3],† and Kwanghee Lee[1,2],‡

[1]Heeger Center for Advanced Materials,
[2]Research Institute for Solar and Sustainable Energies,
[3]Institute of Integrated Technology,
Gwangju Institute of Science and Technology,
Gwangju, Republic of Korea

ABSTRACT

Organic-inorganic hybrid perovskite solar cells (PSCs) have been considered a promising candidate for next-generation photovoltaics due to their sharp increases in power conversion efficiencies (PCEs). Since the

---

* Corresponding Author's Email: gemk@gist.ac.kr.
† Corresponding Author's Email: heejook@gist.ac.kr.
‡ Corresponding Author's Email: klee@gist.ac.kr.

first realisation of efficient PSCs with a PCE of 3.9% in 2009, the record PCE reached 24.2% in 2019. The next step will undoubtedly be developing scale-up techniques for transitioning small-area devices to large-area modules. For the high-throughput and low-cost production of these large-area PSCs, roll-to-roll compatible printing methods such as doctor blading, spray deposition, screen printing and slot-die coating have been developed. Numerous efforts have been made to form a uniform and pinhole-free large-area perovskite films with a minimum drop in the PCE of PSC devices by developing fabrication processes with perovskite precursors; however, achieving fully roll-to-roll-processed large-area PSC modules remains the greatest challenge in the practical applications of PSCs. This book chapter will provide recent studies on the fabrication of PSCs using printing methods and large-area PSC modules, including the module concept, and will discuss various challenges and issues for the commercialisation of PSCs.

Keywords: perovskite solar cells, printing technology, module

1. INTRODUCTION

The commercialisation of photovoltaic technologies demands several requirements, including low cost, mass production with easy fabrication, and high efficiency [1]. Among current developed photovoltaic technologies, perovskite solar cells (PSCs) have been a sensational subject due to their solution processability and marked increase in power conversion efficiency (PCE): compared to those of other types of 3rd generation solar cells, such as organic solar cells, dye-sensitized solar cells, and quantum dot solar cells, the PCEs of PSCs are the highest reported [2]. The remarkable progress in PSCs is attributed to their superior electronic properties, such as high crystallinity (~ micrometer scale), long exciton diffusion lengths (~ 100 nm), small exciton binding energies (~ 20 meV), high absorption coefficients (~ 10^5 cm^{-1}), and low band gaps (onset ≈ 850 nm) [3-7]. Recently, PCEs of over 20% in PSCs have been achieved by the advancement of the deposition process, interfacial engineering, and compositional modifications and the improved crystallinity of perovskite materials [8]. Therefore, a next step toward the commercialisation of PSCs is the demonstration of large-area

PSCs with a module structure. However, the fabrication of high-efficiency large-area PSCs has suffered from a deficiency of high-quality perovskite films and the high sheet resistance of transparent electrodes over a large area, which decreases the PCE in large-area PSCs [9]. Therefore, the next challenge in the research on PSCs is the realisation of the performance of laboratory-scale high-efficiency unit cells (< 1 cm^2) in large-sized modules with a minimal decrease in the PCE.

To fabricate high-quality pinhole-free perovskite films, the development of proper printing methods for perovskite films is required. Fundamentally, the mechanism for the formation of large-area perovskite films with a printing method is completely different from that for the formation of small-size films (< 1 cm^2) with the spin-coating method, even when the same solvents and materials are used. In particular, the formation of a perovskite layer is highly susceptible to experimental conditions, such as humidity, solvent concentration, type of solvent, and surface properties of the underlying substrate (here, the bottom electrode or an interlayer). Therefore, a systematic approach for controlling the coating conditions, including the temperature of the substrate and the perovskite precursor formulation, is required [10]. A high substrate temperature increases the grain size of perovskites in printing processes such as doctor blade coating and slot-die coating. In addition, adding Cl to the perovskite precursor reduces the number of pinholes and increases the grain size, resulting in a uniform large-area printed perovskite layer [11]. Nevertheless, most large-area PSCs prepared with printing methods show lower PCEs than spin-coated devices. Additionally, the device structure is based on high-temperature sintered titanium dioxide, which has been considered a barrier to the realisation of roll-to-roll printing manufacturing.

Even if we could fabricate a large-area pinhole-free perovskite film, the demonstration of a high geometrical fill factor (GFF) in the module structure is also an important issue. Because the current module concept is based on a monolithic geometry to reduce the Ohmic loss caused by the relatively high sheet resistance of transparent electrodes, a resulting module structure comprising a series of connected component subcells can be fabricated by using a delicate stripe patterning process at a regular distance (~ 1 cm) [9-

13]. Although this patterning process ensures the series connections in the module structure, the active area sandwiched by all component layers is inevitably reduced, resulting in a decrease in the ratio between the active area and inactive area (which is called the GFF). Therefore, a significant drop in efficiency has always occurred in PSC modules. To overcome this limitation, a laser scribing technique and delicate pattern-coating method have been introduced, resulting in a high GFF for PSC modules. However, there are still several issues, such as the residue problem with the laser scribing method, high-cost installation, and the difficulty in the delicate patterning method, which can impede market production [13, 14].

In this chapter, we will review the development of printed PSCs using doctor blade coating, slot-die coating, spray coating, inkjet printing, screen printing, bar coating, brush printing, and dip coating. Ultimately, these printing methods differ from each other. In laboratories, the majority of studies on printed PSCs employ the doctor blade coating method because solution waste is very low and the operation is easy. However, continuous solution supply is not available, so this method is not directly compatible with the roll-to-roll process. In this regard, a slot-die coating is compatible with a continuous printing process because it has a reservoir system enabling the continuous supply of solution. Therefore, the slot-die coating method has been considered the most promising printing technique for the commercialisation of PSCs. Both spray coating and inkjet printing use a kind of nozzle to coat the solution onto the substrate. The major difference between these methods is the use of a mask for patterned coating; spray coating requires a mask, whereas inkjet printing can coat a patterned layer without a mask. Unlike the abovementioned noncontact printing methods, screen printing is classified as a contact method, in which solution is squeegeed through a mesh to the substrate. Although screen printing is a very versatile printing method that allows high-throughput processes, there is a requirement for a solution with high viscosity and low volatility [15]. Bar coating is very similar to doctor blade coating; the difference is that bar coating uses a round-shaped blade to form a meniscus on the substrate. Brush printing, which is also called brush painting, is classified as a contact printing method, but only a small number of researchers working on it. Dip coating is also not often used for the fabrication of PSCs, but it is considered a suitable method for mass production. All these printing methods will be reviewed in more detail in this chapter. Furthermore, the development of a new module structure for large-area PSCs will also be reviewed. Finally, in conclusion, the progress and perspective of printed PSCs and PSC modules will be briefly discussed.

2. PRINTED PEROVSKITE SOLAR CELLS

2.1. Doctor Blade Coating

Doctor blade coating is a very useful method for fabricating small-sized devices from solution-processed large-area films (Figure 1 and Figure 2). Because of its simple structure comprising a moving blade and a temperature-controllable plate, this method can be compatible with roll-to-roll printing methods. Furthermore, several advantages, such as the easy fabrication of large-area films and a reduction in the loss of the coating solution, enable this method to be used in the laboratory-scale fabrication of printed devices before applying roll-to-roll printing. Theoretically, the thickness of films can be defined by the following equation:

$d = 1/2(g\ c/\rho)$

Where g is the gap distance between the sharp blade and substrate, c is the concentration of the solid material in the solution (in g/cm^3), and ρ is the density of the material in the film (in g/cm^3). Therefore, the thickness of the thin film can be easily controlled by using these parameters [15]. The first application of the doctor blade coating method in PSCs was reported by H. Hang and coworkers in 2014, with a configuration of FTO/TiO$_2$/ZrO and perovskite/carbon electrode [16]. In this work, the authors used blade coating for fabricating interlayers with submicrometer thickness: ZrO (~ 2 μm) and a carbon film (~ 10 μm). With this configuration, they achieved a high PCE of 12.8%. However, because the perovskite film was deposited by infiltrating a perovskite solution into the porous carbon film by dropping the perovskite precursor solution, further research is required. Meaningful results for PSCs prepared with the doctor blade method have been reported by several research groups. A. Jen and coworkers developed fully printed PSCs with a configuration of ITO/poly(3,4-ethylenedioxythiophene)-poly(styrenesulfonate) (PEDOT:PSS)/perovskite/and phenyl-C_{60}-butyric acid methyl ester ($PC_{60}BM$)/bis-C_{60}/Ag by using the doctor blade coating method [17]. Because they used a simple perovskite composition, methylammonium lead iodide (CH_3NH_3I), as the active layer, they could fabricate all layers of the PSCs without transparent or metal electrodes by using the blade technique under ambient conditions. As a result, this configuration exhibited good PCE values of 10.44% and 7.14% in rigid and flexible PSCs, respectively.

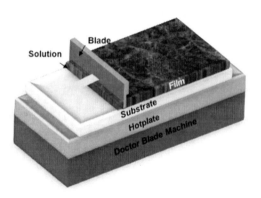

Figure 1. A schematic image of the doctor blade coating process.

Interface engineering of PSCs, such as the introduction of interlayers, is an important issue with doctor blade coating because of the undesirable wettability of perovskite precursor solutions on the bottom electrode. In the inverted structure of PSCs, a configuration of bottom electrode/hole transport layer (HTL)/perovskite photoactive layer/electron transport layer (ETL)/top electrode, this issue is critical for demonstrating a high performance in PSCs. Typically, PEDOT:PSS is used as an HTL material in inverted PSCs. However, because of poor electrostatic interactions between PEDOT:PSS and the perovskite precursor solution, poor coverage of the perovskite layers atop the PEDOT:PSS layer has often been observed. Therefore, in 2016, K. Lee and coworkers introduced a modified PEDOT:PSS layer, which was a mixture of PEDOT:PSS and a small amount of poly(4-styrenesulfonic acid) [18]. With this modified PEDOT:PSS layer, they could fabricate a large-area PSC with a good PCE value of 10.15% and good reproducibility. The high performance of PSCs based on the doctor blade coating method was reported by J. Huang and coworkers in 2017 [11]. By using compositional engineering with the perovskite precursor solution, which is a means of controlling the perovskite grain growth during film deposition by adding Caesium and bromide ions to the precursor solution, they achieved a high PCE value of 19.3%, which is comparable with that of PSCs fabricated by the spin-coating method. Therefore, this result indicates that the printed PSCs are ready to move the next step for printed PSCs with a real printing method, such as a slot-die coating, which is an important component of roll-to-roll printing methods.

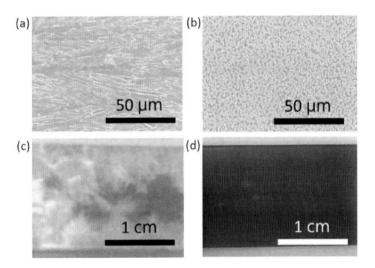

Figure 2. Scanning electron microscopy images of perovskite films coated by a doctor blade machine on (a) room-temperature and (b) 90°C substrates. Photographs of perovskite films coated by a doctor blade machine on (c) room temperature and (d) 90°C substrates.

2.2. Slot-Die Coating

The slot-die coating method is one of the key technologies in roll-to-roll printing methods because they use the same printing head and one-dimensional patterning process. As shown in Figure 3, the slot-die machine comprises a coating head and plate. In principle, the slot-die coating method uses solution to transfer, which is similar to doctor blading. However, the difference between the two printing methods is the continuous and direct supply of solution through a slot-die head by using a pump. Therefore, for the fabrication of thin films, the precursor solution in a stock tank should be injected into the slot-die head. As a second step, a stable meniscus between the shim, which is attached to the slot-die head, and the substrate is formed. Finally, by moving the slot-die head, a wet film of precursor solution is deposited on the substrate. Therefore, the thickness of the thin film can be controlled by the flow rate, web speed, coating width, solid content in the precursor solution, and density of the dried precursor solution [15].

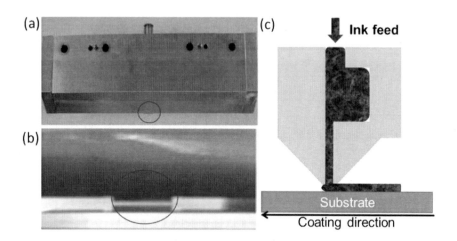

Figure 3. Photography image (a) and enlarged image (b) of a slot-die head with a positive shim mask. (c) Schematic image of the slot-die coating process.

K. Lee and coworkers systematically investigated the slot-die coating method for printed solar cells [19]. In this work, the authors found that the solution viscosity, shim length, and substrate temperature also significantly affect the thickness of the film. However, to fabricate large-area PSCs, more detailed research with various trials is required because of the slow growth of films prepared with the slot-die coating method. Growing perovskite films for different times may cause different crystal growth behaviour from that observed upon depositing a perovskite layer with a spin-coating system. Therefore, significant variation in the PCE of PSCs prepared with the slot-die coating method can be expected. D. Vak and coworkers obtained the first representative result of PSCs fabricated with the slot-die coating method in 2015 [10]. By using a mini-slot-die machine, they fabricated all layers of the PSC except the electrodes. The configuration of the PSC was an ITO/ZnO/perovskite/P3HT doped with lithium-bis(trifluoro methane sulfonyl) imide (Li-TFSI), 4-tert-butylpyridine (TBP))/Ag structure. The perovskite layer was deposited by a two-step process. To fabricate a pinhole-free perovskite layer, the authors used a modified slot-die head consisting of the coating head and an external air blowing system. During the drying process of the deposited PbI_2 layer, high-pressure N_2 gas was sprayed to prevent the overgrowth of perovskite crystals. Finally, methylammonium iodide (CH_3NH_3I) was deposited by the slot-die machine on the completed PbI_2 film. As a result, these PSCs, fully printed with a slot-die system, exhibited a good PCE of 11.96%.

A further enhanced one-step slot-die method for PSCs with a configuration of FTO/TiO$_2$/perovskite/2,2′,7,7′-tetrakis(N,N-di-p-methoxy-phenylamine)-9,9′-spirobifluorene (Spiro-OMeTAD)/V$_2$O$_5$/Au was reported by T. Watson and coworkers in 2017 [20]. In this work, the authors used a mixture of $PbCl_2$ and CH_3NH_3I dissolved in dimethylformamide (DMF) as a precursor solution. To improve the surface coverage of the perovskite layer on the TiO$_2$ substrate, they deposited the precursor solution on a substrate preheated to 65°C for fast solvent evaporation. Furthermore, to improve the crystallization process of the perovskite layer, they used an additional air-knife after finishing the deposition of the precursor solution; thus, the PSCs exhibited a good PCE of 9.2%. Recently, in 2018, Y. Galagan and coworkers demonstrated a large-area (6 in. by 6 in.) PSC module with a configuration of ITO/TiO$_2$/perovskite/Spiro-OMeTAD/Au fabricated by using the slot-die coating method [21]. To fabricate a high-quality perovskite layer with a one-step process, the authors modified the mixed precursor solution containing PbCl and Pb(CH$_3$COO)·3H$_2$O at a molar ratio of 1:4 and CH$_3$NH$_3$I in DMF. The perovskite and HTL layers (here, OMeTAD) were deposited in an N$_2$ gas-filled globe box to prevent undesirable effects by controlling the coating environment. As a result, the authors could fabricate a high-quality perovskite film with an HTL by using the slot-die coating method, resulting in a high PCE value of 16.8%, which is comparable with that of spin-coated small-area PSCs.

2.3. Spray Coating

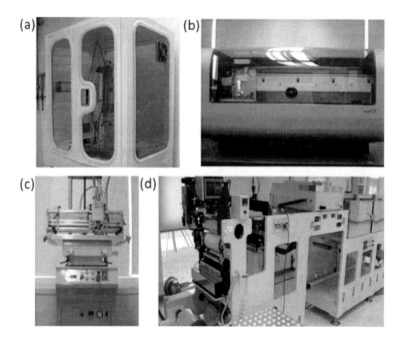

Figure 4. Photographs of (a) spray-coating, (b) inkjet printing, (c) screen printing, and (d) roll-to-roll machines.

Spray coating, which forces printing ink through a nozzle to produce an ultrafine aerosol, has been one of the most promising deposition techniques for the commercialisation of PSCs due to the high-volume production of instruments combined with high-speed, low-cost, and large-area manufacturing (Figure 4a) [22-24]. This coating method enables the perovskite ink to be sprayed over functional and/or irregular surfaces, which makes it highly compatible with multilayer device fabrication with low material consumption [25]. However, the development of highly efficient and stable PSCs fabricated by the spray coating method has been a great challenge because the coating process should be completed by complex sequential steps (i.e., forming ink droplets, merging the droplets into a wet film and drying the thin film), frequently leading to the creation of a number of large pinholes on the surface of the film. This undesirable film morphology may deteriorate interfacial contact between the perovskite film and metal electrode (or organic semiconducting interlayer), thereby reducing the device parameters of PSCs, such as the open-circuit voltage (V_{oc}), short-circuit current density (J_{sc}), and fill factor (FF) [26].

The large pinholes and poor film uniformity of spray-cast PSC films are mainly associated with the high surface tension of the ink (droplet), which determines the large contact angle where the droplet-air interface meets the substrate [27]. Therefore, control of the droplet size and a substrate temperature is key to overcoming these issues. In 2014, D.G. Lidzey and coworkers demonstrated that spray coating with ultrasonic vibrating tips (~35 kHz) allowed minimization of the coalescence of precursor droplets (a mixture of CH_3NH_3I and $PbCl_2$ in a DMF solution) within the nozzle to form a continuous perovskite film (the surface coverage of perovskite was ~85%) on top of a preheated PEDOT:PSS/ITO substrate (75°C) [28]. However, increasing the temperature of the substrate (>75°C) resulted in a dewetting process and decreased surface coverage, which was attributed to the fast drying rate [29]. After an optimised postannealing treatment, the resultant film was used to fabricate planar heterojunction PSCs (ITO/PEDOT:PSS/ perovskite/[6,6]-phenyl C_{61}-butyric acid methyl ester (PCBM)/Ca/Al) exhibiting a PCE of 11.1% (the average PCE ranged from 6 to 8%). Using a similar ultrasonic spray coating method and thermal annealing process as above, K. Xiao and coworkers obtained conventional PSCs (FTO/compact TiO_2/perovskite/spiro-OMeTAD/Au) with a slightly improved PCE of 13% [30]. For further improvement in the PCE of spray-cast PSCs, precisely optimising a blend ratio of perovskite components is required to ensure their high crystallinity and a minimisation of charge trap sites. J. Poortmans and coworkers showed that highly crystalline perovskite layers prepared by the spray-coating method can be obtained from an optimized mixture of 25% $CH_3NH_3PbI_{3-x}Cl_x$ and 75% $CH_3NH_3PbI_{3-x}Ac_x$. As a result, the device exhibited a significantly enhanced PCE of 15.7% (V_{oc} of 0.95, J_{sc} of 22.5 mA·cm^{-2}, and FF of 0.73) [31]. Using the same protocols, the authors also demonstrated large-area perovskite layers (~3.8 cm^2) fabricated with the spray-casting method, exhibiting a PCE of 11.7%. All these spray deposition procedures were conducted in a N_2-filled grove box because perovskite crystallization can be severely limited by humidity and oxygen levels.

Instead of the one-step spray-casting method using a precursor mixture, Q. Meng and coworkers developed a two-step ultrasonic spray deposition technique for large-area and high-quality perovskite layers: 1) PbI_2 in dimethyl sulfoxide (DMSO) was spray-casted to form a thin film on a preheated substrate (60°C); 2) CH_3NH_3I in isopropyl alcohol (IPA) was spray-casted on top of the dried PbI_2 layer [32]. The resultant perovskite layer can be formed through the chemical reaction between PbI_2 and CH_3NH_3I at 100°C for 2 h, affording highly efficient PSCs with a PCE of 16.03% over a small area and 13% over a large area (1 cm^2). More advanced spray deposition techniques with a solvent mixture can lead to further increases in the PCE of PSCs. In 2016, S.H. Im and coworkers developed their own spray-coating steps for balancing the incoming solvent flux and the outgoing evaporation flux by controlling the mixture ratio of DMF and γ-butyrolactone (GBL) [33]. Due to the relatively high boiling point of GBL compared to that of DMF, the transition time from precursor to perovskite film is significantly extended, allowing the formation of large-size perovskite crystal domains. Although the developed method is completely opposite to the conventional spray-casting method, the authors achieved the highest performance of spray-cast PSCs with a PCE of 18.7%, indicating that a dynamic equilibrium in the flux of incoming and outgoing solvent is important for ensuring a well-developed perovskite film.

2.4. Inkjet Printing

Inkjet printing, which can create a layer by propelling ink onto the defined substrate area, has been emerging as an essential research tool and commercialization technique for PSCs due to its many advantages over other deposition methods, including low cost, contactless patterning, high-resolution deposition, and scalability under ambient conditions (Figure 4b) [34-36]. Inkjet printing can also be used to print and/or pattern alternative metal electrodes, such as carbon-based conductors and silver nanowires (AgNWs), without a vacuum process, which can significantly reduce the number of fabrication steps and even increase device performance. In 2014, S. Yang and coworkers demonstrated for the first time inkjet printing-cast planar PSCs with carbon-based electrodes [37]. They developed a mixture of carbon and CH_3NH_3I and then deposited it on top of a PbI_2 film using inkjet printing technology. The authors found that the carbon:CH_3NH_3I blend induced enhanced perovskite crystallinity of perovskite and improved interfacial contact between the carbon electrode and perovskite layer, which are attributed to the simultaneous *in situ* chemical transformation and carbon electrode deposition. Therefore, compared with the reference cell (PCE of 8.51%), the corresponding device showed an increased PCE of 11% (V_{oc} of 0.95 V, J_{sc} of 17.20 mA·cm^{-2}, and FF of 0.71).

The formation of perovskite crystals and their surface coverage on the substrate can also be influenced by the printing table temperature. Y.-L. Song and coworkers demonstrated that the morphology of inkjet-printed $CH_3NH_3PbI_3$ perovskite films on mesoporous TiO_2 can be significantly determined by the printing table temperature ranging from 25 to 60°C [38]. The definition of mesoporous is a layer containing a majority of pores with the size of a few nanometers. In PSCs, the pore sizes of mesoporous TiO_2 are usually below 10 nm [39].

Although an appropriate crystal size and good film coverage of perovskite solution were observed at the optimum table temperature (~ 50°C), the corresponding device showed a relatively low PCE of 7.6%, which was attributed to large pinholes and poor surface morphology. To avoid these issues, the authors used a promising mixed perovskite ink comprising PbI_2, CH_3NH_3I, and CH_3NH_3Cl (molar ratio = 0.4:1:0.6), which resulted in good film morphology and high reproducibility as well as high device performance (V_{oc} of 0.91 V, J_{sc} of 19.55 mA·cm^{-2}, FF of 0.69, and PCE of 12.3%).

For mixed perovskite inks, D. Venkataraman and coworkers used a multichannel inkjet printing method, which enabled the *in situ* mixing of cations (CH_3NH_3I and $HN=CHNH_3^+$) and thus resulted in various compositions of perovskite layers [40]. The multichannel inkjet printing allowed systematic and compositional mapping of mixed cations to find the optimum blend ratio of counterions to achieve highly efficient and stable perovskite solar cells. Therefore, under the optimum condition of a $CH_3NH_3I:HN=CHNH_3^+$ ratio of 2:1, the p-i-n PSCs prepared using their approach provided the highest PCE of 11.1% (V_{oc} of 0.87 V, J_{sc} of 18.77 mA·cm^{-2}, and FF of 0.68).

2.5. Screen Printing

Screen printing, whereby a metal mesh is typically used as a stencil to transfer ink to a target substrate (Figure 4c), has been developed as one of the most efficient printing techniques for large-area and multilayer-structure devices [41]. As the ink fills in the open mesh apertures, a uniform and thin layer can be formed after a blade is moved back and forth across the screen. However, unlike the typical printing technologies for PSCs, which should produce a smooth film, screen printing has widely been employed for establishing nanocrystalline scaffolds in mesoporous PSC architectures [42-44]. M. Wang and coworkers demonstrated that efficient PSCs based on a mesoscopic $TiO_2/Al_2O_3/NiO/carbon$ architecture can be fully achieved by screen printing technology [45]. The authors prepared screen printing pastes by dissolving each nanopowder in an organic solvent such as ethanol. Then, the TiO_2 (400 nm), Al_2O_3 (400 nm) and NiO (800 nm) layers were subsequently screen-printed on an FTO electrode. After thermal annealing of the four layers under the optimum conditions, the mesoscopic $TiO_2/Al_2O_3/NiO/carbon$ structure turned into a scaffold that could be infiltrated by $CH_3NH_3PbI_3$. The screen-printed PSCs showed a high PCE of 15.1% (J_{sc} of 21.62 mA·cm^{-2}, V_{oc} of 0.92, and FF of 0.76).

The screen printing technique is highly beneficial when fabricating both small-area devices and large-area modules because it allows the patterned deposition of unit cells through a designed mask [46]. The screen printing technique can also be combined with other mass production printing technologies, such as slot-die and roll-to-roll printing, to ensure higher throughput for flexible device applications. F.C Krebs and coworkers demonstrated fully printed flexible PSCs including the top electrode and encapsulation [47]. They adopted a planar perovskite solar cell (polyethylene terephthalate (PET)/ITO/ZnO/perovskite/HTL/back electrode) due to its compatibility with continuous and large-area manufacturing. After the required ZnO layer (thickness of ~ 23 nm) and perovskite layer (thickness of ~ 300 nm) was deposited via the slot-die and roll-to-roll printing techniques, respectively, the metal electrode was finally printed by the screen printing method. The resultant flexible PSCs obtained via roll-to-roll mass production exhibited a PCE of ~ 5%, implying that the combination of various printing techniques presents a significant potential for the upscaling of the technology from small laboratory cells to the manufacture of large-area cells via roll-to-roll processes (Figure 4d).

2.6. Other Coating Methods

In addition to the representative coating and printing techniques mentioned above, other methods, such as bar coating, brush painting, and dip coating, have been employed in the fabrication of PSCs. In the bar coating process, the film thickness and quality are determined by the viscosity, concentration, surface energy, bar speed, and the gap distance between coating bar and substrate [48]. T. Wang and coworkers reported efficient planar PSCs prepared by the bar coating of a mixed perovskite solution in GBL, DMSO, and IPA, and they achieved a maximum PCE of 13.0% with very weak hysteresis [49]. Peculiarly enough, an easily purchasable brush was applied to the printing of the perovskite, which has the strong advantages of shear force for the control of the morphology and roll-to-roll compatibility with nonflat and curved surfaces [50]. S.-S. Kim and coworkers first demonstrated PSCs via brush painting of a PEDOT:PSS HTL, a perovskite photoactive layer, and a PCBM ETL, and they particularly investigated the correlation between the perovskite morphology and solvent. The best performance of approximately 9% PCE was obtained using the protic 2-methoxyethanol solvent [51]. The dip-coating method was also used for the fabrication of PSCs. J. Zhang and coworkers successfully coated a sol-gel-derived TiO_2 ETL and a perovskite photoactive layer, which resulted in a PCE of 11% without any hysteresis [52].

3. PEROVSKITE SOLAR MODULES

3.1. Module Concept

Photovoltaic modules comprise a photovoltaic array and panel for the photovoltaic system, which generates solar energy through the photovoltaic effect and supplies it to practical applications such as solar power stations, solar parks, and solar farms [53]. Unlike the silicon solar cell modules, usually fabricated by vacuum processes involving evaporation and sputtering, PSC modules have the advantage of roll-to-roll processability, which allows fast and low-cost production. Before the development of PSC modules, another type of solution-processed solar cell, organic solar cell modules, started to be fabricated by roll-to-roll compatible printing methods in the early 2000s; hence, the F.C. Krebs group led the field of printed organic solar cells fabricated by various printing techniques, especially the slot-die coating method [15]. In the early stage of printed modules, the slot-die coating method was adopted to coat component layers with one-dimensional stripe patterns, thereby ensuring contact areas for the series connection of the subcells of the module [54]. However, the delicate patterned coating of three component layers consisting of the HTL/photoactive layer/ETL is technically very difficult with the slot-die coating method. Thus, the organic solar cell modules showed a poor GFF of approximately 70%, resulting in a drop of approximately 30% in module efficiency over the total area (Figure 5).

Currently, instead of low-cost direct-patterned coating using the slot-die coating method, a laser scribing process is employed for the fabrication of solution-processed solar cell modules for achieving high GFFs [9, 55-58] (Figure 6). In printed organic solar cell modules, a maximum GFF of 98.5% was obtained for the size of 10.4 mm^2 using the delicate laser scribing technique, leading to a low total width of 80 μm for the interconnection region [56]. Similarly, laser patterning has also been applied to the fabrication of PSC modules, and the first efficient PSC modules were reported in 2016 [9]. In this report, the laser scribing technique led to a dead zone of 0.43 mm between adjacent active areas of 4.57 mm, resulting in a high GFF of 91% in the n-i-p-structured PSC modules (total area of 4 cm^2). Although the laser scribing technique has been widely adopted for most photovoltaic modules, it has a residue problem that increases the probability of pinhole formation in the active areas and requires a high-cost installation for industrial mass production. To avoid these disadvantages of the laser scribing technique, a new fabrication architecture for PSC modules was reported; this architecture utilized the intrinsic ion-conducting features of perovskite photoactive materials to create metal-filamentary nanoelectrodes in the series connection regions [59]. By applying a reverse bias to the series connection regions, the metal ions diffused and recombined with electrons from the transparent electrode, thereby forming conductive metal-filamentary nanoelectrodes and ensuring series interconnection between adjacent subcells without the laser scribing technique. With this electrochemical patterning process, the highest GFF of 94.1% was achieved for planar PSC modules, and a high module PCE of 14.0% was obtained for an area of 9.06 cm^2.

In addition to module fabrication using the electrochemical patterning process, pioneering researchers have designed other innovative module structures in the field of organic photovoltaics, which were developed 10 years earlier than PSC modules. However, these new organic solar cell module concepts, such as eliminating the patterning process using a charge recombination feature, introducing seamless interconnections between alternate structures of subcells, and realizing invisible interconnections by the inkjet printing of highly conductive silver lines, have not been applied to the fabrication of PSC modules [60-62]. To fabricate highly efficient PSC modules via a roll-to-roll processing system, the development of not only compact and pinhole-free uniform perovskite layers but also an efficient module structure with a high GFF must occur.

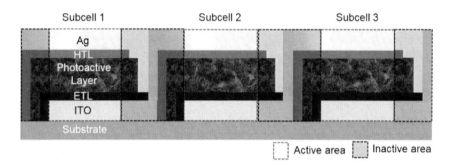

Figure 5. Conventional printed module structure composed of all-patterned layers (ITO/ETL/Photoactive layer/HTL/Ag) to ensure series connection. The area sandwiched between the top and bottom electrodes is denoted as an active area. Otherwise, the area is denoted as an inactive area.

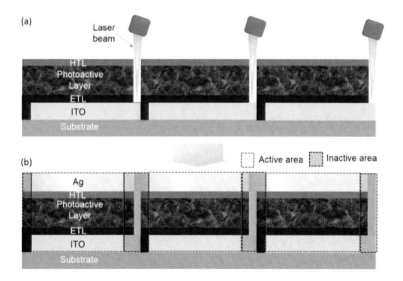

Figure 6. Schematic images of the (a) laser scribing process and (b) completed module structure.

3.2. Printed Perovskite Solar Modules

The ultimate goal of developing the aforementioned techniques is to fabricate efficient PSC modules by integrating various printing methods and module architectures. To achieve high-throughput mass production with a high fabrication yield using the roll-to-roll process, it is necessary to satisfy the following requirements: i) All of the component materials need to be solution-processable to be compatible with continuous printing methods. ii) The general properties of solutions (e.g., viscosity and boiling point) should be appropriate for the coating process. For a solution with too low of a boiling point, maintaining a constant concentration is difficult, while too viscous a solution is more likely to produce non-uniform films. iii) Processing solvents must have no significant harmful effects on the environment and human health. iv) A high-quality (i.e., compact and dense) perovskite film morphology without significant pinholes is beneficial for fabricating highly reproducible perovskite devices. v) During the patterning process, residual byproducts, which have detrimental effects on the device performance and increase device failure probability, should be minimized and completely removed from the PSC modules. Overall, the most critical factors are performance, processability, and reproducibility. In light of these facts, there have been tremendous research efforts in developing 'printed PSC modules'. Due to the difficulty in fabricating PSC modules using printing processes, however, only a few studies have successfully demonstrated perovskite solar modules fabricated by printing techniques.

In 2015, A.D. Carlo and coworkers first pointed out some issues with printed large-area perovskite solar modules [63]. For the successful transition from spin coating to scalable printing techniques, a blade coating method was adopted with the assistance of heated air flow. Considering that the morphology of the PbI_2 precursor layer has a significant impact on the quality of the final perovskite film, the crystallization of the PbI_2 layer was manipulated with delicate control over the solvent evaporation speed. The authors applied a PbI_2 precursor solution with a dilute concentration rather than a supersaturated concentration, which was blade-coated at a speed of 40 mm s^{-1} and a height of 100 µm. Unfortunately, these conditions resulted in non-uniform films with substantial pinholes on FTO/TiO$_2$ substrates. Subsequently, the samples were treated with a hot flow (100°C) of air to quickly remove the residual solvent and thereby form a compact and dense PbI$_2$ film. Finally, perovskite films (CH$_3$NH$_3$PbI$_3$) were obtained by dipping the PbI$_2$ layer in a CH$_3$NH$_3$I solution. To construct a monolithic module structure connected in series, the component layer was patterned using a CO$_2$ laser. The resulting module devices consisted of four subcells with an area of 2.5 cm^2 (total area of 10.1 cm^2), exhibiting a module PCE of 10.3% without considering the GFF.

S.G. Mhaisalkar and coworkers developed PSC modules with an active area of 70 cm^2 fabricated by a modified printing method [64]. In contrast to the conventional deposition process, this modified method involved loading the perovskite precursor solution into a mesoporous scaffold of FTO/TiO$_2$/ZrO$_2$/carbon followed by a drying process. To serve the dual function of the HTL and top electrode, a conductive carbon paste was applied to the FTO/TiO$_2$/ZrO$_2$ scaffold using a screen printing method. The resulting film thicknesses were 500 ± 50 nm for TiO$_2$, 1.4 ± 0.1 μm for ZrO$_2$, and 10-12 μm for the carbon electrode. It was noted that the properties of the carbon electrode determined the performance of the devices; carbon pastes were prepared with different compositional ratios of graphite, carbon nanoparticles, and an organic binder (i.e., ethyl cellulose and terpineol). As a result, the PSC module with compact and dense carbon showed a high performance of 10.74%.

On the other hand, there is a new approach for dealing with scalable PSC module fabrication. To overcome the difficulty in stacking the multilayer structure via a solution process, a lamination process that provides a direct mechanical connection of two cells without chemical deposition has gained considerable research interest in recent years. PSC modules with a device structure of ITO/NiO/CH$_3$NH$_3$PbI$_3$/PCBM/ZnO/polyethyleneimine/ laminated top electrode were demonstrated using a pressure lamination method at a low temperature of 60°C [65]. For a more definite construction, the laminated electrode should be conductive and adhesive, which was accomplished by adopting AgNWs and a transparent conductive adhesive (TCA); the TCA was an organic blend based on a mixture of a highly conductive polymer, PEDOT:PSS, and sorbitol. The pivotal role of PEDOT:PSS is providing sufficient conductivity for efficient charge transport, while sorbitol ensures the adhesive property for securing robust interconnections. Thus, a laminated top electrode was applied as another form of a flexible layered substrate with the configuration of PET/AgNWs/TCA on an original substrate. In addition, module patterning with high resolution and GFF was achieved by using a depth-resolved post-laser-patterning technique, which has the advantage of compatibility with the roll-to-roll process. After careful optimisation of the processing conditions, the laminated PSC modules showed a module PCE of 9.8%.

In addition to the processing procedure, perovskite material engineering has been attempted to be used as a universal method to produce large-scale PSC modules. M.K. Nazeeruddin and coworkers demonstrated highly efficient and stable PSC modules using 2-dimensional (2D)/3D perovskites [66]. This heterojunction material offered the synergistic effect of the enhanced stability of the 2D perovskite and excellent charge transport of the 3D perovskite. By introducing aminovaleric acid iodide (AVAI) as the organic precursor into PbI_2, the 2D perovskite was synthesized. Similar to previous work, the authors demonstrated mesoscopic solar cells with the introduction of hydrophobic carbon electrodes. Consequently, a module efficiency of 11.2% with a large area of 10×10 cm^2 was demonstrated in a fully printable device structure. More importantly, the module exhibited excellent long-term stability under operating conditions (1 sun AM 1.5 G and 55°C), which was mainly attributed to efficient moisture blocking by the 2D perovskite.

Another approach focusing on perovskite material engineering was conducted by changing the chemical composition and processing solvent of the perovskite. K. Zhu and coworkers introduced 30% methylammonium chloride (CH_3NH_3Cl) as an additive into the $CH_3NH_3PbI_3$ precursor system to improve the quality of the perovskite film morphology with a noticeable increase in grain size without concomitant defects, such as pinholes and voids [67]. Because the perovskite morphology is mainly affected by the processing solvent, the authors proposed a new cosolvent system of N-methyl-2-pyrrolidinone (NMP):DMF, which significantly extended the processing window for perovskite film fabrication. As a result, the authors demonstrated a 12.6 cm^2 PSC module consisting of four subcells with a stabilized module PCE of 13.3%.

High-efficiency PSC modules were reported using a new deposition route. L. Han and coworkers developed a pressure processing method for the deposition of high-quality perovskite films [68]. The chemically prepared mixed amine complex precursors (CH$_3$NH$_3$I·3CH$_3$NH$_2$ and PbI$_2$·CH$_3$NH$_2$, 1:1 molar ratio) were cast on the center of the substrate, then a polymer film (polyimide) was placed on the substrate with the precursor solution, and finally uniform pressure was applied to spread the liquid. After the conversion from precursors to perovskite with thermal treatment, the polymer film was gently removed from the substrate. This method offers the following advantageous features: i) Compared with that required by the conventional spin coating process, the amount of materials used can be dramatically reduced. ii) The surface morphology is relatively independent of the properties of the solvents, which expands the solvent choices. iii) This method ensures excellent surface coverage regardless of the surface energy of the underlying substrate. By employing TiO$_2$ and Spiro-OMeTAD as the ETL and HTL, respectively, a regular type of PSC module with a module PCE of 13.9% and aperture area of 36.1 cm^2 was achieved.

Recently, J. Huang and coworkers successfully scaled up PSC modules with minimal PCE loss by adding a small amount of surfactant, which enabled the high-speed deposition of uniform large-area perovskite films with a doctor blade coating method [69]. During the fabrication of PSC devices, the hydrophilic property of the perovskite precursor makes it difficult to coat the precursor on hydrophobic surfaces. Therefore, most printed PSC modules have been fabricated in the structure of hydrophilic TiO$_2$ ETL/perovskite/HTL. In this study, very small amounts of surfactant dramatically increased the adhesion of the perovskite precursor to the underlying hydrophobic organic HTL. Hence, large-area PSC modules were demonstrated using the doctor blade coating method, resulting in a tremendous module PCE of 15.3% with an aperture area of 33.0 cm^2. The other notable result in this work is that every component layer, including the

ETL and HTL, was deposited by a printing method, which increases the area of large-area printed PSC modules.

Finally, the current state-of-the-art results on printable perovskite solar modules are introduced (Figure 7). Primarily, research interests are focused on developing perovskites and optimising the patterning process conditions. We strongly believe that further studies are needed for perfecting perovskite-based photovoltaic technologies, and such efforts will bring a bright future for the successful commercialisation of PSCs.

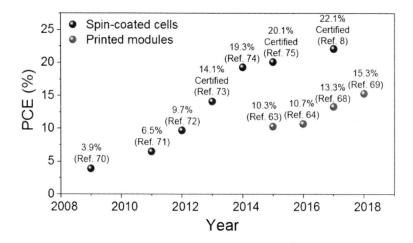

Figure 7. Recently reported PCEs of spin-coated PSCs (blue dots) and printed PSC modules (red dots).

CONCLUSION

In this chapter, we focused on the recent research on printed PSCs prepared using roll-to-roll-compatible printing methods, then discussed the module concept and reviewed PSC modules, which are a practical component of photovoltaic systems. Despite the difficulties in the fabrication of pinhole-free and uniform films via printing techniques, notable results have continuously been reported and have almost caught up with the high efficiencies of spin-coated unit cells by gradually improving the PCE of the printed PSCs. Among the various printing methods, the simple doctor blade coating method has been widely adopted to achieve highly efficient PSCs in studies, reaching PCEs of almost 20% in small-sized unit cells and of 15% in large-sized modules. As the most promising printing method for high-throughput mass production, the slot-die coating method has also been used for the fabrication of printed PSCs. Although the highest PCE of slot-die-coated PSCs is lower than that of doctor blade-coated devices, the PCE is continuously increasing. In addition to these two printing methods, other printing methods are usually specialized for the coating of interlayers, such as the ETL and HTL. Nevertheless, there are many issues for the development of printed PSCs and large-area PSC modules, and it is expected that the collaboration of many researchers and companies in this area will lead to the successful commercialisation of solution-processed photovoltaics in the near future.

Acknowledgments

We thank the Heeger Center for Advanced Materials (HCAM) at the Gwangju Institute of Science and Technology (GIST) of Korea. This work was supported by the GIST Research Institute (GRI) grant funded by GIST in 2019; by the National Foundation of Korea (NRF) grant funded by the Ministry of Science, ICT & Future Planning (MSIP) (NRF-2017R1A2B4 012490 & NRF-2015M1A2A2057510 & No. 2019R1C1C10 07680).

References

[1] Rajagopal, A., Yao, K., Jen, and A. K. Y. (2018). Toward perovskite solar cell commercialization: a perspective and research roadmap based on interfacial engineering. *Adv. Mater.* 30, 1800455.

[2] Green, M. A., Hishikawa, Y., Dunlop, E. D., Levi, D. H., Hohl-Ebinger, J., and Ho-Baillie, A. W. Y. (2018). Solar cell efficiency tables (version 52). *Prog. Photovolt. Res. Appl.* 26, 427-436.

[3] Nie, W., Tsai, H., Asadpour, R., Blancon, J. C., Neukirch, A. J., Gupta, G., Crochet, J. J., Chhowalla, M., Tretiak, S., Alam, M. A., Wang, H. L., and Mohite, A. D. (2015). High-efficiency solution-processed perovskite solar cells with millimeter-scale grains. *Science* 347, 522-525.

[4] Stranks, S. D., Eperon, G. E., Grancini, G., Menelaou, C., Alcocer, M. J. P., Leijtens, T., Herz, L. M., Petrozza, A., and Snaith, H. J. (2013). Electron-hole diffusion lengths exceeding 1 micrometer in an organometal trihalide perovskite absorber. *Science* 342, 341-344.

[5] Yang, Y., Yang, M., Li, Z., Crisp, R., Zhu, K., and Beard, M. C. (2015). Comparison of recombination dynamics in $CH_3NH_3PbBr_3$ and $CH_3NH_3PbI_3$ perovskite films: influence of exciton binding energy. *J. Phys. Chem. Lett.* 6, 4688-4692.

[6] Ziang, X., Shifeng, L., Laixiang, Q., Shuping, P., Wei, W., Yu, Y., Li, Y., Zhijian, C., Shufeng, W., Honglin, D., Minghui, Y., and Qin, G. G. (2015). Refractive index and extinction coefficient of $CH_3NH_3PbI_3$ studied by spectroscopic ellipsometry. *Opt. Mater. Express* 5. 29-43.

[7] Eperon, G. E., Stranks, S. D., Menelaou, C., Johnston, M. B., Herz, L. M., and Snaith, H. J. (2014). Formamidinium lead trihalide: a broadly tunable perovskite for efficient planar heterojunction solar cells. *Energy Environ. Sci.* 7, 982-988.

[8] Yang, W. S., Park, B. W., Jung, E. H. Jeon, N. J., Kim, Y. C., Lee, D. U., Shin, S. S., Seo, J. Kim, E. K., Noh, J. H., and Seok, S. I. (2017). Iodide management in formamidinium-lead-halide–based perovskite layers for efficient solar cells. *Science* 356, 1376-1379.

[9] Qiu, W., Merckx, T., Jaysankar, M., Masse de la Huerta, C., Rakocevic, L., Zhang, W., Paetzold, U. W., Gehlhaar, R., Froyen, L., Poortmans, J., Cheyns, D., Snaith, H. J., and Heremans, P. (2016). Pinhole-free perovskite films for efficient solar modules. *Energy Environ. Sci.* 9, 484-489.

[10] Hwang, K., Jung, Y. S., Heo, Y. J, Scholes, F. H., Watkins, S. E., Subbiah, J., Jones, D. J., Kim, D.-Y., and Vak, D. (2015). Toward large scale roll-to-roll production of fully printed perovskite solar cells. *Adv. Mater.* 27, 1241-1247.

[11] Tang, S., Deng, Y., Zheng, X., Bai, Y., Fang, Y., Dong, Q., Wei, H., and Huang, J. (2017). Composition engineering in doctor-blading of perovskite solar cells. *Adv. Energy Mater.* 7. 1700302.

[12] Yeo, J. S., Lee, C. H., Jang, D., Lee, S., Jo, S. M., Joh, H. I., and Kim, D. Y. (2016). Reduced graphene oxide-assisted crystallization of perovskite via solution process for efficient and stable planar solar cells with module-scales. *Nano Energy* 30, 667-676.

[13] Razza, S., Giacomo, F. D., Matteocci, F., Cinà, L., Palma, A. L., Casaluci, S., Cameron, P., D'Epifanio, A., Licoccia, S., Reale, A., Brown, T. M., and Carlo, A. D. (2015). Perovskite solar cells and large area modules (100 cm^{232}) based on an air flow-assisted PbI$_2$ blade coating deposition process. *J. Power Sources* 277, 286-291.

[14] Röttinger, S., Schwarz, B., Schäfer, S., Gauch, R., Zimmermann, B., and Würfel, U. (2016). Laser patterning of vacuum processed small molecular weight organic photovoltaics. *Sol. Energy Mater. Sol. Cells* 154, 35-41.

[15] Krebs, F. C. (2009). Fabrication and processing of polymer solar cells: A review of printing and coating techniques. *Sol. Energy Mater. Sol. Cells* 93, 394-412.

[16] Mei, A., Li, X., Liu, L., Ku, Z., Liu, T., Rong, Y., Xu, M., Hu, M., Chen, J., Yang, Y., Grätzel, M., and Han, H. (2014). A hole-conductor-free, fully printable mesoscopic perovskite solar cell with high stability. *Science* 345. 295-298.

[17] Yang, Z., Chueh, C. C., Zuo, F., Kim, J. H., Liang, P. W., and Jen, A. K. Y. (2015). High-performance fully printable perovskite solar cells via blade-coating technique under the ambient condition. *Adv. Energy Mater.* 5. 1500328.

[18] Back, H., Kim, j., Kim, G., Kim T. K., Kang, H., Kong, J., Lee, S. H., and Lee, k. (2016). Interfacial modification of hole transport layers for

efficient large-area perovskite solar cells achieved via blade-coating. *Sol. Energy Mater. Sol. Cells* 144, 309-315.

[19] Hong, S., Lee, J., Kang, H., and Lee, K. (2013). Slot-die coating parameters of the low-viscosity bulk-heterojunction materials used for polymer solar cells. *Sol. Energy Mater. Sol. Cells* 112, 27-35.

[20] Cotella, G., Baker, J., Worsley, D., Rossi, F. D., Pleydell-Pearce, D., Carnie, M., and Watson, T. (2017). One-step deposition by slot-die coating of mixed lead halide perovskite for photovoltaic applications. *Sol. Energy Mater. Sol. Cells* 159, 362-369.

[21] Giacomoa, F. D., Shanmugama, S., Fledderusa, H., Bruijnaersb, B. J., Verheesc, W. J. H., Dorenkamperc, M. S., Veenstrac, S. C., Qiud, W., Gehlhaard, R., Merckxd, T., Aernoutsd, T., Andriessena, R., and Galagan, Y. (2018). Up-scalable sheet-to-sheet production of high efficiency perovskite module and solar cells on 6-in. substrate using slot die coating. *Sol. Energy Mater. Sol. Cells* 181, 53-59.

[22] Tedde, S. F., Kern, J., Sterzl, T., Furst, J., Lugli, P., and Hayden, O. (2009). Fully spray coated organic photodiodes. *Nano Lett.* 9, 980-983.

[23] Girotto, C., Moia, D., Rand, B. P., and Heremans, P. (2011). High-performance organic solar cells with spray-coated hole-transport and active layers. *Adv. Funct. Mater.* 21, 64-72.

[24] Colsmann, A., Reinhard, M., Kwon, T. H., Kayser, C., Nickel, F., Czolk, J., Lemmer, U., Clark, N., Jasieniak, J., Holmes, A. B., and Jones, D. (2012). Inverted semi-transparent organic solar cells with spray coated, surfactant free polymer top-electrodes. *Sol. Energy Mater. Sol. Cells*, 98, 118-123.

[25] Jung, Y. S., Hwang, K., Scholes, F. H., Watkins, S. E., Kim, D. Y., and Vak, D. (2016). Differentially pumped spray deposition as a rapid screening tool for organic and perovskite solar cells. *Sci. Rep.* 6, 20357.

[26] Dualeh, A., T´etreault, N., Moehl, T., Gao, P., Nazeeruddin M. K., and Grätzel, M. (2014). Effect of annealing temperature on film morphology of organic-inorganic hybrid pervoskite solid-state solar cells, *Adv. Funct. Mater.* 24, 3250-3258.

[27] Bishop, J. E., Routledge, T. J., Lidzey, D. G. (2018). Advances in spray-cast perovskite solar cells. *J. Phys. Chem. Lett.* 9, 1977-1984.

[28] Barrows, A. T., Pearson, A. J., Kwak, C. K., Dunbar, A. D. F., Buckley, A. R., and Lidzey, D. G. (2014). Efficient planar heterojunction mixed-halide perovskite solar cells deposited via spray-deposition. *Energy Environ. Sci.* 7, 2944-2950.

[29] Eslamian, M. (2014). Spray-on thin film PV solar cells: advances, potentials and challenges. *Coatings* 4, 60-84.

[30] Das, S., Yang, B., Gu, G., Joshi, P. C., Ivanov, I. N., Rouleau, C. M., Aytug, T., Geohegan, D. B., and Xiao, K. (2015). High-performance flexible perovskite solar cells by using a combination of ultrasonic spray coating and low thermal budget photonic curing. *ACS Photonics* 2, 680-686.

[31] Tait, J. G., Manghooli, S., Qiu, W., Rakocevic, L., Kootstra, L., Jaysankar, M., Huerta, C. A. M., Paetzold, U. W., Gehlhaar, R., Cheyns, D., Heremans, P., and Poortmans, J. (2016). Rapid composition screening for perovskite photovoltaics via concurrently pumped ultrasonic spray coating. *J. Mater. Chem. A* 4, 3792-3797.

[32] Huang, H., Shi, J., Zhu, L., Li, D., Luo, Y., and Meng, Q. (2016). Two-step ultrasonic spray deposition of $CH_3NH_3PbI_3$ for efficient and large-area perovskite solar cell. *Nano Energy* 27, 352-358.

[33] Heo, J. H., Lee, M. H., Jang, M. H., Im, S. H. (2016). Highly efficient $CH_3NH_3PbI_{3-x}Cl_x$ mixed halide perovskite solar cells prepared by re-dissolution and crystal grain growth via spray coating. *J. Mater. Chem. A* 4, 17636-17642.

[34] Singh, M., Haverinen, H. M., Dhagat, P., Jabbour, G. E. (2010). Inkjet printing-process and its applications. *Adv. Mater.* 22, 673-685.

[35] Franeker, J. J., Voorthuijzen, W. P., Gorter, H., Hendriks, K. H., Janssen, R. A. J., Hadipour, A., Andriessen, R., and Galagan, Y. (2013). All-solution-processed organic solar cells with conventional architecture. *Sol. Energy Mater. Sol. Cells* 117, 267-272.

[36] Peng, X., Yuan, J., Shen, S., Gao, M., Chesman, A. S. R., Yin, H., Cheng, J., Zhang, Q., and Angmo, D. (2017). Perovskite and organic

solar cells fabricated by inkjet printing: progress and prospects. *Adv. Funct. Mater.* 27, 1703704.
[37] Wei, Z., Chen, H., Yan, K., and Yang, S. (2014). Inkjet printing and instant chemical transformation of a CH$_3$NH$_3$PbI$_3$/nanocarbon electrode and interface for planar perovskite solar cells. *Angew. Chem. Int. Ed.* 53, 13239-13243.
[38] Li, S. G., Jiang, K. J., Su, M. J., Cui, X. P., Huang, J. H., Zhang, Q. Q., Zhou, X. Q., Yang, L. M., and Song, Y. L. (2015). Inkjet printing of CH$_3$NH$_3$PbI$_3$ on a mesoscopic TiO$_2$ film for highly efficient perovskite solar cells. *J. Mater. Chem. A* 3, 9092-9097.
[39] Sarkar, A., Jeon, N. J., Noh, J. H., and Seok, S. I. (2014). Well-organized mesoporous TiO$_2$ photoelectrodes by block copolymer-induced sol-gel assembly for inorganic-organic hybrid perovskite solar cells. *J. Phys. Chem. C* 118, 16688-16693.
[40] Bag, M., Jiang, Z., Renna, L. A., Jeong, S. P., Rotello, V. M., and Venkataraman, D. (2016). Rapid combinatorial screening of inkjet-printed alkyl-ammonium cations in perovskite solar cells. *Materials Letters* 164, 472-475.
[41] Razza, S., Castro-Hermosa, S., Carlo, A. D., and Brown, T. M. (2016). Research Update: Large-area deposition, coating, printing, and processing techniques for the upscaling of perovskite solar cell technology. *APL Materials* 4, 091508.
[42] Rong, Y., Ku, Z., Mei, A., Liu, T., Xu, M., Ko, S., Li, X., and Han, H. (2014). Hole-conductor-free mesoscopic TiO$_2$/CH$_3$NH$_3$PbI$_3$ heterojunction solar cells based on anatase nanosheets and carbon counter electrodes. *J. Phys. Chem. Lett.* 5, 2160-2164.
[43] Zhang, L., Liu, T., Liu, L., Hu, M., Yang, Y., Mei, A., and Han, H. (2015). The effect of carbon counter electrodes on fully printable mesoscopic perovskite solar cells, *J. Mater. Chem. A* 3, 9165-9170.
[44] Ku, Z., Rong, Y., Xu, M., Liu, T., and Han, H. (2013). Full printable processed mesoscopic CH$_3$NH$_3$PbI$_3$/TiO$_2$ heterojunction solar cells with carbon counter electrode. *Sci. Rep.* 3, 3132.
[45] Cao, K., Zuo, Z., Cui, J., Shen, Y., Moehl, T., Zakeeruddin, S. M., Grätzel, M., and Wang, M. (2015). Efficient screen printed perovskite

solar cells based on mesoscopic $TiO_2/Al_2O_3/NiO$/carbon architecture. *Nano Energy* 17, 171-179.

[46] Giacomo, F. D., Zardetto, V., D'Epifanio, A., Pescetelli, S., Matteocci, F., Razza, S., Carlo, A. D., Licoccia, S., Kessels, W. M. M., Creatore, M., Brown, T. M. (2015). Flexible perovskite photovoltaic modules and solar cells based on atomic layer deposited compact layers and UV-irradiated TiO_2 scaffolds on plastic substrates. *Adv. Energy Mater.* 5, 1401808.

[47] Schmidt, T. M., Larsen-Olsen, T. T., Carlé, J. E., Angmo, D., and Krebs, F. C. (2015). Upscaling of perovskite solar cells: fully ambient roll processing of flexible perovskite solar cells with printed back electrodes. *Adv. Energy Mater.* 5, 1500569.

[48] Khim, D., Han, H., Baeg, K. J., Kim, J., Kwak, S. W., Kim, D. Y., and Noh, Y. Y. (2013). Simple bar-coating process for large-area, high performance organic field-effect transistors and ambipolar complementary integrated circuits. *Adv. Mater.* 25, 4302-4308.

[49] Liu, X., Xia, X., Cai, Q., Cai, F., Yang, L., Yan, Y., and Wang, T. (2017). Efficient planar heterojunction perovskite solar cells with weak hysteresis fabricated via bar coating. *Sol. Energy Mater. Sol. Cells* 159, 412-417.

[50] Kim, S. S., Na, S. I., Jo, J., Tae, G., Kim, D. Y. (2017). Efficient polymer solar cells fabricated by simple brush painting. *Adv. Mater.* 19, 4410-4415.

[51] Lee. J. W., Na, S. I., Kim, S. S., (2017). Efficient spin-coating-free planar heterojunction perovskite solar cells fabricated with successive brush-painting. *J. Power Sources* 339, 33-40.

[52] Huang, L., Li, C., Sun, X., Xu, R., Du, Y., Ni, J., Cai, H., Li, J., Hu, Z., and Zhang, J. (2017). Efficient and hysteresis-less pseudo-planar heterojunction perovskite solar cells fabricated by a facile and solution-saving one-step dip-coating method. *Org. Electron.* 40, 13-23.

[53] Krebs, F. C., Espinosa, N., Hösel, M., Søndergaard, R. R., and Jørgensen, M. (2014). 25th Anniversary article: rise to power – OPV-based solar parks. *Adv. Mater.* 26, 29-39.

[54] Krebs, F. C., Tromholt, T. and Jørgensen, M. (2010). Upscaling of polymer solar cell fabrication using full roll-to-roll processing. *Nanoscale* 2, 873-886.

[55] Spyropoulos, G. D., Kubis, P., Li, N, Baran, D., Lucera, L., Salvador, M., Ameri, T., Voigt, M. M., Krebs, F.C., and Brabe, C.J. (2014). Flexible organic tandem solar modules with 6% efficiency: combining roll-to-roll compatible processing with high geometric fill factors. *Energy Environ. Sci.* 7, 3284-3290.

[56] Lucera, L., Machui, F., Kubis, P., Schmidt, H. D., Adams, J., Strohm, S., Ahmad, T., Forberich, K., Egelhaa, H. J., and Brabec, C. J. (2016). Highly efficient, large area, roll coated flexible and rigid OPV modules with geometric fill factors up to 98.5% processed with commercially available materials. *Energy Environ. Sci.* 6, 89-94.

[57] Yang, M., Li, Z., Reese, M. O., Reid, O. G., Kim, D. H., Siol, S., Klein, T. R., Yan, Y., Berry, J. J., Hest, M. F. A. M., and Zhu, K. (2017). Perovskite ink with wide processing window for scalable high-efficiency solar cells. *Nat. Energy* 2, 17038.

[58] Deng, Y., Zheng, X., Bai, Y., Wang, Q., Zhao, J., and Huang, J. (2018). Surfactant-controlled ink drying enables high-speed deposition of perovskite films for efficient photovoltaic modules. *Nat. Energy* 3, 560-566.

[59] Hong, S., Lee, J., Kang, H., Kim, G., Kee, S., Lee, J. H., Jung, S., Park, B. Kim, S., Back, H., Yu, K., and Lee, K. (2018). High-efficiency large-area perovskite photovoltaic modules achieved via electrochemically assembled metal-filamentary nanoelectrodes. *Sci. Adv.* 4, eeat3604.

[60] Hong, S., Kang, H., Kim, G., Lee, S., Kim, S., Lee, J. H., Lee, J., Yi, M., Kim, J., Back, H., Kim, J. R., and Lee, K. (2016). A series connection architecture for large-area organic photovoltaic modules with a 7.5% module efficiency. *Nat. Commun.* 7, 10279.

[61] Lee, J., Back, H., Kong, J., Kang, H., Song, S., Suh, H., Kang, S. O., and Lee, K. (2013). Seamless polymer solar cell module architecture built upon self-aligned alternating interfacial layers, *Energy Environ. Sci.* 6, 1152-1157.

[62] Maisch, P., Tam, K. C., Schilinsky, P., Egelhaaf, H. J., and Brabec, C. J. (2018). Shy organic photovoltaics: digitally printed organic solar modules with hidden interconnects. *Sol. RRL* 2, 1800005.

[63] Razza, S., Giacomo, F. D., Matteocci, F., Cinà, L., Palma, A. L., Casaluci, S., Cameron, P., D'Epifanio, A., Licoccia, S., Reale, A., Brown, T. M., and Carlo A. D. (2015). Perovskite solar cells and large area modules (100 cm^2) based on an air flow-assisted Pbi$_2$ blade coating deposition process. *J. Pow. Sources* 277, 286-291.

[64] Priyadarshi, A., Haur, L. J., Murray, P., Fu, D., Kulkarni, S., Xing, G., Sum, T. C., Mathews, N., and Mhaisalkar, S. G. (2016). A large area (70 cm^2) monolithic perovskite solar module with a high efficiency and stability. *Energy Environ. Sci.* 9, 3687-3692.

[65] Spyropoulos, G. D., Ramirez Quiroz, C. O., Salvador, M., Hou, Y., Gasparini, N., Schweizer, P., Adams, J., Kubis, P., Li, N., Spiecker, E., Ameri, T., Egelhaaf, H.-J., and Brabec, C. J. (2016). Organic and perovskite solar modules innovated by adhesive top electrode and depth-resolved laser patterning. *Energy Environ. Sci.* 9, 2302-2313.

[66] Grancini, G., Roldán-Carmona, C., Zimmermann, I., Mosconi, E., Lee, X., Martineau, D., Narbey, S., Oswald, F., De Angelis, F., Graetzel, M., and Nazeeruddin, M. K. (2017). One-year stable perovskite solar cells by 2D/3D interface engineering. *Nat. Commun.* 8, 15684.

[67] Yang, M., Li, Z., Reese, M. O., Reid, O. G., Kim, D. H., Siol, S., Klein, T.R., Yan, Y., Berry, J. J., Hest, M. F. A. M., and Zhu, K. (2017). Perovskite ink with wide processing window for scalable high-efficiency solar cells. *Nat. Energy* 2, 17038.

[68] Chen, H., Ye, F., Tang, W., He, J., Yin, M., Wang, Y., Xie, F., Bi, E., Yang, X., Grätzel, M., and Han, L. (2017). A solvent-and vacuum-free route to large-area perovskite films for efficient solar modules. *Nature* 550, 92-95.

[69] Deng, Y., Zheng, X., Bai, Y., Wang, Q., Zhao. J., and Huang, J. (2018). Surfactant-controlled ink drying enables high-speed deposition of perovskite films for efficient photovoltaic modules. *Nat. Energy* 3, 560-566.

[70] Kojima, A., Teshima, K., Shirai, Y., and Miyasaka, T. (2009). Organometal halide perovskites as visible-light sensitizers for photovoltaic cells. *J. Am. Chem. Soc.* 131, 6050-6051.

[71] Im, J. H., Lee, C. R., Lee, J. W., Park, S. W., and Park, N. G. (2011). 6.5% efficient perovskite quantum-dot-sensitized solar cell. *Nanoscale* 3, 4088-4093.

[72] Kim, H. S., Lee, C. R., Im, J. H., Lee, K. B., Moehl, T., Marchioro, A., Moon, S. J., Humphry-Baker, R., Yum, J. H., Moser, J. E., Grätzel, M., and Park, N. G. (2012). Lead iodide perovskite sensitized all-solid-state submicron thin film mesoscopic solar cell with efficiency exceeding 9%. *Sci. Rep.* 2, 591.

[73] Burschka, J., Pellet, N., Moon, S. J., Humphry-Baker, R., Gao, P., Nazeeruddin, M.K., and Grätzel. M. (2013). Sequential deposition as a route to high-performance perovskite-sensitized solar cells. *Nature* 499, 316-319.

[74] Zhou, H., Chen, Q., Li, G., Luo, S., Song, T., Duan, H. S., Hong, Z., You, J., Liu, Y., and Yang, Y. (2014). Interface engineering of highly efficient perovskite solar cells. *Science* 345, 542-546.

[75] Yang, W. S., Noh, J. H., Jeon, N. J., Kim, Y. C., Ryu, S., Seo, J., and Seok, S. I. (2015). High-performance photovoltaic perovskite layers fabricated through intramolecular exchange. *Science* 346, 1234-1237.

In: Perovskite Solar Cells
Editor: Murali Banavoth
ISBN: 978-1-53615-858-8
© 2019 Nova Science Publishers, Inc.

Chapter 6

FLEXIBLE PEROVSKITE SOLAR CELLS (FPSCS)

Banavoth Murali[*], T. Swetha, Sachin G. Ghugal and Ranadeep Raj Sumukam
School of Chemistry, University of Hyderabad,
Hyderabad, Telangana

ABSTRACT

The demand for cost-effective, flexible optoelectronic devices has increased rapidly over the recent years owing to their numerous practical applications. Perovskites, due to their phenomenal performances have marked a new era in the class of efficient photovoltaics. Albeit the conventional substrate configurations accomplishing the skyrocketing efficiencies in par with silicon, their flexible substrate configurations are yet to be realised for widespread utilisation. Flexible perovskite solar cells (FPSCs) is undoubtedly is a most attractive option their large-area, roll-to-roll processing, conformability, light-weight, easy bendability, stretching, wearability and high-power conversion efficiencies (PCEs). Selection of proper materials, processing techniques and long-term durability have

---

[*] Corresponding Author's Email: murali.banavoth@uohyd.ac.in.

played a vital role in commercialisation of the f-PSC. In this chapter, we have discussed the achievements of the FPSCs with their device structure, operating principle and various intrinsic properties. The most commonly used flexible substrates are also discussed in detail. Finally, the processability of FPSCs with encapsulation are summarised.

Keywords: flexible perovskite solar cells (FPSCs), roll-to-roll processing, power conversion efficiencies, flexible substrates and durability

1. INTRODUCTION

Flexible perovskites solar cells (FPSCs) are emerging technology towards commercialisation of pervoskites solar cells (PSCs), owing to their intrinsic advantages such as light weight and bendability with low-temperature solution processed deposition on various flexible substrates. This makes them suitable for transportation, installation, and integration with architectures and wearable electricity-generating devices. FPSCs exhibits additional advantages compared to the rigid PSCs as: (1) produce higher power conversion efficiency (PCE) with low weight, (2) proper stretching and bending, (3) easy device architectures modifications on to the other substrates, and (4) possibility of roll-to-roll processes in a large scale. Generally, the flexible substrates include transparent conductive materials and interfacial materials. The high symmetry cubic perovskites exhibit better photovoltaic properties compared to the other forms of perovskite. The organic-inorganic perovskites are more flexible compared to the inorganic perovskites; a) lower phase transition temperatures; ($CH_3NH_3PbI_3$; 327 K) ($CsPbI_3$; 598 K) [1], b) compact and smooth layers formed at low annealing temperatures by spin coating method; c) favourable mechanical properties; d) good solubility in organic polar solvents [2] gives better thin film formation using simple solution processes like ink jet printing [3], spray coating [4], spin coating [5], blade coating [6] slot-die coating [7], two-step deposition process and chemical vapour deposition (CVD) [8]. Compared to different techniques, spray coating is cost-effective, important area technique used for roll-to-roll processing [9]. The morphology and crystallinity of the perovskite layer play a vital role to enhance the PCEs of FPSCs [10]. Two-step deposition process (spin-spin [11] and spin-dip [12]) improves the morphology of the perovskite layer with better crystallinity. Critical parameters and efficiency trend for FPSCs is shown in Figure 1.

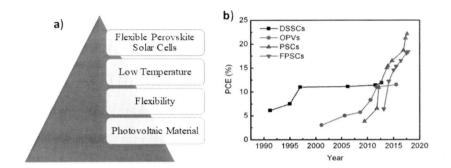

Figure 1. a) A summary of the critical parameters, which are critically important to achieving high-performance FPSCs. b) Efficiency progress of DSSCs, OPVs, PSCs, and FPSCs. The values for the efficiency of PSCs and FPSCs are obtained from published reports in the literature, and the efficiencies of the OPVs and DSSCs have certified values obtained from the National Renewable Energy Laboratory.

The PSCs have already delivered high PCEs on rigid substrates at low temperatures (< 150°C). The perovskite layer was processed at a temperature compatible with plastic substrates and the top contact layer was deposited with lower temperatures compared to the deposition of active layer temperature. In contrast, the bottom metal oxide layer was processed with high temperature used to avoid the recombination with the substrate and collect the carriers. Thus, more research efforts have devoted to developing FPSCs with alternative substrates or low-temperature processing for those layers.

A strategic alternative to overcome the temperature is to use substrates having excellent barrier properties like metal substrates, polymeric substrates etc., which will influence the processing of the PSC devices. For example, no need to use high-temperature processing for a polymeric film with a mandatory semi-transparent top contact.

2. DEVICE STRUCTURE AND OPERATING PRINCIPLE OF FPSCs

A typical FPSC consist of flexible substrate (metal foils and plastics etc.,) a perovskite absorber, opaque/transparent conductive electrodes, a hole transport layer (HTL) and an electron transport layer (ETL). Upon light illumination, the perovskite layer absorbs the light and undergoes excitation; the excited charge carriers diffuse to the interface of perovskite/HTL or perovskite/ETL (the holes or electrons injected into the valence band of the HTL or conduction band of the ETL) and holes/electrons collected by the respective electrodes.

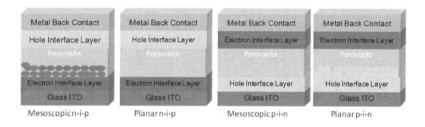

Figure 2. Schematic representation of perovskite solar cells heterojunctions according to the sequence.

Compared to several modified device architectures (e.g., ETL free, HTL free), the planar and mesoporous structures are mostly adopted device fabrication in FPSCs, due to their low-temperature and simple solution processable features. The mesoporous architecture consists of a metal oxide (Al_2O_3, TiO_2) frame with the perovskite layer; these metal oxide frames requires high-temperature sintering. The HTL and ETL are deposited front and rear side of perovskite/metal oxide layer respectively. In the planar device architecture, the perovskite layer is sandwiched between ETL and the HTL. HTL and ETL plays a vital role, ensures the hole/electron extraction at the interface and thereby collecting at respective electrodes. According to the sequence of the heterojunctions, the perovskite solar cells are further classified into two type's p-i-n and n-i-p as shown in Figure 2.

3. FLEXIBLE PEROVSKITE LAYERS: CRYSTALLINITY AND MORPHOLOGY

The FPSCs can give good device performance with proper morphology and better crystallinity of the perovskite layer. The prepared perovskite ($CH_3NH_3PbI_3$) using two-step spin dip solution process [13] in which PbI_2 spin-coated followed by a dip in CH_3NH_3I solution at room temperature, exhibited tetragonal crystals with a crystal size of 75 nm. This crystalline size resulted in improvement of the carrier mobility generated in the perovskite layer further reduce the trap state recombination with an increment in the effective optical path length. The fabricated FPSCs with this perovskite layer exhibited a PCE of 10.3% with V_{oc} of 1.03 V, J_{sc} of 13.4 $mA^{-1}cm^{-2}$ and FF of 73.9%, whereas the counterpart on rigid-ITO exhibited a PCE of 15.7%. The quality of the perovskite crystals obtained from the spin-dip processing technique depends upon the immersion time [14]. The best photovoltaic and kinetic parameters are observed at a dipping time of 40 s because total PbI_2 converted to perovskite layer with coverage of the large area, further perovskite layer is washed away after increasing the dipping time to 50 s and resulted in reduced device performance.

In FPSCs, a mixed-halide perovskite layer also has been exhibited excellent performance. The chloride-iodide perovskite layer ($CH_3NH_3PbCl_xI_{3-x}$) was known high lifetime of carrier mobility and excellent electrical properties, the synthesized perovskite layer [15] by combining 2.4 M CH_3NH_3I with 0.8 M of $PbCl_2$ in the ratio of 3:1 by one step deposition process found be uniform with some pin holes and having a thickness of 340 nm, the thicker the perovskite layer better will be the absorption further the XRD characterization revealed that $CH_3NH_3PbCl_xI_{3-x}$ crystals are along the direction of (110) and same as that of $CH_3NH_3PbI_3$ crystals. The X-ray photoelectron spectroscopy (XPS) reveals that the amount of Cl was found to be very minute in the $CH_3NH_3PbCl_xI_{3-x}$. The percentage composition of Cl was found to be 2.2% in (Cl+I). The $CH_3NH_3PbCl_xI_{3-x}$ layer having a band gap of 1.5 eV exhibited a PCE of 9.2% [16]. Similarly $CH_3NH_3PbCl_xI_{3-}$

x synthesised using one-step deposition process of $PbCl_2$ and CH_3NH_3I at 1:3 ratios on a TCO-free flexible substrate exhibited a PCE of 4.9% [17].

The $CH_3NH_3PbI_3$ layer synthesised using fast-deposition-crystallisation processing [18] method (1:1 ratio of CH_3NH_3I and PbI_2 in anhydrous DMF, stirred for overnight at 60°C in the glove box, then the reaction mixture was filtered using 0.45 μm polyvinylide fluoride. The formed perovskite crystals are with no pinholes on the surface, with a plate-like shape of diameter from 200-700 nm range. This perovskite film crystallized in tetragonal planes of (110), (211), (220) and (310) on single ITO substrate and ITO-PEDOT: PSS substrate. The high-resolution transmission electron microscope, selected-area-electron diffraction (SAED) pattern reveals that a 0.31 nm inter-planar spacing and tetragonal crystal structure of the synthesized perovskite.

A record efficiency of 12.7% [19] on PET-ITO substrate was obtained using $Cl-FAPbI_3$ layer; synthesized by the reaction of CH_3NH_3Cl and dual akylammoniumhalides of formamidinium iodide solutions $(CH(NH_2)_2I)$ in the ratio of 5:50, spin coated on PbI_2 (annealed) at 3000 rpm for the 30s, further an FPSC fabricated with modified α-$HC(NH_3)_2PbI_3$ (α-$FAPbI_3$) [20] exhibited a PCE of 13.03%; the perovskite layer synthesized by deposition of PbI_3 on flexible substrate (PEN-ITO/PEDOT:PSS) under a pressure of 10^{-5} mbar and heated at temperature 60 ^0C for 1 min in the glove box, 0.063 mol L^{-1} of MABr, FAI or their mixture (1:1, 1:2, 1:3, 1:4, 1:0) was spin coated at 2000 rpm for the 20s in isopropanol solution. The moisture in the perovskite layer was removed by heating at 90°C, 120°C and 150°C for 10 min, respectively, finally annealed for 5 min at 120°C. The crystal layer was changed to cubic from trigonal concerning the applied temperature.

The FPSC on a willow-glass substrate fabricated [21], the perovskite layer was deposited using a sequential two-step thermal evaporation deposition method. It was carried out by deposition of quartz crucibles having CH_3NH_3I and PbI_2 on the metal oxide (ITO/ZnO). The desired thickness was achieved using a calibrated quartz sensor. The best device performance observed with a film thickness of 340 nm. Similarly, gas assisted processing [22] technique was employed to synthesize the perovskite layer on a flexible substrate (PET-ZNO). During this method, an even perovskite layer is observed with fast solvent evaporation under the blowing of inert gas during spin coating, and it shows a tetragonal crystal of perovskite layer.

4. FLEXIBLE SUBSTRATES

4.1. Polymer Substrates

PSCs fabricated on the rigid substrates exhibited higher PCEs. These substrates are known to have corrosion resistance, heat resistance, good contact with the TCO films and good optical transmittance. The main drawbacks; frequently carried out at a very high temperature, and fragile. Polyethylene naphthalate (PEN) and polyethylene terephthalate (PET) films are known to be promising flexible substrates are owing to their low-cost, deactivation to common solvents, bendability, and excellent optical transparency [23]. The glass transition temperature (T_g) of PET and PEN are 80°C and 125°C respectively, the sheet resistance (R_{sh}) of PET/ITO and PEN/ITO conductive substrate have a significant increase after heating to a temperature 200°C and 240°C about 30 min respectively. Alternate to PET and PEN an intensive research studies are going on a transparent polyimide (CPI) substrates. The CPI/ITO exhibited better properties than PET/ITO (low Rsh of 57.8 Ω sq^{-1} and high transmittance of 83.6%). The outstanding flexible PSC developed with an annealing temperature of 300°C which balances the high-temperature processing and flexibility with a PCE of 15.5% on 60 μm thick CPI with roll-to-roll sputtered ITO films [24].

4.2. Metal Foils

Metal substrates can also be utilized as an alternative to polymer substrates, owing to their moderate flexibility, remarkable thermal stability, ultralow R_{sh}, higher permeation barrier properties and higher conducting properties compared to plastic substrates. The major drawback of these substrates is processing of transparent top electrode without damaging the perovskite layer. For industrial or practical use of metal foils one should connect single cells, or else, these can be used as carriers because it is difficult to develop a series connection of a uniform module on metal [25]. The substrate is coated with a metallic thin film electrode after insulator layer deposition, easy to etch with photo (laser) ablation. Itavoids the necessary process, mechanical cutting of the metal foil to electrically isolate different areas which will affect the mechanical stability of the foil. In industrial production of dye-sensitized solar cells, mechanical cutting is successfully employed on Ti-foil, because in that scenario counter electrode (conductive foil) is deposited on Ti-foil to ensure the mechanical stability [26].

The similar procedure adopted for the FPSCs to deposit conductive and transparent top electrode. Still, FPSCs on metal foil limited to a single cell as the bottom electrode. The metal foils utilized as bottom electrode roughness of the metal foil is an issue, indeed enhanced the electropolishing pre-treatment observes device performance on the metal foil. The FPSC of 25 µm Ti-foil as photocathode (using electrochemical anodization highly ordered TiO_2-nanotube arrays grown on Ti foil) and carbon nanotube (CNT) network covered with spiro-OMeTAD as the transparent photo anode exhibited a PCE of 5.68%, further increased to 8.31% by reduced recombination and efficient passivation treatment of $TiCl_4$ after the Ti anodization [27]. Laminated Ti-foil (PEDOT: PSS covered nickel mesh on top) as a transparent electrode, insulating Al_2O_3 nanoparticles as a mesoporous scaffold with bilayer HTL (vanadium pentoxide (V_2O_5)-doped spiro-OMeTAD) showed a PCE of 10.3% with good bending stability [28]. The PCE further improved and exhibited a record value of 13.07% on metal foil-based FPSCs with the Ti/TiO_2 nanowire as a cathode and a PEN/ITO/PEDOT: PSS as an anode [29]. The critical issue of metal substrates is their opaque nature to serve as a top electrode with later conductivity and high transparency. Apart from this, silver nanowires, Ag-embedded ITO and ultrathin metal films are also successfully employing in FPSCs with Ti foils [30-32]. In addition to Ti foil, Cu foils with a device structure of $Cu/CuI/MAPbI_3/ZnO/Ag$ nanowires exhibited a PCE of 12.80%. Here, the Cu foil was also acting as Cu source and reacts with iodine vapour to produce CuI. Similar to Ti foil characteristics, collecting the electrons in the device, Cu has the similar work function of ITO, facilitates to collect holes extracted from the CuI layer [33].

4.3. Fibers

Self-powering merged e-textiles (wearable electronics) is one of the research areas of FPSCs, which can be functionalized with wearable and fiber features. In the OLEDs and other photovoltaic technologies, the metal substrate is also be replaced in the form of meshes and fiber. Impossible to adopt standard coating methods requires flat substrates. The fabrication of the device is entirely different compared to the flexible and rigid substrate. Nowadays common coating method is used dip-coating because one can coat uneven surfaces. Stainless steel or titanium can be transformed into fiber. Apart from the dip-coating method electrodeposition or chemical bath deposition also employed. Electrodeposition method is proved to be an effective method for lead salt deposition for the second step synthesis, resulting in a significant enhancement in the performance [34]. The low-cost fiber-shaped PSC (a mesoporous TiO_2, the perovskite absorber and HTM deposited on stainless-steel fiber finally a CNT sheet coiled as anode electrode) exhibited a PCE of 3.3% with high flexibility [35]. Later a more flexible CNT fiber, constructed a double-twisted solar cell architecture [36] exhibited a PCE of 3.03%, showed excellent bending stability and long-term ambient stability even more than 1000 bending cycles. Further, the coaxial fiber-shaped PSC optimized by electrochemically anodized Ti wires to induce dimpled compact TiO_2 and Ag nanowires were deposited by spray method and used as a top anode electrode, showed a PCE of 3.85% [37]. Generally fiber shaped devices made by dip-coating deposition resulting in low-quality perovskite layer and low PCE. To overcome this problem, Peng et al., developed high-quality perovskite layers on Ti wires via the indirect process [38]. This indirect solution-fabrication process resulted in a highly ordered crystalline PbI_2 and reacted with MAI to form perovskite nanocrystals. The corresponding fiber PSC with CNT sheets as photoanode exhibited a PCE of 7.1%. The 1D ribbon-like shape fiber with device structure of PEN/ITO/TiO$_2$/perovskite/CNT achieved a PCE of 9.49% [39]. Till now these fiber PSCs performances lower than the FPSCs but these are proven to be an application of FPSCs in wearable electronics.

5. TRANSPARENT CONDUCTIVE ELECTRODES

5.1. Indium Oxide Based Flexible Substrates

Tin-doped indium oxide (ITO) and zinc-doped indium oxide have shown better usage in a variety of optoelectronic applications like solar cells and displays as transparent electrodes. ITO is a wide band gap highly transparent n-type semiconducting material with the excellent light transmission in the visible and near-infrared region and good conductive electrical properties [40, 41]. ITO thin film is highly degenerate due to the existence of its Fermi level (EF) above the conduction band, have carrier concentration between 10^{20}-10^{21} cm^3 and low resistivity in the range 2-4×10^{-4} Ω cm [41]. ITO deposited on various polymer substrates (Polystyrene, willow-glass, PET and PEN) have been used as flexible substrates. PET-IZO FPSC with a R$_{sh}$ of 15 Ω sq^{-1} of exhibited a PCE of 12.3% [42]. The UV-Vis absorption of ITO-based substrates (PET-ITO and PEN-ITO, the FPSC device showed a PCE of 15.3%, with a good transmittance in the visible and near-infrared region [43].

5.2. Non-Indium-Oxide Based Flexible Substrates

Metal-doped indium oxide semiconductors are expensive, and indium is less abundant material. Hence, Non-indium oxide-based substrates have mainly been used in the fabrication of FPSCs to achieve large-scale, low-cost commercialization. Roldan-Carmona et al. [44] developed a multilayered conductive flexible substrate using DC magnetron sputtering inside a batch coater (pressure of 5×10^{-3} mbar) by depositing 30 nm thick aluminium-doped zinc oxide (AZO with two wt% Al), 9 nm thick Ag and 30 nm thick AZO on 50 μm PET substrate. The PET-AZO/Ag/AZO electrode has an R_{sh} of 7.5 Ω at the transparency of 81%. The FPSCs using this substrate exhibited a PCE of 7% [44-46].

Highly conductive PEDOT:PSS (HC-PEDOT) electrode is also an excellent alternate owing to its exhibiting low R_{sh}, high conductivity with treatment, favorable mechanical properties, high transmission coefficient, high surface coverage, good film processability [47] is synthesized by Poorkazem et al. [45] A mixture of PEDOT:PSS (Clevios PH1000), 0.5% (v/v) Zonyl F-300 fluorosurfactant and 5% (v/v) dimethyl sulfoxide (DMF) were filtered through a 0.45 mm polyvinylidene difluoride (PVDF) syringe filter. A pre-clean PET substrate (2.54 cm×2.54 cm) was spin coated (1000 rpm, 60 s followed by 2000 rpm 60 s) with a 300 mL of the above mixture further annealed for 30 min at 120°C. A thin strip of silver paste formed an electrical contact with PET-PEDOT: PSS electrode. The electrode exhibited transmittance of 85% at 800 nm corresponds to the onset absorption of the perovskite layer, therefore transmitting a large amount of light to the perovskite layer. An FPSC exhibited a PCE of 3.29% employing PET-Ag NW-ZnO: F (FZO) flexible substrate with R_{sh} of 17 Ω sq^{-1} [48], further PCE increased to 7.92% using PET/Ag NW-GO substrate [49]. A highly flexible and mechanically recoverable FPSC (HC- PEDOT: PSS on a shape-recoverable polymer, Noland Optical Adhesive 63 (NOA 63)) retained performance less than 10.4% of its original, even at a bend radius of 1 mm) [50]. HC-PEDOT: PSS FPSCs exhibited excellent bending tolerance employing as both the bottom and top electrodes on PET and ultrathin CPI [51].

A metal-mesh embedded HC-PEDOT: PSS was exhibiting good mechanical durability and high conductivity by overcoming the rough surface issue with enhancing the lateral conductivity of the mesh [52-54]. Ultrathin FPSC (PET (57 μm)/Ag-mesh/HC-PEDOT:PSS; annealed at 120°C for 20 min exhibited a transmittance of 82% to 86% in the visible region and had a low R_{sh} of ~ 3 Ω sq^{-1}) [55] showed a PCE of 14.2%, with a superior tolerance against mechanical bending at a radius of 5 mm [46]. An air-stable, ultra flexible, ultra lightweight, FPSC of 1.4 μm PET/PEDOT:

PSS and chromium oxide-metal contacts, (PET/PEDOT: PSS/perov skite/PCBM/PTCDI/Cr$_2$O$_3$/Cr/Au) [55] at a power-per-weight of 23 W g^{-1} exhibited a PCE of 12% during bending. The ultra-lightweight solar cells are magnificently attached on a dried leaf skeleton and used to power aviation models [55].

Li et al. developed a high-quality perovskite film by the two-step interdiffusion method showing efficient charge extraction and better interfacial contact [56]. An hydrophilic fullerene acceptor((([6,6]-phenyl-C61-butyricacid-(3,4,5-tris(2-(2-(2-ethoxyethoxy)ethoxy)ethoxy)phenyl) methanol ester (PCBB-OEG)) in perovskite solution, improved the electronic coupling and band alignment at the interface by forming an vertical gradient distribution with an ultrathin layer on the top during annealing process and also provided a lower driving force, less nucleus density by generating a soft-templating effect in the film. Therefore, it gave a better crystalline perovskite film with fewer trap states and good morphology on hydrophilic (PEDOT: PSS) or hydrophobic (PTAA) HTLs. The PSCs with this PTAA HTL exhibited a PCE of 20.2% on a rigid substrate, and a record PCE of 18.1% is achieved for FPSC on hybrid electrode compared to other FPSCs [57-62]. Compared to ITO based flexible substrate using PEN-graphene electrode formed a better flexible and efficient FPSC, showed a comparable PCE of 16.8% with no hysteresis and exhibited better stability against deformation of bending. The graphene-based device retained >90% and 85% efficiency even after 1000 and 5000 cycles at a bent radius 2 or 4 mm [63]. Further, FPSCs based on graphene as electrode showed excellent bending tolerance [64]. This outstanding bending stability shows the importance of graphene electrodes in the applications of wearable and flexible photovoltaic applications.

6. MECHANICAL TOLERANCE

FPSCs can be easily produced on a commercial scale by the printable roll-to-roll process. To commercial use in society, these products in the electronic community should be bendable, portable wearable, comfortable and lightweight. These should have excellent mechanical tolerance and retain their device parameters properties even after subjected to physical deformation of stretching, straining, bending, twisting and compressing. The device durability of FPSC was carried out by various research groups with a bending test, established that the perovskite layer withstands even at rigorous bending test. The decrease in the performance of FPSCs after some bending cycles at a particular radius have been endorsed cracks in the indium-oxide based substrates resulting carrier leakage increase in series resistance.

Using the metal counter electrode/ indium oxide-based electrodes, the FPSCs rate of degradation depending on the direction of bending [65, 66]. In FPSCs the radius of bending is essential criteria. High recombination rate, enhanced R_{sh}, more charge traps observed during the smaller bending radius via a faster generation of more micro-cracks.

Compared to Indium-oxide based electrodes non-indium-oxide based electrodes are known to exhibit better mechanical robustness. Therefore, non-indium-oxide based electrodes are proven to be a better replacement for metal-based and indium-oxide based electrodes with a high mechanical resistant FPSCs under severe mechanical stress for a long time. Many research groups estimated the bending test with device performance of FPSCs after and before bending at various radius with certain cycles of bending.

Towards stretchable electronics of PSCs, the mechanical durability of FPSCs under compression and stretching was demonstrated [55], using a pre-stretched acrylic elastomer (3 M VHB). The FPSC suffered a 40% compressive strain of 100 to 300 μm with a maximum amplitude of 140 μm due to the sinusoidal folds on the FPSC and a 3-D morphology. There is no significant decrement in the photovoltaic performance of FPSC even after 50% compressive strain. A slight decrease in the J_{SC} due to smaller active area illumination with no change in the V_{oc} and FF. The minimal decrement in J_{SC} due to the light trapping effect exhibited by the folds and wrinkles. The molar extinction coefficient of the device without the Au electrode increased with compressive strain. The EQE improved due to an increase in the formation of microtexture; results increase in photocurrent by more trapping of light.

To evaluate the tolerance effect, FPSC on Noland Optical Adhesive 63 (NOA 63) substrate retained its original shape from the distorted device after annealing 10 s at a temperature of 80°C [50]. The recovered device exhibited a decrement in the PCE from 10.2% to 6.1%. The crumpled FPSC exhibited two forms of bending strains, high strain and low strain regions with a radius less than 1 mm and greater than 1 mm respectively. Further, it showed better mechanical tolerance even after 50 bending cycles with a PCE of 5.88% (decreased less than 50% of its original PCE).

A fiber FPSC, with a device structure PEN-ITO-TiO_2-perovskite-CNT exhibited a PCE of 9.49% and retained 90% of its initial PCE even after 500 bending cycles. The SEM image reveals that there is no noticeable change in the photoactive [67]. Further, demonstrated the mechanical tolerance of double twisted fiber shape FPSC to complex deformation [36] extended to the typical wearable application after and before light illumination.

As fiber-shaped FPSCs known to have applications in wearable electronics and opened a way to power fabrics due to their withstanding nature during deformation twisting and retained their 90% of its initial performance [38]. Due to the interaction between the CNT electrode and the perovskite layer, FF was improved. The device showed a 10% reduction in the PCE after 400 cycles due to a decrease in J_{sc}.

Piezo-phototropic effect on FPSC [68] observed by employing ZnO microwaves (modulate the transport, dissociation, generation and recombination of charge carriers) on polystyrene substrate. This effect increased the PCE of the FPSC under compressive strain. The Piezo-phototronic effect of devices examined on the orientation of c-axis of the ZnO perovskite compound, shows ZnO microwave is directed away from the perovskite crystals, increases the photovoltaic properties (V_{oc} and J_{sc}) under compressive strains and decreases with increase in tensile strain, and FF, PCE increases under both tensile and compressive strains [68]. In contrast, V_{oc} and J_{sc} decrease when the c-axis of ZnO microwave oriented in the direction of perovskite [68].

7. PROSPECTS OF ROLL-TO-ROLL PROCESSING FOR FPSCS

Fabrication of solar cells for next-generation photovoltaics can be divided 1) single device 2) batch processing and 3) roll-to-roll processing [69, 70]. In laboratory scale, single device fabrication is employed (active areas ≤1 cm^2), both on flexible and rigid substrates for optimization and research purpose. The batch process devices usually manufactured in modules, (usually in the range ~ 1 m^2 and 100 cm^2) and particularly on rigid (glass) substrates, those are significantly bigger than laboratory scale devices, generally developed as a batch or single unit. The roll-to-roll processing is utilized for the bulk production of long flexible substrates with a typical continuous web-based fabrication process (till the flexible substrate unrolled). The roll-to-roll process is cost effective and large area processing in the organic photovoltaic field. As mentioned, various processing techniques (ink-jet printing, spray coating, doctor blade etc.), compatible with bulk production are employed for fabrication of PSCs. Till now some of these techniques are utilizing for PSCs on rigid (glass) substrate only, they opened the way to transfer these manufacturing techniques for FPSCs, even for roll-to-roll processing.

The ultrasonic spray coating technique was used to develop a planar heterojunction PSC device [71] on pre-patterned ITO rigid substrates with greater than 80% coverage of perovskite layer by various processing parameters like the volatility of solvents, annealing conditions and temperature of the substrate for perovskite layer. The FPSC (active area <1 cm^2) with a roll-to-roll processed perovskite layer exhibited a PCE of ≈11% having perovskite layer (processed by roll-to-roll method). As the bottom and top polymer layers (PEDOT: PSS and PCBM) were spin-coated, limits the large-scale fabrication of the device. PSCs employing carbon as insulating layer [72] without an HTM is promising as their device fabrication is compatible with roll-to-roll processing. The first carbon-based PSC [73] developed using perovskite/C bilayer by the inkjet printing method, employing in situ generated perovskite ($CH_3NH_3PbI_3$) from CH_3NH_3I/C ink formulation with PbI$_2$ and exhibited a PCE of ≈11.6% in the laboratory scale rigid based PSCs. A modified 3-D printing technology employed for large-scale PSCs. A slot-die coating with a 3-D printer to a solution processed PSCs exhibited a PCE of ≈ 4.6% in large area modules of ITO substrates (≈ 50 cm^2) and 11.6% for laboratory-scale devices. Further, PCE improved up to ≈ 12% (area of < 1 cm^2) by improving the perovskite layer quality using slot-die, gas-quenching assisted slot-die coating; HTM, perovskite, and ETL layers were printed using slot-die method whereas the top contact layer was deposited in vacuum. Similarly, A fully printable PSCs developed on a glass substrate using doctor blade at low-temperature processing, even though processed at low temperature, (except ETL layer is annealed); so it can be extended to flexible devices [6]. In continuation, the PSC devices with vacuum processed perovskite layer [74] also proved its compatibility with batch processing but limited to the rigid substrates. First screen printing and UV- assisted batch processing of flexible large area PSC device developed by Giacomo et al. [75] exhibited a PCE of ≈3%.

In continuation Schmidt et al. [76] developed fully printed FPSC by the roll-to-roll process on ITO-PET substrate and showed a PCE of ≈ 5%, this PCE is approximately half of the device developed by spin-coating on ITO/rigid substrate, a similar trend observes too that of printed OPVs. Combined photonic curing and spray-coating processing manufactured a FPSCs exhibited a PCE of 8% on PET substrate, can be applied to roll-to-roll processing which can produce a better quality of pin-hole free perovskite layers.

All the reported methods, annealing of the perovskite layer carried out for ≈ one hr, which might hinder the roll-to-roll processing. To overcome this, a rapid near-infrared processing [77] of perovskite precursor was developed which requires only 2.5s with no significant decrease in the PCE. Further, the processing time decreased to milliseconds range by photonic curing [78].

8. ENCAPSULATION

Towards development for commercialization and stable photovoltaic performance, a long-time photovoltaic technology is an essential criterion. Instead of high PCEs of PSCs, these have been concerned about the long-term durability of the device. The degradation of their performance is not only due to their intrinsic properties (interaction with transport layers and structural instability etc.), but also with extrinsic properties (temperature, UV-light, moisture, slow charge dynamics, hysteresis due to light soaking and photon dose). FPSCs are more challenging to be encapsulated because these are more effective for the above-mentioned factors. FPSCs are usually developed on polymeric substrates (PEN, PET etc.), which are more sensitive to oxygen and moisture compared to rigid substrates, liable to an effect of UV exposure and high-temperature cycling, especially outdoor installed photovoltaic device faces. Therefore, durability of the FPSCs ensured by the development of more stable HTMs, ETLs, perovskites, effective penetration barriers, sealing and contact material combinations to reduce the effect of extrinsic factors.

Material section plays a vital role to develop stable and durable FPSCs. Hydrophobic materials avoid the degradation of the perovskite layer, whereas hydrophilic materials readily react with moisture and destabilise the device. Reduced graphene oxide has efficiently hindered the degradation of the device to the moisture [79]. Counter electrode is also playing an important role to reduce the rate of degradation of the device, metal electrodes undergo oxidation with redox electrolyte and form corresponding halides. Buffer layers and carbon-based electrodes proved to be a better choice for the durability of PSC devices.

To improve the durability of the device's encapsulation is one of alternative for commercialization of FPSCs [80]. Epoxy as encapsulating for FPSC enhanced the device performance and durability [68]. To estimate the effect encapsulation on durability, developed three sets (entirely, partially and non-encapsulated) of FPSCs (PET-ITO-TiO$_2$-perovskite-SM-Au) [81]. A plastic encapsulant with a thickness of 240 µm, water vapour transmission rate of 10^{-3} gm^{-2}day^{-}.[1] Moreover, 90% of transparency in the visible light sealed on the metal electrode with acrylic adhesive. The percentage of moisture content of the device was reduced to less than 1 ppm by drying under the vacuum.

The partially encapsulated device retained 75% of the initial efficiencies for 400 hr, and a rapid decrease in the performance observed beyond 400 hr. Whereas, the encapsulated FPSC retained the performance over 500 hr. Further, they investigated the degradation of the device via encapsulation of calcium sensor with the device. The Ca film reacted with oxygen or moisture became transparent, for wholly encapsulated FPSCs moisture entered via copper wire electrical connections whereas partially encapsulated devices through edges.

Conclusion

Perovskite solar cells (PSCs) are an emerged class of photovoltaic research field, which have attracted research community due to their remarkable optoelectrical properties, including high carrier mobilities, long carrier diffusion lengths, high absorption coefficients, tunable bandgaps, low cost, and facile fabrication. PSCs have reached efficiencies of 22.70% and 18.36% on rigid fluorine-doped tin oxide (FTO) and poly (ethylene terephthalate) (PET) substrates, respectively. Each layer is essential for the performance of PSCs; hence, we discuss achievements in flexible perovskite solar cells (FPSCs) includes device structure, crystallinity and morphology of flexible substrates, mechanical tolerance, roll-to-roll processing techniques and encapsulation of the FPSCs. The use of roll-to-roll production is cost-effective for making this device more attractive to the market. Finally, the stretchability and flexibility of these devices lead to portable and lightweight self-powered devices. This chapter will provide the essential knowledge for the researcher community to further development of the FPSCs.

ACKNOWLEDGMENTS

The authors acknowledge the Indo-Korea grant (INT/Korea/P – 40) and early Career Research Award (ECR/2017/003092), from Department of Science and Technology (DST), SERB, India for their financial support.

REFERENCES

[1] Stoumpos, C. C., Malliakas, C. D. and Kanatzidis, M. G. (2013). Semiconducting tin and lead iodide perovskites with organic cations: phase transitions, high mobilities, and near-infrared photoluminescent properties. *Inorganic Chemistry,* 52 (15): 9019-9038.

[2] Long, M., Zhang, T., Xu, W., Zeng, X., Xie, F., Li, Q., Chen, Z., Zhou, F., Wong, K. S., Yan, K. and Xu, J. (2017). Large-grain formamidinium $pbi_{3-x}br_x$ for high-performance perovskite solar cells via intermediate halide exchange. *Advanced Energy Materials,* 7 (12): 1601882-1601889.

[3] Li, S. G., Jiang, K. J., Su, M. J., Cui, X. P., Huang, J. H., Zhang, Q. Q., Zhou, X. Q., Yang. L. M. and Song, Y. L. (2015). Inkjet printing of $CH_3NH_3PbI_3$ on a mesoscopic TiO_2 film for highly efficient perovskite solar cells. *Journal of Materials Chemistry. A,* 3 (17): 9092-9097.

[4] Barrows, A. T., Pearson, A. J., Kwak, C. K., Dunbar, A. D., Buckley, A. R. and Lidzey, D. G. (2014). Efficient planar heterojunction mixed-halide perovskite solar cells deposited *via* spray-deposition. *Energy and Environmental Science,* 7 (9): 2944-2950.

[5] Zhang, H., Cheng, J., Li, D., Lin, F., Mao, J., Liang, C., Jen, A. K. Y., Grätzel, M. and Choy, W. C. (2017). Toward all room-temperature, solution-processed, high-performance planar perovskite solar cells: a new scheme of pyridine-promoted perovskite formation. *Advanced Materials,* 29 (13): 1604695-1604701.

[6] Yang, Z., Chua, C. C., Zuo, F., Kim, J. H., Liang, P. W. and Jen, A. K. Y. (2015). High-Performance Fully Printable Perovskite Solar Cells

via Blade-Coating Technique under the Ambient Condition. *Advanced Energy Materials*, 5(13): 1500328-1500333.
[7] Hwang, K.; Jung, Y. S.; Heo, Y. J.; Scholes, F. H.; Watkins, S. E.; Subbiah, J.; Jones, D. J.; Kim, D. Y. and Vak, D. (2015). Toward large-scale roll-to-roll production of fully printed perovskite solar cells. *Advanced Materials*, 27 (7): 1241-1247.
[8] Malinkiewicz, O., Yella, A., Lee, Y. H., Espallargas, G. M., Grätzel, M., Nazeeruddin, M. K. and Bolink, H. J. (2014). Perovskite solar cells employing organic charge-transport layers. *Nature Photonics*, 8 (2): 128-132.
[9] Das, S., Yang, B., Gu, G., Joshi, P. C., Ivanov, I. N., Rouleau, C. M., Aytug, T., Geohegan, D. B. and Xiao, K. (2015). High-performance flexible perovskite solar cells by using a combination of ultrasonic spray-coating and low thermal budget photonic curing. *ACS Photonics*, 2 (6): 680-686.
[10] Luo, Y., Meng, F., Zhao, E., Zheng, Y. Z., Zhou, Y. and Tao, X. (2016). Fine control of perovskite-layered morphology and composition via sequential deposition crystallization process towards improved perovskite solar cells. *Journal of Power Sources*, 311: 130-136.
[11] Song, T. B., Chen, Q., Zhou, H., Jiang, C., Wang, H. H., Yang, Y. M., Liu, Y., Yu, J. and Yang, Y. (2015). Perovskite solar cells: film formation and properties. *Journal of Materials Chemistry A, 3* (17): 9032-9050.
[12] Habibi, M., Zabihi, F., Ahmadian-Yazdi, M. R. and Eslamian, M. (2016). Progress in emerging solution-processed thin film solar cells-Part II: Perovskite solar cells. *Renewable and Sustainable Energy Reviews*, 62: 1012-1031.
[13] Liu, D. and Kelly, T. L. (2013). Perovskite solar cells with a planar heterojunction structure prepared using room-temperature solution processing techniques solution processing techniques. *Nat Photonics*, 8:133–138.
[14] Luo, Y., Meng, F., Zhao, E., Zheng, Y., Zhou, Y., and Tao X. (2016). Fine control of perovskite-layered morphology and composition via

sequential deposition crystallization process towards improved perovskite solar cells. *J Power Sources,* 311:130-136.
[15] You, J., Hong, Z., Yang Y. M, Chen, Q., Cai, M., Song, T. B., Chen, C. C., Lu, S., Liu, Y., Zhou, H. and Yang. (2014) Y. Low-Temperature Solution-Processed Perovskite Solar Cells with High Efficiency and Flexibility. *ACS Nano, 8* (2):1674-1680
[16] Kim, H. S., Lee, C. R., Im, J. H., Lee, K.B., Moeh, T., Marchioro, A., Moon, S. J., Humphry-Baker, R., Yum, J. H., Moser, J. E., Grätzel, M and Park, N. G. (2012). Lead Iodide Perovskite Sensitized All-Solid-State Submicron Thin Film Mesoscopic Solar Cell with Efficiency Exceeding 9%. *Scientific Reports,* 2:591.
[17] Dianetti, M., Giacomo, F. D., Polino, G., Ciceroni, C., Liscio, A., D'Epifanio, A., Licoccia, S., Brown, T. M., Di Carlo, A. and Brunettia, F. (2015). TCO-free flexible organo metal trihalide perovskite planar-heterojunction solar cells. *Sol Energy Mater& Sol Cells,* 140:150–157.
[18] Zhang, Y., Hu, X., Chen, L., Huang, Z., Fu, Q., Liu, Y., Zhang, L., and Chen, Y. (2016). Flexible, hole transporting layer-free and stable $CH_3NH_3PbI_3$/PC61BM planar heterojunction perovskite solar cells. *Org Electron,* 30:281-288.
[19] Xu, X., Chen, Q., Hong, Z., Zhou, H., Liu, Z., Chang, W., Sun, P., Chen, H., Marco, N. D., Wang, M. and Yang, Y. (2015). Working Mechanism for Flexible Perovskite Solar Cells with Simplified Architecture. *Nano Lett,* 15 (10): 6514–6520.
[20] Xi, J., Wu, Z., Xi, K., Dong, H., Xia, B., Lei, T., Yuan, F., Wu, W., Jiao, B. and Hou, X. (2016). Initiating crystal growth kinetics of α-$HC(NH_2)_2PbI_3$ for flexible solar cells with long-term stability. *Nano Energy* 26:438–445.
[21] Tavakoli, M. M., Tsui, K. H., Zhang, Q., He, J., Yao, Y., Li, D. and Fan, Z. (2015). Highly Efficient Flexible Perovskite Solar Cells with Antireflection and Self-Cleaning Nanostructures. *ACS Nano, 9* (10):10287-10295.
[22] Dkhissi, Y., Huang, F., Rubanov, S., Xiao, M. and Bach, U. (2015). Low-temperature processing of flexible planar perovskite solar cells with efficiency over 10%. *Journal of Power Sources* 278:325-331.

[23] Zardetto, V., Brown, T. M., Reale, A. and Di Carlo, A. (2011). Substrates for flexible electronics: A practical investigation on the electrical, film flexibility, optical, temperature, and solvent resistance properties. *Journal of Polymer Science Part B: Polymer Physics,* 49 (9): 638-648.

[24] Park, J. I., Heo, J. H., Park, S.-H., Hong, K. I., Jeong, H. G., Im, S. H. and Kim, H. K. (2017). Highly flexible InSnO electrodes on thin colourless polyimide substrate for high-performance flexible $CH_3NH_3PbI_3$ perovskite solar cells. *Journal of Power Sources,* 341: 340-347.

[25] Kessler, F. and D. Rudmann. (2004). Technological aspects of flexible CIGS solar cells and modules *Sol. Energy,* 77 (6): 685-695.

[26] J. Ryan, (2011). Methods of scoring for fabricating interconnected photovoltaic cells, US 7932464 B2.

[27] Wang, X., Li, Z., Xu, W., Kulkarni, S. A., Batabyal, S. K., Zhang, S., Cao, A. and Wong, L. H. (2015). TiO_2 nanotube arrays based flexible perovskite solar cells with transparent carbon nanotube electrode. *Nano Energy,* 11:728-735.

[28] Troughton, J., Bryant, D., Wojciechowski, K., Carnie, M. J., Snaith, H., Worsley, D. A. and Watson, T. M. (2015). Highly efficient, flexible, indium-free perovskite solar cells employing metallic substrates. *Journal of Materials Chemistry A,* 3 (17): 9141-9145.

[29] Xiao, Y., Han, G., Zhou, H. and Wu, J. (2016). An efficient titanium foil based perovskite solar cell: using a titanium dioxide nanowire array anode and transparent poly (3, 4-ethylene dioxythiophene) electrode. *RSC Advances,* 6 (4): 2778-2784.

[30] Lee, M., Jo, Y., Kim, D. S., Jeong, H. Y. and Jun, Y. (2015). Efficient, durable and flexible perovskite photovoltaic devices with Ag-embedded ITO as the top electrode on a metal substrate. *Journal of Materials Chemistry A,* 3 (28): 14592-14597.

[31] Lee, M., Ko, Y., Min, B. K. and Jun, Y. (2016). Silver nanowire top electrodes in flexible perovskite solar cells using titanium metal as substrate. *ChemSusChem,* 9 (1): 31-35.

[32] Lee, M., Jo, Y., Kim, D. S. and Jun, Y. (2015). Flexible organo-metal halide perovskite solar cells on a Ti metal substrate. *Journal of Materials Chemistry A*, 3 (8): 4129-4133.

[33] Nejand, B. A., Nazari, P., Gharibzadeh, S., Ahmadi, V. and Moshaii, A. (2017). All-inorganic large-area low-cost and durable flexible perovskite solar cells using copper foil as a substrate. *Chemical Communications*, 53 (4): 747-750.

[34] Deng, J., Qiu, L., Lu, X., Yang, Z., Guan, G., Zhang, Z. and Peng, H. (2015). Elastic perovskite solar cells. *Journal of Materials Chemistry A*, 3 (42), 21070-21076.

[35] Qiu, L., Deng, J., Lu, X., Yang, Z. and Peng, H. (2014). Integrating perovskite solar cells into a flexible fiber. *Angewandte Chemie International Edition*, 53 (39): 10425-10428.

[36] Li, R., Xiang, X., Tong, X., Zou, J. and Li, Q. (2015). Wearable Double-Twisted Fibrous Perovskite Solar Cell. *Advanced Materials*, 27 (25): 3831-3835.

[37] Lee, M., Ko, Y. and Jun, Y. (2015). Efficient fiber-shaped perovskite photovoltaics using silver nanowires as the top electrode. *Journal of Materials Chemistry A*, 3 (38): 19310-19313.

[38] Qiu, L., He, S., Yang, J., Deng, J. and Peng, H. (2016). Fiber-Shaped Perovskite Solar Cells with High Power Conversion Efficiency. *Small*, 12 (18): 2419-2424.

[39] Qiu, L., He, S., Yang, J., Jin, F., Deng, J., Sun, H., Cheng, X., Guan, G., Sun, X. and Zhao, H. (2016). An all-solid-state fiber-type solar cell achieving 9.49% efficiency. *Journal of Materials Chemistry A*, 4 (26): 10105-10109.

[40] Park, Y., Choong, V., Gao, Y., Hsieh, B. R. and Tang, C. W. (1996). Work function of indium tin oxide transparent conductor measured by photoelectron spectroscopy. *Applied Physics Letters*, 68 (19): 2699-2701.

[41] Kim, H., Gilmore, A. C., Pique, A., Horwitz, J., Mattoussi, H., Murata, H., Kafafi, Z. and Chrisey, D. (1999). Electrical, optical, and structural properties of indium-tin-oxide thin films for organic light-emitting devices. *Journal of Applied Physics*, 86 (11): 6451-6461.

[42] Dkhissi, Y., Huang, F., Rubanov, S., Xiao, M., Bach, U., Spiccia, L., Caruso, R. A. and Cheng, Y.-B. (2015). Low-temperature processing of flexible planar perovskite solar cells with efficiency over 10%. *Journal of Power Sources*, 278: 325-331.

[43] Shin, S. S., Yang, W. S., Noh, J. H., Suk, J. H., Jeon, N. J., Park, J. H., Kim, J. S., Seong, W. M. and Seok, S. II. (2015). High-performance flexible perovskite solar cells exploiting Zn_2SnO_4 prepared in solution below 100 °C. *Nature communications,* 6:1-6.

[44] Roldán-Carmona, C.; Malinkiewicz, O.; Soriano, A.; Espallargas, G. M.; Garcia, A.; Reinecke, P.; Kroyer, T.; Dar, M. I.; Nazeeruddin, M. K. and Bolink, H. J. (2014). Flexible high-efficiency perovskite solar cells. *Energy and Environmental Science*, 7 (3): 994-997.

[45] Poorkazem, K., Liu, D. and Kelly, T. L. (2015). Fatigue resistance of a flexible, efficient, and metal oxide-free perovskite solar cell. *Journal of Materials Chemistry A 3* (17): 9241-9248.

[46] Li, Y., Meng, L., Yang, Y. M., Xu, G., Hong, Z., Chen, Q., You, J., Li, G., Yang, Y. and Li, Y. (2016) High-efficiency robust perovskite solar cells on ultrathin flexible substrates. *Nat Commun*, 7:10214-10223.

[47] Xia Y and Ouyang J. (2012). Significant Different Conductivities of the Two Grades of Poly (3,4-ethylene dioxythiophene): Poly(styrene sulfonate), Clevios P and Clevios PH1000, Arising from Different Molecular Weights. *ACS Appl Mater Interfaces*,4(8):4131–4140.

[48] Han J, Yuan S, Liu L, Qiu X, Gong H, Yang X, Li, C., Haob, Y., and Cao, B. (2015). Fully indium-free flexible Ag nanowires/ZnO: F composite transparent conductive electrodes with high haze. *Journal of Material Chemistry A,* 3 (10):5375-5384.

[49] Lu, H., Sun, J., Zhang, H., Lu, S. and Choy, W. C. (2016). Room-temperature solution-processed and metal oxide-free nano-composite for the flexible transparent bottom electrode of perovskite solar cells. *Nanoscale*, 8 (11): 5946-5953.

[50] Park, M., Kim, H. J., Jeong, I., Lee, J., Lee, H., Son, H. J., Kim, D. E. and Ko, M. J. (2015). Mechanically recoverable and highly efficient

perovskite solar cells: investigation of intrinsic flexibility of organic-inorganic perovskite. *Advanced Energy Materials,* 5 (22): 1501406.

[51] Zhang, Y., Wu, Z., Li, P., Ono, L. K., Qi, Y., Zhou, J., Shen, H., Surya, C. and Zheng, Z. (2018). Fully Solution-Processed TCO-Free Semitransparent Perovskite Solar Cells for Tandem and Flexible Applications *Advanced Energy Materials,* 8(1): 1701569.

[52] Xu, G., Shen, L., Cui, C., Wen, S., Xue, R., Chen, W., Chen, H., Zhang, J., Li, H. and Li, Y. (2017). High-Performance Colorful Semitransparent Polymer Solar Cells with Ultrathin Hybrid-Metal Electrodes and Fine-Tuned Dielectric Mirrors. *Advanced Functional Materials.* 27 (15): 1605908.

[53] Zhang, J., Xue, R., Xu, G., Chen, W., Bian, G. Q., Wei, C., Li, Y. and Li, Y. (2018). Self-Doping Fullerene Electrolyte-Based Electron Transport Layer for All-Room-Temperature-Processed High-Performance Flexible Polymer Solar Cells. *Advanced Functional Materials,* 28 (13): 1705847.

[54] Li, Y.; Xu, G.; Cui, C. and Li, Y. (2018). Flexible and semi-transparent organic solar cells. *Advanced Energy Materials,* 8 (7): 1701791.

[55] Kaltenbrunner, M., Adam, G., Głowacki, E. D., Drake, M., Schwödiauer, R., Leonat, L., Apaydin, D. H., Groiss, H., Scharber, M. C. and White, M. S. (2015). Flexible high power-per-weight perovskite solar cells with chromium oxide-metal contacts for improved stability in air. *Nature Materials,* 14 (10): 1032-1039.

[56] Xu, G., Xue, R., Chen, W., Zhang, J., Zhang, M., Chen, H., Cui, C., Li, H., Li, Y. and Li, Y. (2018). New Strategy for Two-Step Sequential Deposition: Incorporation of Hydrophilic Fullerene in Second Precursor for High-Performance p-i-n Planar Perovskite Solar Cells. *Advanced Energy Materials,* 8 (12): 1703054.

[57] Han, J., Yuan, S., Liu, L., Qiu, X., Gong, H.; Yang, X., Li, C., Hao, Y. and Cao, B. (2015). Fully indium-free flexible Ag nanowires/ZnO: F composite transparent conductive electrodes with high haze. *Journal of Materials Chemistry A,* 3 (10): 5375-5384.

[58] Roldán-Carmona, C., Malinkiewicz, O., Soriano, A., Espallargas, G. M., Garcia, A., Reinecke, P., Kroyer, T., Dar, M. I., Nazeeruddin, M.

K. and Bolink, H. J. (2014). Flexible high-efficiency perovskite solar cells. *Energy and Environmental Science,* 7 (3): 994-997.

[59] Kim, A., Lee, H., Kwon, H. C., Jung, H. S., Park, N. G., Jeong, S. and Moon, J. (2016). Fully solution-processed transparent electrodes based on silver nanowire composites for perovskite solar cells. *Nanoscale,* 8 (12): 6308-6316.

[60] Dong, H., Wu, Z., Jiang, Y., Liu, W., Li, X., Jiao, B., Abbas, W. and Hou, X. (2016). A flexible and thin graphene/silver nanowires/polymer hybrid transparent electrode for optoelectronic devices. *ACS applied materials and interfaces,* 8 (45): 31212-31221.

[61] Xu, M., Feng, J., Fan, Z. J., Ou, X. L., Zhang, Z. Y., Wang, H. Y. and Sun, H. B. (2017). Flexible perovskite solar cells with ultrathin Au anode and vapour-deposited perovskite film. *Solar Energy Materials and Solar Cells,* 169: 8-12.

[62] Lu, H., Sun, J., Zhang, H., Lu, S. and Choy, W. C. (2016). Room-temperature solution-processed and metal oxide-free nano-composite for the flexible transparent bottom electrode of perovskite solar cells. *Nanoscale,* 8 (11): 5946-5953.

[63] Yoon, J., Sung, H., Lee, G., Cho, W., Ahn, N., Jung, H. S. and Choi, M. (2017). Super flexible, high-efficiency perovskite solar cells utilizing graphene electrodes: towards future foldable power sources. *Energy & Environmental Science,* 10 (1): 337-345.

[64] Liu, Z., You, P., Xie, C., Tang, G. and Yan, F. (2016). Ultrathin and flexible perovskite solar cells with graphene transparent electrodes. *Nano Energy,* 28: 151-157.

[65] Giacomo, F. Di., Zardetto, V., Epifanio, A. D., Pescetelli, S., Matteocci, F., Razza, S., Carlo, A. D., Licoccia, S., Kessels, W. M. M., Creatore, M. and Brown, T. M. (2015). Flexible Perovskite Photovoltaic Modules and Solar Cells Based on Atomic Layer Deposited Compact Layers and UV-Irradiated TiO_2 Scaffolds on Plastic Substrates. *Advanced Energy Materials,* 5:1401808.

[66] Jung, K., Lee, J., Kim, J., Chae W. and Lee M. (2016). Solution-processed flexible planar perovskite solar cells: A strategy to enhance efficiency by controlling the ZnO electron transfer layer, PbI_2 phase,

and CH$_3$NH$_3$PbI$_3$ morphologies. *Journal of Power Sources*, 324 (30): 142-149.

[67] Qiu, L., He, S., Yang, J., Jin, F., Deng, J., Sun, H., Cheng, X., Guan, G., Sun, X., Zhao, H. (2016). An all-solid-state fiber-type solar cell achieving 9.49% efficiency. *Journal of Materials Chemistry A*, 4 (26): 10105-10109.

[68] Hu, G., Guo, W., Yu, R., Yang, X., Zhou, R., Pan, C. and Wang, Z. L. (2016). Enhanced performances of flexible ZnO/perovskite solar cells by piezo-phototronic effect. *Nano Energy*, 23: 27-33.

[69] Krebs, F. C., Tromholt, V. and M. Jorgensen. (2010) Upscaling of polymer solar cell fabrication using full roll-to-roll processing. *Nanoscale*, 2 (6): 873-886.

[70] Barrows, A. T., Pearson, A. J., Kwak, C. K., Dunbar, A. D. F., Buckley, A. R. and Lidzey, D. G. (2014). Efficient planar heterojunction mixed-halide perovskite solar cells deposited *via* spray-deposition. *Energy Environmental Scicence*, 7 (9):2944-2950.

[71] Mei, A., Li, X., Liu, L., Ku, Z., Liu, T., Rong, Y., Xu, M., Hu, M., Chen, J., Yang, Y., Grätzel, M. and Han, H. (2014). A hole-conductor-free, fully printable mesoscopic perovskite solar cell with high stability. *Science*, 345(6194): 295-298.

[72] Wei, Z., Chen, H., Yan, K. and Yang, S. (2014). Inkjet printing and instant chemical transformation of a CH$_3$NH$_3$PbI$_3$/Nano carbon electrode and interface for planar perovskite solar cells. *Angewandte Chemie International Edition*, 53 (48): 13239-13243.

[73] Vak, D., Hwang, K., Faulks, A., Jung, Y. S., Clark, N., Kim, D. Y., Wilson, G. J. and Watkins, S. E. (2015). 3D Printer Based Slot-Die Coater as a Lab-to-Fab Translation Tool for Solution-Processed Solar Cells. *Advanced Energy Materials*, 5:1401539.

[74] Fakharuddin, A., Palma, A. L., Di Giacomo, F., Casaluci, S., Matteocci, F., Wali, Q., Rauf, M., Di Carlo, A., Brown, T. M. and Jose, R. (2015). Solid state perovskite solar modules by vacuum vapor assisted sequential deposition on Nd: YVO$_4$ laser patterned rutile TiO$_2$ nanorods *Nanotechnology*, 26: 494002.

[75] Di Giacomo, F., Zardetto, V., D'Epifanio, A., Pescetelli, S., Matteocci, F., Razza, S., Di Carlo, A., Licoccia, S., Kessels, W. M. M., Creatore, M. and Brown, T. M. (2015). Flexible Perovskite Photovoltaic Modules and Solar Cells Based on Atomic Layer Deposited Compact Layers and UV-Irradiated TiO_2 Scaffolds on Plastic Substrates. *Advanced Energy Materials,* 2015, 5:1401808.

[76] Schmidt, T. M., Larsen-olsen, T. T. Carle, J. E., Angmo, D. and Krebs, F. C. (2015). Upscaling of Perovskite Solar Cells: Fully Ambient Roll Processing of Flexible Perovskite Solar Cells with Printed Back Electrodes. *Advanced Energy Materials,* 5:1500569.

[77] Troughton, J., Charbonneau, C., Carnie, M. J., Davies, M. L., Worsley, D. A. and Watson, T. M. (2015). Rapid processing of perovskite solar cells in under 2.5 seconds. *Journal of Materials Chemistry A,* 3(17): 9123-9127.

[78] Troughton, J., Carnie, M. J., Davies, M. L., Charbonneau, C., Jewell, E. H., Worsley, D. A. and Watson, T. M. (2016). Photonic flash-annealing of lead halide perovskite solar cells in 1 ms. *Journal of Materials Chemistry A,* 4(9): 3471-3476.

[79] Bouclé, J. and Herlin-Boime, N. (2016). The benefits of graphene for hybrid perovskite solar cells. *Synthetic Metals,* 222:3-16.

[80] Wang, D., Wright, M., Elumalai, N. K. and Uddin, A. (2016). Stability of perovskite solar cells. Sol Energy Mater Sol Cells, 147: 255-275.

[81] Weerasinghe, H. C.; Dkhissi, Y.; Scully, A. D.; Caruso, R. A. and Cheng, Y.-B. (2015). Encapsulation for improving the lifetime of flexible perovskite solar cells. *Nano Energy,* 18: 118-125.

In: Perovskite Solar Cells
Editor: Murali Banavoth
ISBN: 978-1-53615-858-8
© 2019 Nova Science Publishers, Inc.

Chapter 7

BISMUTH AND ANTIMONY BASED PEROVSKITE, PEROVSKITE-LIKE, AND NON-PEROVSKITE MATERIALS FOR LEAD-FREE PEROVSKITE SOLAR CELLS

Ashish Kulkarni[1,*] *and Trilok Singh*[2,†]
[1]Graduate School of Engineering, Toin University of Yokohama, Aoba, Yokohama, Kanagawa, Japan
[2]School of Energy Science and Engineering, Indian Institute of Technology Kharagpur, Kharagpur, India

ABSTRACT

Group 15 metals such as bismuth (Bi^{3+}) and antimony (Sb^{3+}) have been considered as the potential materials with similar electronic configuration and optoelectronic properties suitable for solar cell applications. Here, in this chapter, we limelight the recent trends and developments made by the research community on the zero-, two-, and three-dimensional Bi^{3+} and

---

[*] Corresponding Author's Email: ashish.kulkarni786@gmail.com
[†] Corresponding Author's Email: trilok@iitkgp.ac.in.

Sb^{3+} based non-toxic perovskite/non-perovskite materials and discuss the possibilities to improve the efficiency of the same. This book chapter covers the designing of three dimensional (3D) double perovskite having a structure of A$_2$MM'X$_6$ (A = MA, Cs; M = Bi, Sb; M' = Ag, Cu; X = I, Cl, Br) which was the first step towards development of Bi and Sb-based perovskite materials. These materials possess strong photoluminescence in the visible region and long recombination lifetime similar to that of lead-perovskites. However, they possess wide bandgap limiting their absorption. Subsequently, ternary Bi and Sb-based lower dimensional perovskite materials emerged as a potential candidate due to its high moisture stability, in addition to the tunability of bandgap ranging from 1.8 to 2.1 eV. To date, PCE of ~3% has been achieved with lower dimensional ternary metal based perovskite materials. The chapter covers bismuth iodide and antimony iodides, which are used as precursor materials for the preparation of perovskite materials, based solar cells as they have bandgap suitable for single and tandem junction devices with suitable optoelectronic properties.

Further, the development of 3D Bi-based non-perovskite materials in a combination of a monovalent cation such as silver (Ag) and copper (Cu) and their photovoltaic developments towards efficient lead-free materials will be discussed. Finally, we will discuss further direction and perspective towards the development of Bi and Sb-based materials for non-toxic solar cells.

1. INTRODUCTION

Enhancement in power conversion efficiency (PCE) of organic-inorganic hybrid lead (Pb) halide perovskite solar cells (PSCs) have witnessed the fastest growth in photovoltaic technologies compare to the already established one [1, 2]. Ever since the seminal works on organic-inorganic hybrid lead halide perovskite solar cells (PSCs) lead to prodigious rise of PCE from 0.37% (in solid state device) [3] and 3.8% (in liquid state) [4] to certified PCE of 23% [5] in just 10 years (from its initial attempt in solid junction device). In addition to cost-effective simple solution processability, [2, 6] which makes facile fabrication route for up-scaling the production, lead perovskites possess number of remarkable optoelectronic properties such as direct band gap of ~1.6 eV, which is near to Schockley-Queisser gap (1.43 eV) for single junction solar cell [4], high absorption coefficient (5 x 10^{4}cm^{-1}) than silicon and GaAs, sharp absorption onset, long carrier diffusion length as long as 1μm, less electron and hole effective masses leading to ambipolar charge transport property, photon recycling capability, to name few [1], making them "first high-quality halide semiconductors" [7].

Despite becoming an *"affluential material"* of photovoltaic technologies, lead-based perovskites suffers from longevity issue [8]. The cation A (from AMX$_3$) gets eliminated from the perovskite crystal structure after exposing to moisture [9], continuous illumination under sunlight [10, 11], and with heating; [8] leaving behind the degraded product lead iodide (PbI$_2$). Development of robust encapsulation strategies can help the perovskite module to last long [12]. However, it is still believed that lead toxicity can stands as a major obstacle in a way to commercialization of this star rising material [13]. The maximum permissible amount of Pb, by the U.S. EPA, is 15 μg/L and 0.15 μg/L in water and air, respectively, which are much less than the amount of Pb (0.4 g) estimated in a 1m^2 solar panel with 300 nm thick perovskite layer [14]. Additionally, the solubility product (K$_{sp}$) of lead is on the order of 10^{-8} which is much higher than K$_{sp}$ of cadmium (Cd^{2+}) in CdTe. This indicates a higher possibility of environment toxication, after exposure, due to Pb^{2+} than Cd^{2+} [14]. Moreover, it is reported that exposure of lead to human can be catastrophic effecting nervous and reproductive systems [13]. This undesired lead toxicity issue, regardless of high efficiency, motivated the research communities to pave the way towards the development of lead-free perovskite materials.

Initially, tin (Sn) based perovskites were considered as a promising candidate to replace Pb because of its similar ionic radii [15]. Moreover, Sn-based perovskites possess exceptional optoelectronic properties such as bulk *n*-type electrical conductivity, long carrier diffusion length, high electron mobility than traditional semiconductors, such as Si and CdTe. Additionally, Sn-perovskites have shown a narrow optical gap with the potential to absorb up to 1000 nm, and higher charge carrier mobility of 10^2-10^3 cm^2/V.s compared to their lead analogous [16]. Despite several advantages, Sn^{2+} in Sn-perovskite rapidly oxidizes to Sn^{4+} limiting the efficiency [16]. To note, the best efficiency so far reported for Sn-perovskite devices is 9% and the majority of reported works have demonstrated PCE of 4-5% [17, 18]. The oxidized Sn^{4+} causes self-doping effect leading to fast carrier recombination within the bulk of the material and thus low device efficiency.[16] Concurrently, germanium (Ge) also found its place in perovskite family and among all the Ge-perovskites, only $CsGeI_3$ possess bandgap of 1.6 eV and incorporating any other organic/inorganic cation leads to the formation of perovskite with bandgap >2.2 eV [19]. Furthermore, a similar phenomenon of rapid oxidation was also observed for germanium (Ge) based perovskite due to its $4s^2$ orbital, thus limiting device efficiency [19]. Numerous theoretical and experimental efforts have been carried out to replace lead with Cu^{2+}, Mn^{2+}, Ni^{2+}, Co^{2+}, Mg^{2+} but their device efficiency has been far behind that of lead-perovskites [20]. Recently, bismuth and antimony-based materials have attracted lots of research attention and here in this book chapter, we discuss some of the important findings and developments of Bi and Sb-based perovskites for lead-free solar cells.

2. BISMUTH AND ANTIMONY BASED LOWER DIMENSIONAL PEROVSKITE MATERIALS

Group 15 metals, especially, bismuth (Bi) and antimony (Sb) attracted recent attention in perovskite family as they possess similar electronic configurations and comparable ionic radii with that of Pb [21]. Optical and electronic properties of various lower dimensional bismuth perovskites have been investigated previously [22, 23, 24, 25, 26], however, efforts to incorporate them into photovoltaic devices was lately addressed. Dating back to 2016, Park et al. [27] employed three different bismuth perovskite materials and demonstrated PCE of 1%, 0.1% and 0.01% for $Cs_3Bi_2I_9$, $MA_3Bi_2I_9$, and $MA_3Bi_2I_{9-x}Cl_x$ respectively. The absorption coefficient for Bi-perovskites was estimated to be 1 x 10^5 cm^{-1} at 450 nm, which is less than $MAPbI_3$ at the same wavelength. Additionally, Wannier-Mott exciton binding energy of 70 meV, 150 meV, and 300 meV was observed for $MA_3Bi_2I_9$, $Cs_3Bi_2I_9$ and $MA_3Bi_2I_{9-x}Cl_x$ respectively. Interestingly, the exciton binding energy of MA$_3$Bi$_2$I$_9$ (70 meV) is slightly higher than MAPbI$_3$ (25-50 meV) indicating its promising application for a photovoltaic cell. This initial report triggered the development of group 15 metal based lower and higher dimensional perovskite materials. Hoye et al. deposited MA$_3$Bi$_2$I$_9$ by a conventional two-step deposition method and studied its crystal properties and stability. Phase pure MA$_3$Bi$_2$I$_9$ showed an indirect optical gap of 2.04 eV. Further, they compared the long-term stability of MA$_3$Bi$_2$I$_9$ and MAPbI$_3$ by exposing them to a high humidity atmosphere (60%). It has been observed that the colour of MAPbI$_3$ changed from brown to yellow in 5 days while MA$_3$Bi$_2$I$_9$ showed no sign of degradation even after 13 days with a slight change to bright colour after ~25 days owing to the formation of Bi$_2$O$_3$ or BiOI layer on the surface [28]. To note, Bi$_2$O$_3$ or BiOI layer can help in charge transport process across the interfaces and also acts as a passivation layer to reduce the recombination. Single crystal structure of MA$_3$Bi$_2$I$_9$ was systematically studied by Eckhardt et al. in which the author observed that three equivalent symmetrical I$^-$ help to combine every two Bi^{3+} and these I$^-$ are situated on the other sides of mirror planes as shown in Figure 1 [29].

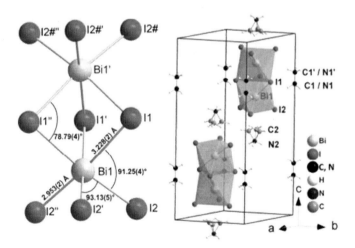

Figure 1. MA$_3$Bi$_2$I$_9$ crystal structure showing the local structure of Bi$_2$I$_9^{3-}$ anion and cation and anion positions in the unit cell.

Further, Oez et al. deposited $MA_3Bi_2I_9$ perovskite in inverse structure and observed hexagonal morphology contrasting the first report in which Park et al. observed dendrite-like morphology. Despite this, there was no difference in device performance [30]. However, it is very interesting to note that with the change in device architecture, the morphology of the $MA_3Bi_2I_9$ perovskite changes significantly. To further understand the role of different architecture Singh et al. performed a systematic study by depositing $MA_3Bi_2I_9$ perovskite on TiO_2 based planar architecture, and on brookite and anatase TiO_2 mesoporous layer. The difference between the anatase and brookite mesoporous is that the later one has inter-necking particles with a particle size of around 10 nm in contrast to anatase (particle size 18 nm). Interestingly, dendrite-like morphology of $MA_3Bi_2I_9$ showed uniform capping layer only on anatase TiO_2 mesoporous layer. This uniform capping layer further facilitated better interfacial connection with neighboring charge transport layer while on brookite mesoporous and TiO_2 compact layer (CL) $MA_3Bi_2I_9$ showed poor capping layer hampering the efficient charge transport across the interface. As a result, PCE of 0.25%, 0.1%, and 0.09% was achieved for anatase TiO_2, planar and brookite TiO_2 mesoporous based device respectively. Moreover, the devices were reasonably stable for 10 weeks after exposing to relative humidity atmosphere of ~60% [31]. It is noted here that despite a change in the ETL layer, $MA_3Bi_2I_9$ demonstrated dendrite-like morphology owing to rapid crystallization during the spin coating process. To address this issue, Kulkarni et al. employed solvent-engineering technique by introducing N-methyl-2-pyrrolidone (NMP) as an additive to reduce the crystallization process of $MA_3Bi_2I_9$. Different concentration of NMP was incorporated as an additive and at a particular concentration of 25 µL in 1 mL of $MA_3Bi_2I_9$-DMF solution the morphology was much enhanced showing almost complete coverage over TiO_2 mesoporous layer (Figure 2). However, the device performance was not much enhanced and best device efficiency showed 0.32% Vs 0.15% (without NMP) [20]. This indicated that tuning the morphology does not necessarily help to enhance the performance of $MA_3Bi_2I_9$ perovskite and tuning intrinsic optoelectronic properties are of vital importance. Further, Shin et al. [32] studied the effect of solvent engineering technique by introducing a mix solvent approach (DMF-DMSO, DMF-TBP) for $MA_3Bi_2I_9$, $Cs_3Bi_2I_9$, and $FA_3Bi_2I_9$. With this, in addition to anti-solvent dripping, $MA_3Bi_2I_9$ films showed compact, uniform morphology (without pinholes) with high coverage and the devices with spiro-OMeTAD (as HTM) demonstrated PCE of 0.5% (Figure 3). Further, they employed a new HTM, PIF8-TTA, having a deeper HOMO level and the device demonstrated enhanced V_{oc} up to 0.85 V (Vs 0.69 V) and thus the PCE improved to 0.7% [31].

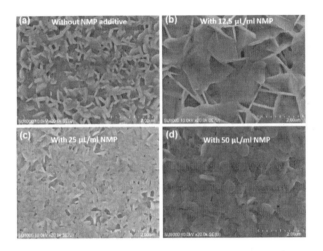

Figure 2. Top surface SEM image of $MA_3Bi_2I_9$ (a) without and (b, c, d) with different concentration of NMP.

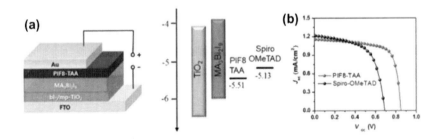

Figure 3. (a) Device structure and their corresponding energy-band diagram, (b) J-V curves of best-performing devices incorporating Spiro-OMeTAD and PIF8-TAA.

In addition to solvent engineering technique (as discussed above), efforts have also been made to deposit MA$_3$Bi$_2$I$_9$ perovskite by a various deposition technique. For instance, Ran et al. deposited MA$_3$Bi$_2$I$_9$ perovskite in an inverse structure by novel two-step evaporation-spin coating method in which initially BiI$_3$ layer was deposited by thermal evaporation followed by converting it to MA$_3$Bi$_2$I$_9$ via spin coating the methylammonium iodide (MAI) solution. As a result, highly compact MA$_3$Bi$_2$I$_9$ perovskite layer was obtained which demonstrated open-circuit voltage (V$_{oc}$) of 0.8 V. Despite of this, the device exhibited PCE of 0.39% mainly attributing to low short-circuit current (J$_{sc}$) (1.3 mA/cm^2) and poor fill factor (FF) (0.33). Such low FF also indicates that poor charge transport and high series resistance across the interfaces. Further, they calculated the diffusion length of MA$_3$Bi$_2$I$_9$ film (having a thickness of 120 nm) with the help of PL decay (Figure 4) and observed electron and hole diffusion length of 46 and 26 nm respectively, which are less than MAPbI$_3$ (>100 nm).

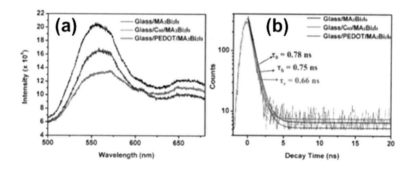

Figure 4. (a) PL and (b) TRPL decay spectra of MA$_3$Bi$_2$I$_9$ on different substrates.

Additionally, the trap-state density within the MA$_3$Bi$_2$I$_9$ film was reported to be 5.29 x 10^{17} cm^{-3} using space charge limit current (SCLC) method which is approximately one order of magnitude higher than that of the lead-perovskites thin film [33]. Also, as mentioned above [26] that the exciton binding energy of MA$_3$Bi$_2$I$_9$ is higher than MAPbI$_3$ which would make charge carriers to recombine easily. All these above-mentioned factors can cause inefficient charge transport resulting in low FF. To address this low FF issue, Zhang et al. deposited MA$_3$Bi$_2$I$_9$ film by dual source evaporation method in which the BiI_3 was deposited under high vacuum and was converted to $MA_3Bi_2I_9$ by evaporating MAI under low vacuum. The resulting perovskite layer showed large submicron-micron grains, with compact and pin-hole free layer; as a result of which, the device showed enhanced J_{sc} (2.95 mA/cm^2), V_{oc} (0.81 V), improved FF (0.79), with the PCE of 1.64% [34]. In addition to the above-mentioned issues, it is reported that a metallic form of bismuth (Bi^0) is present in BiI_3 which affects the charge transport properties and thus device performance [35]. To investigate and improve this issue Jain et al. employed vapour-assisted solution process by exposing BiI_3 films to MAI vapour at the different time interval. They further observed that long-time exposure resulted in mitigation of bismuth metal defect sites and improvement in the crystallinity in $MA_3Bi_2I_9$. Moreover, with exposure of MAI vapour for 25 minutes, $MA_3Bi_2I_9$ showed uniform morphology resulting in enhanced PCE up to 3.17% which is to date the highest efficiency reported for $MA_3Bi_2I_9$ based solar cells [36].

Even though in the very first report of bismuth perovskite Park et al. [27] demonstrated PCE of 1% for $Cs_3Bi_2I_9$ based device, very few studies have been carried out compared to $MA_3Bi_2I_9$ as subsequent reports have shown its poor reproducibility in the device performance. Despite of reasonable optical gap (~2 eV) and absorption coefficient [27], Ghosh et al. found the limitations in achieving high PCE for $Cs_3Bi_2I_9$ based devices mainly owing to its zero-dimensional molecular crystal structure. Moreover, they observed that adding excess BiI_3 into the precursor solution shows beneficial effect on the performance due to suppression of intrinsic defects, however, the efficiency was still below 0.3% and concluded that poor charge transport can be one of the possible reason for such low performance and suggested to move towards development of three-dimensional bismuth materials for better charge transport properties and device performance [37]. On the contrary, recently Bai et al. employed a dissolution-re-crystallization method to fabricate high-quality $Cs_3Bi_2I_9$ perovskite nanosheets films. In this method, after spin coating the solution and annealing, the $Cs_3Bi_2I_9$ coated substrates were again spin-coated with few solvent drops resulting in the formation of $Cs_3Bi_2I_9$ nanosheets. Further, they employed three different HTM, namely, spiro-OMeTAD, PTAA and copper iodide (CuI) and the devices showed PCE of 1.77%, 2.30%, and 3.20% respectively with high stability [38]. This indicates that developing such novel deposition techniques in addition to suitable neighbouring charge transport layers can further help to enhance the efficiency of $Cs_3Bi_2I_9$ perovskite solar cells. From the above discussion, we speculate that wide bandgap and intrinsic optoelectronic properties such as high exciton binding energy are the major limiting factors for $MA_3Bi_2I_9$ and $Cs_3Bi_2I_9$. Although, recent reports have demonstrated efficiency over 3%, intense study in terms of understanding the material at a fundamental level as well as in device perspective is highly required to further explore the potential of ternary bismuth perovskite materials for single and tandem junction devices.

3. BANDGAP TUNING OF LOWER-DIMENSIONAL BISMUTH PEROVSKITE MATERIALS

In the above section, we have discussed a suitable charge transport layer and deposition techniques for bismuth perovskite materials, however, $Cs_3Bi_2I_9$ and $MA_3Bi_2I_9$ possess indirect large optical gap of ~2.1 eV which are not suitable for single junction device and also the efficiency has been far behind of lead-perovskites [26]. Hence, various efforts have also been made to tune the bandgap of $MA_3Bi_2I_9$ and $Cs_3Bi_2I_9$ and some of the important findings will be discussed in this section. Firstly, Vigneshvaran et al. demonstrated the bandgap tuning of $MA_3Bi_2I_9$ by sulfur doping [39]. In this report, ethyl xanthate precursor was employed to dope sulfur at relatively low temperature (~120°C) resulting in narrowing of bandgap to 1.45 eV (lower than $MAPbI_3$ (1.56 eV)). Moreover, the Hall-effect measurement revealed that resultant perovskite material behaves as a *p*-type semiconductor with enhanced carrier concentration and mobility compared to un-doped $MA_3Bi_2I_9$. However, no report on device performance is available so far. Further, Hong et al. [40] theoretically suggested that bandgap of $Cs_3Bi_2I_9$ can be reduced by applying dual trivalent metal cation to form $MA_3Bi_2M^{III}I_9$ and very recently Gu et al. verified it experimentally by doping Ru^{3+} into $Cs_3Bi_2I_9$ [41]. With the incorporation of Ru^{3+}, by hydrothermal synthesis, they observed significant changes in the optoelectronic properties such as narrowing of the optical bandgap, induced shallow defect states, and more radiative recombination centers. Moreover, doping of Ru^{3+} into $Cs_3Bi_2I_9$ upshifts the overall band structure with a high work function. Furthermore, they observed enhanced stability of $Cs_3Bi_{2-x}Rb_xI_9$ against moisture and heat-stress but attempts to incorporate into the device was not demonstrated. In addition to the doping efforts to tune the bandgap, changing the stoichiometric ratio of the precursor components tunes the bandgap of the material. In this regards, Johansson et al. synthesized $CsBi_3I_{10}$ from solution process and optical properties and the crystal structure was compared with $Cs_3Bi_2I_9$ perovskite [42]. Interestingly, the XRD pattern suggested that $CsBi_3I_{10}$ possess a layered structure with a different dominating crystal growth direction than $Cs_3Bi_2I_9$. Moreover, with a change in the stoichiometry, the bandgap tuned from 2.03 eV for $Cs_3Bi_2I_9$ to 1.77 eV for $CsBi_3I_{10}$ with extension in visible light absorption spectrum up to 700 nm. Also, the morphology was highly uniform in contrast to $Cs_3Bi_2I_9$, which showed dendrite-like morphology. The device incorporating $CsBi_3I_{10}$ showed PCE of 0.4% Vs 0.07% (for $Cs_3Bi_2I_9$). Several other attempts have also been made to tune the bandgap by incorporating new organic ligands, for instance, Li et al. [43] synthesized two novel iodobismuthate compounds, [py][BiI_4] and [mepy][BiI_4], having a band gap of 1.78 eV and 2.38 eV respectively and studied their structural, electronic characterization and photovoltaic performance in mesoscopic hole-conductor free architecture. Interestingly, the aforementioned compounds feature one-dimensional $[BiI_4]^-$ infinite anionic chains while the protonated aromatic organic cations play an active role in promoting intermolecular interactions leading to enhanced pseudo-three-dimensional charge carrier transport ability and the devices (hole-conductor free) demonstrated the efficiency of 0.9%. Here the device efficiency was not reproducible but this study provides a new perspective in designing lead-free materials for photovoltaic applications. Further, more novel iodobismuthate compounds are reported using thiazolium, aminothiazolium and imidazolium organic species forming $[C_3H_4NS]_3[Bi_2I_9]$, $[C_3H_4N_2]_3[Bi_2I_9]$ and $[C_3H_5N_2S][BiI_4]$, respectively. The first two compounds showed zero-dimensional structure, whereas one-dimensional edge-sharing chain structure of BiI_6-octahedra was observed in $[C_3H_5N_2S][BiI_4]$. Moreover, red-shift in the optical gap was observed by moving from $[C_3H_4NS]_3[Bi_2I_9]$, $[C_3H_4N_2]_3[Bi_2I_9]$ and $[C_3H_5N_2S][BiI_4]$ with bandgap of 2.08 eV, 2.0 eV and 1.78 eV respectively. Further, they fabricated devices using $[C_3H_5N_2S][BiI_4]$ as an active layer in hole-conductor free architecture and demonstrated PCE of 0.47% with good reproducibility [44]. From the above discussion, it can be concluded that changes/replacement in the organic cation and metal cation significantly changes the bandgap. However, to date, the efficiency of the bandgap tuned bismuth perovskites are lower and much more efforts are required to obtain suitable low bandgap bismuth perovskite materials for efficient solar cells. Additionally, research on the fundamental aspects of the low bandgap bismuth perovskites cannot be ruled out.

3.1. Antimony Based Lower Dimensional Perovskites

Similar to bismuth, antimony (Sb) also found its place in the perovskite family due to similar electronic configuration, comparable electronegativity, and +3 oxidation state. They also exhibit suitable optoelectronic properties for photovoltaic cells. Additionally, $A_3Sb_2I_9$ (A = MA^+, FA^+, Cs^+) based perovskite materials have also shown exceptional stability against humidity and heat-stress. Initial reports on $MA_3Sb_2I_9$ have shown relatively poor PCEs (below 0.5%) (Figure 5), [45] mainly due to poor morphology.

To address the poor morphology issue, Boopathi et al. performed hydroiodic acid (HI) additive approach to deposit $MA_3Sb_2I_9$ perovskite with which the morphology of perovskite was highly controlled and the inverted planar architecture cell showed PCE of 2.04% [46]. In another report from the same group, different hydrophobic HTM such as perylene, pyrene was employed and interestingly $MA_3Sb_2I_9$ crystallized with larger grain size on pyrene in contrast to PEDOT:PSS. As a result, $MA_3Sb_2I_9$ showed enhanced crystallinity and the device demonstrated PCE of 2.8% efficiency which is to date highest efficiency ever reported for $MA_3Sb_2I_9$ based solar cells in inverted architecture [47]. Another interesting work was reported by Jiang et al. who incorporated chlorine (Cl⁻) into the MA$_3$Sb$_2$I$_9$ and observed the formation of a layered phase resulting in improved charge transport properties. Further, they employed layered MA$_3$Sb$_2$I$_9$ perovskite in regular TiO$_2$ mesoporous based structure and obtained PCE of 2.2% which is the best PCE obtained so far in a regular structure [48].

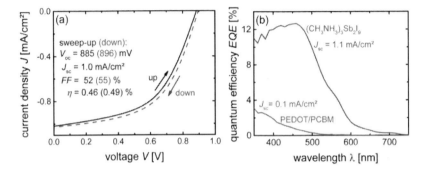

Figure 5. J-V characteristic curve of MA$_3$Sb$_2$I$_9$ solar cells measured with up and down sweep with a rate of 0.1 V/s and (b) its corresponding EQE spectrum.

In addition to MA$_3$Sb$_2$I$_9$, Cs$_3$Sb$_2$I$_9$ have also been explored, however, in contrast to Cs$_3$Bi$_2$I$_9$, Cs$_3$Sb$_2$I$_9$ exist in (dimer) zero- and two-dimensional (layered) form and show direct bandgap of 2.05 eV similar to MA$_3$Bi$_2$I$_9$. However, when deposited by a solution process, the zero-dimer structure is preferentially formed which hinders the charge transport properties leading to poor efficiency (less than 0.5%) [49]. To address this issue, Singh et al. performed SbI$_3$ dissolved DMF solvent vapour annealing in which the Cs$_3$Sb$_2$I$_9$ deposited substrates were covered with Petri dish with few SbI$_3$ dissolved solvent drops. As a result, layered Cs$_3$Sb$_2$I$_9$ perovskite was formed showing improved exciton lifetime and binding energy of 6 ns and 100 meV respectively [50]. Moreover, the Urbach energy of layer-form of Cs$_3$Sb$_2$I$_9$ was less than that of the dimer form of Cs$_3$Sb$_2$I$_9$. And, the device incorporating the layered Cs$_3$Sb$_2$I$_9$ perovskite (in inverse structure) showed PCE of 2.18%. Another strategy to obtain a layered form of antimony perovskite was reported by Harikesh et al. who replaced Cs⁺ with rubidium (Rb⁺) cation which templated the formation of layered phase (2D) irrespective of processing condition. However, they observed slight blue-shift with an optical gap of 2.24 eV for $Rb_3Sb_2I_9$ and the devices demonstrated PCE of 0.66% [51].

To summarize, bismuth perovskite-based devices showed enhanced efficiency with novel deposition (mostly based on 2-step deposition) techniques and also the performance showed high dependence on the choice of suitable HTM. Few efforts were also made to tune the bandgap, however, the efficiency was below 1% and hence much research direction is necessary towards finding novel strategies to tune the bandgap as well as material properties which can help to enhance PCE and absorption spectrum in the visible region. In a combination of already established deposition techniques and/or developing new deposition routes can also help to enhance the photocurrent/voltage and performance of the cell. In the case of antimony perovskites, the layered phase of perovskite-based devices showed improvement in PCE compared to the dimer phase. Employing hydrophobic and inorganic HTM also played an important role in enhancing the PCE of bismuth and antimony perovskite and hence exploring more suitable neighboring charge transport materials can further pave the path towards efficiency improvement.

4. 3D LEAD-FREE PEROVSKITE AND PEROVSKITE-LIKE MATERIALS

4.1. Tuning the Dimensionality by Mixing with Monovalent Cation: The Family of Double Perovskites

Bismuth or antimony based double perovskites has been explored as a class of three-dimensional materials for lead-free perovskite solar cells. The possibilities of heterovalent substitution of Pb^{2+} by trivalent metals and monovalent metals such as silver (Ag), copper (Cu), potassium (K) etc. into the perovskite structure forms double perovskite with molecular structure of $A_2MM'X_6$, where A = Cs, MA; M = Bi, Sb; M' = Ag, Au, Cu, K; X = I, Cl,

Br. Investigation of $(CH_3NH_3)_2KBiCl_6$, theoretically and experimentally, was the first step towards the development of lead-free double perovskite materials, however, its bandgap was reported to be ~3.08 eV which is not suitable for photovoltaic applications [52]. And, among various double perovskite materials explored, most of the investigations have been performed with $Cs_2BiAgBr_6$ and $Cs_2BiAgCl_6$ due to its acceptable bandgap (~2 eV). Firstly, Slavney et al. synthesized $Cs_2BiAgBr_6$ which showed strong photoluminescence in the visible region and long recombination lifetime, however, it showed an indirect nature of bandgap [53]. In a subsequent report, McClure et al. synthesized both $Cs_2BiAgBr_6$ and $Cs_2BiAgCl_6$ and observed the bandgap of 2.19 eV and 2.77 eV respectively. The difference in the bandgap of $Cs_2BiAgBr_6$ compared to the first report can be due to the crystallinity and different processing condition. Further, they studied the stability of $Cs_2BiAgBr_6$ and $Cs_2BiAgCl_6$ thin films and observed no significant changes in $Cs_2BiAgCl_6$ while $Cs_2BiAgBr_6$ showed dramatic changes after two weeks of simultaneous exposure to light and ambient air [54, 55]. Recently, Docampo et al. demonstrated a double perovskite-based device using $Cs_2BiAgBr_6$ as an absorber layer. The resulting thin film featured bandgap comparable to that of single crystal and polycrystalline powder. Very interestingly, the device showed PCE of 2.5% with V_{oc} exceeding 1V with high stability when compared to $MAPbI_3$ under constant illumination, demonstrating its suitability for optoelectronic applications [56]. It is very interesting to note that according to the McClure et al. report, the synthesized $Cs_2BiAgBr_6$ showed poor stability with simultaneous exposure of light and air while $Cs_2BiAgBr_6$ based device (according to Docampo et al.) showed much-enhanced stability which we believe can be due to the effect of neighboring charge transport layers and variation in the processing condition from lab to lab [55, 56]. Concurrently, Wu et al. incorporated $Cs_2AgBiBr_6$ using SnO_2 and P3HT as electron and hole transport layers respectively and showed 1.44% PCE, showing its compatibility with different charge transport layers [57]. Another important study was performed by Ning et al. who employed and deposited $Cs_2AgBiBr_6$ by one-step deposition process from single-crystal solution in TiO_2 based planar architecture. Despite the obtained PCE of 1.22%, which was lower than that of previous reports, they observed a long diffusion length of 110 nm [58]. To note, Docampo et al. [56] employed mesoporous architecture while Ning et al. [58] incorporated $Cs_2AgBiBr_6$ in TiO_2 planar architecture which we believe to be the reason for the variation in the PCE. Although double perovskite materials have shown promising optoelectronic properties, compatibility to incorporate in the device and long-term stability. Recently, Savory et al. reported its limitations owing to the wide indirect bandgap of these class of materials and high carrier effective masses and suggested to explore materials beyond Ag-Bi combinations. To overcome this, further, they performed theoretical calculations to design novel double perovskite materials bismuth-indium (Bi-In) and bismuth-thallium (Bi-Tl) which possess a direct bandgap of 2.0 eV [59]. Dong et al. investigated this experimentally by synthesizing $(CH_3NH_3)_2BiTlBr_6$ and they also observed direct bandgap of 2.0 eV. However, it is well known that Tl is even more toxic than Pb which limits its consideration as an eco-friendly material for lead-free perovskite solar cells. Although the family of double perovskites is promising alternatives as lead-free perovskites, their wide optical gap limits its performance and hence efforts need to be directed towards developing low bandgap double perovskite suitable for single junction device. Additionally, high bandgap double perovskites have demonstrated V_{oc} around 1V, however, considering the bandgap (~2 eV) the V_{oc} can be improved to beyond 1.5 eV indicating lots of room for further improvements.

4.2. Three Dimensional Perovskite and Perovskite-Like Materials Beyond AMX_3 and $A_2MM'X_6$

Sun et al. firstly performed theoretical studies, using split-anion approach, to design Bi-perovskite materials in combination with chalcogenide compounds to form chalcogenide-halide mix perovskite having a structure of $MABiXY_2$ compounds (X = chalcogen; Y = halogen). Interestingly, the atomic structure of these compounds is similar to that of $MAPbI_3$ and the bandgap lies in between 1.3 eV to 1.4 eV (smaller than

MAPbI$_3$ and similar to FAPbI$_3$) which are quite suitable for single junction solar cells [60]. Recently, Seok et al. performed an experimental effort to obtain a thin film of MASbSI$_2$ and calculated its tolerance factor to be 0.99 which satisfies a perfect cubic structure. This also indicates that MASbSI$_2$ perovskite materials have fewer defects related to crystal structure, unlike, MAPbI$_3$ which possess tetragonal structure and has defect centers. Further, they assessed its photovoltaic property by fabricating a device with TiO$_2$ and PCPDTBT as ETM and HTM respectively and the cells demonstrated PCE of 3.06% with good reproducibility and enhanced stability against humidity exposure, continuous light illumination, and heat-stress [61]. On a contrary, through a combination of experimental and computational analysis, Mitzi et al. observed no evidence of the formation of MASbSI$_2$ and instead observed a mixture of binary and ternary compounds (Sb$_2$S$_3$ and MA$_3$Sb$_2$I$_9$). Moreover, density functional theory (DFT) calculation suggests that MASbSI$_2$ types of perovskite phase are thermodynamically unstable (Figure 6) [62]. We believe that more investigation in terms of crystal structure as well as in terms of device perspective is required to validate this as only two papers have been reported so far.

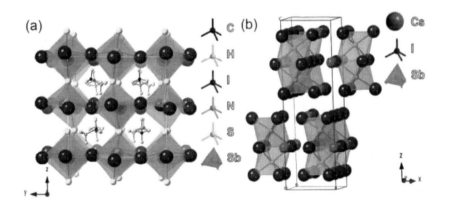

Figure 6. Proposed structure of MASbSI$_2$ from DFT relaxation and (b) crystal structure of Cs$_3$Sb$_2$I$_9$ (a member of A$_3$Sb$_2$I$_9$ family).

4.3. Silver-Bismuth Halide Materials: A Family of Three-Dimensional Perovskite-Like Materials

Various silver bismuth halide materials such as $AgBi_2I_7$, $AgBiI_4$, Ag_2BiI_5, and Ag_3BiI_6 emerged as promising candidates for lead-free perovskite solar cells. Firstly, Sargent et al. reported a cubic phase of $AgBi_2I_7$ by mixing silver iodide (AgI) and bismuth iodide (BiI_3) in 1:2 molar ratios and achieved a conversion efficiency of 1.2%. Additionally, the devices showed enhanced stability against humidity and light soaking [63]. However, in the subsequent reports, the efficiency was not reproducible, for instance, Johansson et al. and Shao et al. reported 0.4% and 0.5% respectively [64, 65]. Regarding the crystal structure, with the help of theoretical studies, Mitzi et al. reported that $AgBi_2I_7$ is not preferentially formed and instead a sub-stoichiometric $AgBiI_4$ is formed which can be the reason for the non-reproducible performance of the cells [66]. On the other hand, Jung et al. performed experimental studies by mixing AgI and BiI_3 in different compositions and observed that $AgBiI_4$ does not form and instead $AgBi_2I_5$ is formed which is the most highly stable phase of silver-bismuth halide family. Also, they have explored many other combinations of silver-bismuth halide materials and among all, the cells with $AgBi_2I_5$ showed the best PCE of 1.7% (with spiro-OMeTAD as HTM) [67]. Concurrently, Johansson et al. employed P3HT as HTM and fabricated devices with $AgBi_2I_5$ as an absorber layer and achieved PCE of 2.1%.[65] Further, they also reported that $AgBi_2I_5$ is slightly sensitive to the humid atmosphere as less intense XRD peaks were observed for the $AgBi_2I_5$ thin film (after exposing to sir) and with P3HT upper layer, the XRD peak intensity enhanced further indicating that P3HT upper layer is protecting $AgBi_2I_5$ from degradation [65]. In addition to the above mentioned compounds, Turkevych et al. reported Ag_3BiI_6 thin film based solar cells. They deposited Ag_3BiI_6 by solution process followed by anti-solvent (toluene) drenching with electrode deposited TiO_2 and PTAA layers acting as ETM and HTM respectively. As a result, the device demonstrated PCE of 4.3% [68] which is to date the highest efficiency reported for silver-bismuth halide class of materials.

In summary, bismuth-based double perovskite has been explored mainly focusing on $Cs_2AgBiBr_6$ and the devices demonstrated 2% efficiency, however, wide and indirect bandgap limits its light absorbing capability and hence tuning the bandgap of double perovskite materials (by doping) and simultaneously maintaining the long diffusion length can be an effective strategy towards development of double perovskite materials. On the other hand, intense studies are required on understanding the formation, crystal structure of chalcogenide-halide based perovskite materials as well as $AgBi_2I_7$ and $AgBiI_4$ based silver-bismuth halide materials. As mentioned above, $AgBi_2I_5$ has been reported to be sensitive to air but it can be noted that there are no hygroscopic components and both AgI and BiI_3 are not sensitive to moisture, hence further understanding is required on the materials instability against ambient atmosphere. Strategies to explore various device architecture and neighboring charge transport layers and its suitability for double perovskite, chalcogenide-halide perovskite, and silver-bismuth halide materials also cannot be ruled out.

4.4. Bismuth and Antimony Based Non-Perovskite Materials

It can be noted that in the synthesis of bismuth perovskite, and perovskite-like materials, BiI_3 is used as one of the precursor components and interestingly BiI_3 has a layered structure and possess interesting optical properties such as higher optical absorption than Si and GaAs, the large static dielectric constant, electron diffusion length of 4.9 µm [69, 70]. Additionally, $6s^2$ outer shell electronic configuration of BiI_3 results in dispersed valance band, high dielectric constant, shallow intrinsic point defects; all of these are serviceable properties of defect-tolerant material. Buonassisi et al. observed room temperature photoluminescence (PL) of BiI_3 at 550 nm which further endorse its promising application in optoelectronic devices [71]. Initially, BiI_3 was employed as an HTM in organic solar cells with fullerene-based light absorber as an active layer [72] and in later reports, Lehner et al. and Hamdeh et al. employed it in TiO_2 based planar architecture and reported an efficiency of 0.3% and 1% respectively [73, 74].

Kulkarni et al. employed it in mesoporous architecture and studied the effect of different annealing methods such as thermal annealing, solvent vapour annealing, and Petri dish covered recycled vapour annealing and at different temperatures ranging from 90-180°C. Interestingly with an increase in temperature, the grain size increased and film uniformity improved and the device based on BiI_3 films processed at 150°C with Petri dish covered recycled vapour annealing demonstrated PCE of 0.5% which is so far the best efficiency reported for BiI_3 solar cells employing organic HTM (in mesoporous architecture). Moreover, the devices were stable against 50% humidity atmosphere, 100°C heat-stress, and 20 min light soaking test [75]. In addition to BiI_3 based cells, few other antimony based semiconducting materials have been reported such as Sb_2S_3, $Cs_2Sb_8S_{13}$, and $MA_2Sb_8S_{13}$ which possess bandgap of 1.72, 1.85, and 2.08 eV respectively [76].

In summary, bismuth halide and other non-perovskite semiconducting materials have been explored, however, their device efficiency lags far behind the lead-perovskite. One of the critical issues is controlling their morphology which can be improved by combining the already known deposition technique with some new novel methodologies. Theoretical calculations are further required to understand the reason for such low performance and experimental efforts are required to understand it from the fundamental point of view. Additionally, TiO_2 ETL is not suitable for these non-perovskite semiconducting materials and thus suitable ETL such as ZnO, SnO_2 should be explored. Moreover, dopants used in widely employed Spiro-OMeTAD degrade these materials and exploring and developing dopant-free HTM cannot be ruled out.

REFERENCES

[1] Miyasaka, T. (2016). Perovskite photovoltaics: Rare functions of organo lead halide solar cells and optoelectronic devices. *Chem. Lett.*, 44(6): 720-729.

[2] Singh, T., and Miyasaka, T. (2018). Stabilizing the efficiency beyond 20% with a mixed cation perovskite solar cell fabricated in ambient air under controlled humidity. *Adv. Energy Mater.*, 8(3): 1700677.

[3] Kojima, A., Teshima, K., Miyasaka, T., and Shirai, Y. (2016). Novel photoelectrochemical cell with mesoscopic electrodes sensitized by lead-halide compounds. *210th ECS Meeting Abstracts*, 397.

[4] Kojima, A., Teshima, k., Shirai, Y., and Miyasaka, T. (2009). Organo-metal halide perovskites as visible-light sensitizers for photovoltaic cells. *J. Am. Chem. Soc.*, 131 (17): 6050-6051.

[5] Best solar cell efficiency chart, https://www.nrel.gov/pv/assets/pdfs/pv-efficiencies-07-17-2018.pdf, accessed on 14th Sept. 2018.

[6] Kulkarni, A., Jena, A., K., Chen, H.-W., Sanehira, Y., Ikegami, M., and Miyasaka, T. (2016). Revealing and reducing the possible recombination loss in TiO_2 compact layer by incorporating MgO layer in perovskite solar cells. *Sol. Energy*, 136: 379-384.

[7] Yablonovitch, E. (2016). Lead halides join the top optoelectronic league. *Science*, 351: 1401-1401.

[8] Legitens, T., Bush, K., Chaecharoen, R., Beal, R., Bowring, A., and McGehee, M., D. (2017). Towards enabling stable lead halide perovskite solar cells; interplay between structural, environmental and thermal syability. *J. Mater. Chem. A.*, 5: 11483-11500.

[9] Chaudhary, B., Kulkarni, A., Jena, A., K., Ikegami, M., Udagawa, Y., Kunugita, H., Ema, K., and Miyasaka, T. (2017). Poly(4-vinyl-pyridine)-based interfacial passivation to enhance voltage and moisture stability of lead halide perovskite solar cells. *Chem Sus Chem*, 10: 2473-2479.

[10] Singh, T., Oz, S., Sasinska, A., Frohnhoven, R., Mathur, S., and Miyasaka, T. (2018). Sulfate-assisted interfacial engineering for high yield and efficiency of triple cation perovskite solar cells with alkali-doped TiO_2 electron-transport layers. *Adv. Fun. Mater.*, 28(14): 1706287.

[11] Pinpithak, P., Chen, H. W., Kulkarni, A., Sanehira, Y., Ikegami, M., and Miyasaka, T. (2017). Low-temperature and ambient air processes

of amorphous SnO$_x$-based mixed halide perovskite planar solar cell. *Chem. Lett.*, 46(3): 382-384.

[12] Matteocci, F., Cina, L., Lamanna, E., Cacovich, S., Divitini, G., Midgley, P., A., Ducati, C., and Carlo, A., D. (2016). Encapsulation for long-term stability enhancement of perovskite solar cells. *Nano Energy*, 30: 162-172.

[13] Babayigit, A., Ethirajan, A., Muller, M., and Conings, B. (2016). Toxicity of organometal halide perovskite solar cells. *Nature Materials*, 15: 247-251.

[14] Turkevych, I., Kazaoui, S., Ito, E, Urano, T., Yamada, K., Tomiyasu, H., Yamagishi, H., Kondo, M., and Aramaki, S. (2017). Photovoltaic rudorffites: lead-free silver bismuth halides alternatives to hybrid lead halide perovskites. *Chem Sus Chem*, 10 (19): 3754-3759.

[15] Hao, F., Stoumpos, C., C., Cao, D., H., Chang, R., P., H., and Kanatzidis, M., G. (2014). Lead-free solid-state organic-inorganic halide perovskite solar cells. *Nature Photonics*, 8: 489-494.

[16] Konstantaku, M., and Stergiopoulous, T. (2017). A critical review on tin halide perovskite solar cells. *J. Mater. Chem. A*, 5: 11518-11549.

[17] Shao, S., Liu, J., Portale, G., Fang, H. H., Blake, G., R., Brink, G., H., Koster, L., J., A., and Loi, M., A. (2018). Highly reproducible Sn-based hybrid perovskite solar cells with 9% efficiency. *Adv. Energy Mater.*, 8: 1702019.

[18] Konstantakou, M., Stergiopoulos, T. (2017). A critical review on tin halide perovskite solar cells. *J. Mater. Chem. A*, 5: 11518-11549.

[19] Krishnamoorthy, T., Ding, H., Yan, C., Leong, W., L., Baike, T., Zhang, Z., Sherburne, M., Li, S., Asta, M., Mathews, N., and Mhaisalkar, S., G. (2015). Lead-free germanium iodide perovskite materials for photovoltaic applications. *J. Mater. Chem. A.*, 3: 23829-23832.

[20] Hoefler, S., F., Trimmel, G., and Rath, T. (2017). Progress on lead-free metal halide perovskites for photovoltaic applications: a review. *Monatsh. Chem.*, 148 (5): 795-826.

[21] Kulkarni, A., Singh, T., Ikegami, M., and Miyasaka, T. (2017). Photovoltaic enhancement of bismuth halide hybrid perovskite by N-

methyl pyrrolidone-assisted morphology conversion. *RSC Adv.*, 7: 9456-9460.

[22] Kawai, T., Ishii, A., Kitamura, T., Shimanuki, S., Iwata, M., and Ishibashi, Y. (1996). Optical absorption in band-edge region of (CH$_3$NH$_3$)$_3$Bi$_2$I$_9$ single crystals. *J. Phys. Soc. Jpn.*, 65:1464-1468.

[23] Kawai, T., and Shimanuki, S. (1993). Optical studies of (CH$_3$NH$_3$)$_3$Bi$_2$I$_9$ single crystals. 177: K43.

[24] Lindqvist, O. (1968). The crystal structure of cesium bismuth iodide, Cs$_3$Bi$_2$I$_9$. *Acta Chem. Scand.*, 22: 2943-2952.

[25] Arakcheeva, A., V., Bonin, M., Chapuis, G., and Zaitsev, A., I. (1999). The phase of Cs$_3$Bi$_2$I$_9$ between RT and 190 K. *Z. Kristallogr.*, 214: 279-284.

[26] Arakcheeba, A., V., Chapuis, G., Meyer, M. (2001). The LT phase of Cs$_3$Bi$_2$I$_9$. *Z. Kristallogr.*, 216: 199-205.

[27] Park, B. W., Philippe, B., Zhang, X., Ransmo, H., Boschloo, G., and Johansson, E., M., J. (2015). Bismuth based hybrid perovskites A$_3$Bi$_2$I$_9$ (A: metnylammonium or cesium) for solar cell applications. *Adv. Mater.*, 27:6806-6813.

[28] Hoye, R., Brandt, R., E., Osherov, A., Stevanovic, V., Stranks, S., D., Wilson, M., W., B., Akey, A., J., Kurchin, R., C., Poindexter, J., R., Wang, E., N., Bawendi, M., G., Bulovic, V., and Buonassisi, T. (2016). Methylammonium bismuth iodide as a lead-free, stable hybrid organic-inorganic solar absorber. *Chem. Eur. J.*, 22: 2605-2610.

[29] Echkardt, K., Bon, V., Getzschmann, J., Grothe, J., Wisser, F., and Kaske, S. (2016). Crystallographic insights into (CH$_3$NH$_3$)$_3$Bi$_2$I$_9$: a new lead-free hybrid organic-inorganic material as a potential absorber for photovoltaics. *Chem. Comm.*, 52: 3058-3060.

[30] Oz, S., Hebig, J., C., Jung, E., Singh, T., Lepcha, A., Olthof, S., Flohre, J., Gao, Y., German, R., van Loosdrecht, P., H., M., Meerholz, K., Kirchartz, T., and Mathur, S. (2016). Zero-dimensional (CH$_3$NH$_3$)$_3$Bi$_2$I$_9$ perovskite for optoelectronic applications. *Sol. Energy Mater Sol. Cells*, 158(2): 195-201.

[31] Singh, T., Kulkarni, A., Ikegami, M., and Miyasaka, T. (2016). Effect of electron transporting layer on bismuth based lead free perovskite

($CH_3NH_3)_3Bi_2I_9$ for photovoltaic applications. *ACS Appl. Mater. Interfaces*, 8(23): 14542-14547.

[32] Shin, S., S., Baena, J., P., C., Kurchin, R., C., Polizzotti, A., Jungwan, J., Y., Weighold, S., Bawendi, M., G., and Buonassisi, T. (2018). Solvent-engineering method to deposit compact bismuth based thin films: mechanism and application to photovoltaics. *Chem. Mater.*, 30(2): 336-343.

[33] Ran, C., Wu, Z., Xi, J., Yuan, F., Dong, H., Lei, T., He, X., and Hou, X. (2017). Constructing of compact methylammonium bismuth iodide film promoting lead-free inverted heterojunction organohalide solar cells with open-circuit voltage over 0.8 V. *J. Phys. Chem. Lett.*, 8(2): 394-400.

[34] Zhang, Z., Li, X., Xia, X., Wang, Z., Huang, Z., Lei, B., and Gao, Y. (2017). High-quality $(CH_3NH_3)_3Bi_2I_9$ film-based solar cells: pushing efficiency up to 1.64%. *J. Phys. Chem. Lett.*, 8: 4300-4307.

[35] Hamdeh, U., M., Nelson, R., D., Ryan, B., J., Bhattacharjee, U., Petrich, J. W., and Panthani, M., G. (2016). Solution-processed BiI_3 thin films for photovoltaic applications: improved carrier collection via solvent annealing. *Chem. Mater.*, 28 (18): 6567-6574.

[36] Jain, S., M., Phuyal, D., Davies, M., L., Li, M., Philippe, B., De Castro, C., Qiu, Z., Kim, J., Watson, T., Tsoi, W., C., Karis, O., Rensmo, H., Boschloo, G., Edvinsson, T., and Durrant, J., R. (2018). An effective approach of vapour assisted morphological tailoring for reducing metal defect sites in lead-free, $(CH_3NH_3)_3Bi_2I_9$ bismuth-based perovskite solar cells for improved performance and long-term stability. *Nano Energy*, 49: 614-624.

[37] Ghosh, B., Wu, B., Mulmudi, H., M., Guet, C., Weber, K., Sun, T., C., Mhaisalkar, S., G., and Mathews, N. (2018). Limitations of $Cs_3Bi_2I_9$ as lead-free photovoltaic absorber materials. *ACS Appl. Mater. Interfaces*, doi: 10.1021/acsami.7b14735.

[38] Bai, F., Hu, Y., Hu, Y., Qui, T., Miao, X., and Zhang, S. (2018). Lead-free, air-stable untrathin $Cs_3Bi_2I_9$ perovskite nanosheets for solar cells. *Sol. Energy Mater Sol. Cells*, 184: 15-21.

[39] Vigneshwaran, M., Ohta, T., Iikubo, S., Kapil, G., Ripolles, T., S., Ogomi, Y., Ma, T., Pandey, S., S., Shen, Q., Toyoda, T., Yoshino, K., Minemoto, T., and Hayase, S. (2016). Facile synthesis and characterization of sulfur doped low bismuth based perovskites by soluble precursor route. *Chem. Mater.*, 28 (18), 6436-6440.

[40] Hong, K. H., Kim, J., Debbichi, L., Kim, H., and Im, S., H. (2017). Band gap engineering of $Cs_3Bi_2I_9$ perovskites with trivalent atoms using a dual metal cation. *J. Phys. Chem. C*, 121 (1): 969-974.

[41] Gu, J., Yan, G., Lian, Y., Mu, Q., Jin, H., Zhang, Z., Deng, Z., and Pend, Y. (2018). Bandgap engineering of lead-free defect perovskite $Cs_3Bi_2I_9$ through trivalent doping of Ru^{3+}. *RSC Adv.*, 8: 25802.

[42] Johansson, M., B., Zhu, H., and Johansson, E., M., J. (2016). Extended photo-conversion spectrum in low-toxic bismuth halide perovskite solar cells. *J. Phys. Chem. Lett.*, 7 (17): 3467-3471.

[43] Li, T., Hu, Y., Morrison, A., A., Wu, W., Han, H., and Robertson, N. (2017). Lead-free pseudo-three-dimensional organic-inorganic iodobismuthates for photovoltaic applications. *Sustainable Energy Fuels*, 1: 308-316.

[44] Li, T., Wang, Q., Nichol, G., S., Morrison, C., A., Han, H., Hu, Y., and Robertson, N. (2018). Extending lead-free hybrid photovoltaic materials to new structures: thiazolium, aminothiazolium, and imidazoliumiodobismuthates. *Dalton Trans.*, 47: 7050-7058.

[45] Hebig, J. C., Kuehn, I., Flohre, J., Kirchartz, T. (2016). Optoelectronic properties of $(CH_3NH_3)_3Sb_2I_9$ Thin Films for photovoltaic applications. *ACS Energy Lett.*, 1: 309-314.

[46] Boopathi, K. M., Karuppuswamy, P., Singh, A., Hanmandlu, C., Lin, L., Abbas, S. A., Chang, C. C., Wang, P. C., Li, G., Chu, C. W. (2017). Solution-processable antimony-based light-absorbing materials beyond lead halide perovskites. *J. Mater. Chem. A*, 5: 20843-20850.

[47] Karuppuswamy, P., Boopathi, K., M., Mohapatra, A., Chen, H. C., Wong, K. T., Wang, P. C., Chu, C. W. (2018). Role of a hydrophobic scaffold in controlling the crystallization of methylammonium antimony iodide for efficient lead-free perovskite solar cells. *Nano Energy*, 45: 330-336.

[48] Jiang, F., Yang, D., Jiang, Y., Liu, T., Zhao, X., Ming, Y., Luo, B., Qin, F., Fan, J., Han, H., Zhang, L., Zhou, Y. (2018). Chlorine-incorporation-induced formation of layered phase for antimony-based lead-free perovskite solar cells. *J. Am. Chem. Soc.*, 140 (3): 1019-1027.

[49] Saparov, B., Hong, F., Sun, J.-P., Duan, H.-S., Meng, W., Cameron, S., Hill, I. G., Yan, Y., Mitzi, D. B. (2015). Thin film preparation and characterization of $Cs_3Sb_2I_9$: A lead-free layered perovskite semiconductor. *Chem. Mater.*, 27 (16): 5622-5632

[50] Singh, A., Boopathi, K., M., Mohapatra, A., Chen, Y., F., Li, G., Chu, C. W. (2018). Photovoltaic performance of vapour-assisted solution-processed layer polymorph of $Cs_3Sb_2I_9$. *ACS Appl. Mater. Interfaces*, 10 (3): 2566-2573.

[51] Harikesh, P., C., Mulmudi, H., K., Ghosh, B., Goh, T., W., Teng, Y., T., Thirumal, K., Lockrey, M., Weber, K., Koh, T., M., Li, S., Mhaisalkar, S., Mathews, N. (2016). Rb as an alternative cation for templating inorganic lead-free perovskites for solution processed photovoltaics. *Chem Mater.*, 28 (20): 7496-7504.

[52] Wei, F., Deng, Z., Sun, S., Xie, F., Kieslich, G., Evans, D., M., Carpenter, M., A., Bristowe, P., D., Cheetham, A., K. (2016). The synthesis, structure and electronic properties of a lead-free hybrid inorganic-organic double perovskite $(MA)_2KBiCl_6$ (MA = methylammonium). *Mater. Horiz.*, 3: 328-332.

[53] Slavney, A., H., Hu, T., Lindenberg, A., M., Karunadasa, H., I. (2016). A bismuth-halide double perovskite with long carrier recombination lifetime for photovoltaic applications. *J. Am. Chem. Soc.*, 138 (7): 2138-2141.

[54] McClure, E., T., Ball, M., R., Wind, W., Woodward, P., M. (2016). Cs_2AgBiX_6 (X = Br, Cl): New visible light absorbing, lead-free halide perovskite semiconductors. *Chem. Mater.*, 28 (5): 1348-1354.

[55] Giustino, F., Snaith, H., J. (2016). Towards lead-free perovskite solar cells. *ACS Energy Lett.*, 1(6): 1233-1240.

[56] Greul, E., Petrus, M., L., Binek, A., Docampo, P., Bein, T. (2017). Highly stable, phase pure Cs$_2$AgBiBr$_6$ double perovskite thin films for optoelectronic applications. *J. Mater. Chem. A*, 5:19972-19981.

[57] Wu, C., Zhang, Q., Liu, Y., Luo, W., Huo, X., Huang, Z., Ting, H., Sun, W., Zhong, X., Wei, S., Wang, S., Chen, Z., Xiao, L. (2018). The dawn of lead-free perovskite solar cells: highly stable double perovskite Cs$_2$AgBiBr$_6$ film. *Adv. Sci.*, 5:1700759.

[58] Ning, W., Wang, F., Wu, B., Lu, J., Yan, Z., Liu, X., Tao, Y., Liu, J.-M., Huang, W., Fahlman, M., Hultman, L., Sun, T., C., Gao, F. (2018). Long electron-hole diffusion length in high quality lead-free double perovskite films. *Adv. Mater.*, 30 (20): 1706246.

[59] Savory, C., N., Walsh, A., Scanlon, D., O. (2016). Can Pb-free halide double perovskites support high-efficiency solar cells? *ACS Energy Lett.*, 1(5): 949-955.

[60] Sun, Y. Y., Shi, J., Lian, J., Gao, W., Agiorgousis, M., L., Zhang, P., Zhang, S. (2016). Discovering lead-free perovskite solar materials with a split-anion approach. *Nanoscale*, 8:6284-6289.

[61] Nie, R., Mehta, A., Park, B.-W., Kwon, H.-W., Im, J., Seok, S., I. (2018). Mixed sulfur and iodide-based lead-free perovskite solar cells. *J. Am. Chem. Soc.*, 140(3): 872-875.

[62] Li, T., Wang, X., Yan, Y., Mitzi, D., B. (2018). Phase stability and electronic structure of peospective Sb-based mixed sulfide and iodide 3D perovskite (CH$_3$NH$_3$)SbSI$_2$. *J. Phys. Chem. Lett.*, 9(14): 3829-3833.

[63] Kim, Y., yang, Z., Jain, A., Voznyy, O., Kim, G.-H., Liu, M., Quan, L., N., Arquer, F., P., G., Comin, R., Fan, J., Z., Sargent, E., H. (2016). Pure phase cubic hybrid iodobismuthates AgBi$_2$I$_7$ for thin-film photovoltaics. *Angew. Chem.*, 55 (33): 9586-9590.

[64] Zhu, H., Pan, M., Johansson, M., B., Johansson, E., M. (2017). High photon-to-current conversion in solar cells based on light absorbing silver bismuth iodide. *ChemSusChem.* 10 (12): 2592-2596.

[65] Shao, Z., Mercier, T., L., Madec, M., B., Pauporte, T. (2018). Exploring AgBi$_x$I$_{3x+1}$ semiconductor thin films for lead-free perovskite solar cells. *Materials and Design*, 141: 81-87.

[66] Xiao, Z., Meng, W., Mitzi, D., B., Yan, Y. (2016). Crystal structure of AgBi$_2$I$_7$ thin films. *J. Phys. Chem. Lett.*, 7(19): 3903-3907.

[67] Jung, K., W., Sohn, M., R., Lee, H., M., Yang, I., S., Sung, S., D., Kim, J., Diau, E., W.-G., Lee, W., I. (2018). Silver bismuth iodides in various compositions as potential Pb-free light absorbers for hybrid solar cells. *Sustainable Energy Fuels*, 2: 294-302.

[68] Turvevych, I., Kazaoui, S., Ito, E., Urano, T., Yamada, K., Tomiyasu, H., Yamagishi, H., Kondo, M., Aramaki, S. (2017). Photovoltaic rudorffites: lead-free silver bismuth halides alternative to hybrid lead halide perovskites. *ChemSusChem*. 10: 3754-3759.

[69] Green, M., A., Keevers, M., J. (1995). Optical properties of intrinsic silicon at 300 K. *Prog. Photovoltaics*, 3: 189-912.

[70] Blakemore, J., S. (1982). Semiconducting and other major properties of gallium arsenide. *J. Appl. Phys.* 53: R123-R181.

[71] Brandt, R., E., Kurchin, R., C., Hoye, R., L., Z., Poindexter, J., R., Wilson, M., W., B., Sulekar, S., Lenahan, F., Yen, P., X., T., Stevanovic, V., Nino, J., C., Bawedi, M., G., Buonassisi, T. (2015). Investigation of bismuth halide (BiI$_3$) for photovoltaic applications. *J. Phys. Chem. Lett.*, 6: 4297-4302.

[72] Boopathi K., M., Raman, S., Mohanraman, R., Chou, F. C., Chen, Y., Y., Lee, C. H., Chang, F. C., Chu, C. W. (2014). Solution-processable bismuth iodide nanosheets as hole transport layers for organic solar cells. *Sol. Energy Mater Sol. Cells*, 121:35-41.

[73] Lehner, A., J., Wang, H., Fabini, D., H., Liman, C., D., Hebert, C.-A., Perry, E., E., Wang, M., Bazon, G., C., Chabinyc, M., L., Seshadri, R. (2015). Electronic structure and photovoltaic application of BiI$_3$. *Appl. Phys. Lett.*, 107: 131109.

[74] Hamdeh, U., M., Nelson, R., D., Ryan, B., J., Bhattacharjee, U., Petrich, J., W., Panthani, M., G. (2016). Solution-processed BiI$_3$ thin films for photovoltaic applications: improved carrier collection via solvent annealing. *Chem. Mater.*, 28(18): 6567-6574.

[75] Kulkarni, A., Singh, T., Jena, A., K., Pinpithak, P., Ikegami, M., Miyasaka, T. (2018). Vapour annealing controlled crystal growth and photovoltaic performance of bismuth triiodide embedded in

mesostructcured configuration. *ACS Appl. Mater. Interfaces*, 10(11): 9547-9554.

[76] Yang, R., X., Butler, K., T., Walsh, A. (2015). Assessment of hybrid organic-inorganic antimony sulfides for earth-abundant photovoltaic applications. *J. Phys. Chem. Lett.*, 6(24): 5009-5014.

In: Perovskite Solar Cells
Editor: Murali Banavoth
ISBN: 978-1-53615-858-8
© 2019 Nova Science Publishers, Inc.

Chapter 8

EFFECTS OF NH₄CL OR PbI₂ ADDITIONS TO CH₃NH₃PbI₃ PEROVSKITE SOLAR CELLS

Takeo Oku[*], *Yuya Ohishi and Naoki Ueoka*
Department of Materials Science, The University of Shiga Prefecture,
Hikone, Shiga, Japan

ABSTRACT

Effects of NH₄Cl or PbI₂ addition to perovskite CH₃NH₃PbI₃ precursor solutions on photovoltaic properties were investigated. TiO₂/CH₃NH₃PbI₃(Cl)-based photovoltaic devices were fabricated by spin-coating and air blowing techniques, and the microstructures of the devices were investigated by X-ray diffraction, optical microscopy and scanning electron microscopy. Current density voltage characteristics were improved by a small amount of Cl-doping, which resulted in an improvement of the efficiencies of the devices. The structure analysis indicated the formation of a homogeneous microstructure by NH₄Cl addition to the perovskite phase, and formation of PbI₂ was suppressed by the NH₄Cl or PbI₂ addition. The microstructure analysis also indicated that the perovskite layer contained dense grains with strong (100) orientation, as a result of NH₄Cl addition and air blowing. The ratio of the (100)/(210)

---

[*] Corresponding Author's Email: oku@mat.usp.ac.jp.

reflection intensities for the perovskite crystals was 2000 times higher than that of randomly oriented grains. The devices were stable when stored in ambient air for two weeks.

Keywords: perovskite, $CH_3NH_3PbI_3$, solar cell, NH_4Cl, PbI_2, air blow

1. INTRODUCTION

Since the development of perovskite solar cells [1–3], thin film solar cells containing methylammonium trihalogenoplumbate(II) ($CH_3NH_3PbI_3$) compounds with perovskite structures have been widely studied [4–9]. This has been due to their easy fabrication processes and high photoconversion efficiencies compared with conventional fullerene based organic solar cells. Conversion efficiencies for these conventional cells have reached 15% [10], but higher efficiencies have been achieved for various perovskite compounds and device structures [11–17]. Conversion efficiencies above 20% have been reported [18–32].

The photovoltaic properties of perovskite-based solar cells strongly depend on the composition and crystal structure of the perovskite compound [33–35]. The basic crystal structure of $CH_3NH_3PbI_3$ shown in Figure 1(a) [34]. Introducing metal atoms such as tin (Sn) [36–40], antimony (Sb) [41–45], copper (Cu) [46–51], arsenic (As) [52–54], germanium (Ge) [55–57], zinc (Zn) [57, 58], manganese (Mn) [59, 60], yttrium (Y) [60], strontium (Sr) [61], indium (In) [56], cobalt (Co) [31] or thallium (Tl) [56] at lead (Pb) sites has been performed. The optical absorption range of perovskite compounds has been extended by Sn or Tl doping [36, 37, 56].

Introducing cesium [22, 62–65], rubidium [23, 66, 67], potassium (K) [26, 68], sodium (Na) [68], formamidinium ($NH=CHNH_3$, FA) [20, 66, 69], ethylammonium ($CH_3CH_2NH_3$, EA) [70] or guanidinium (CH_6N_3, GA) [71] at methylammonium (CH_3NH_3, MA) sites can also improve conversion efficiencies. Studies on doping with bromine (Br) [20, 72, 73] or chlorine (Cl) [74–77] at iodine (I) sites of perovskite crystals have been reported.

Doping with Cl reportedly increases the diffusion length, which improves the conversion efficiency [74, 75]. Various elemental and molecular dopants at Pb, I, and/or MA sites reportedly affect the photovoltaic properties and microstructures of perovskite-based solar cells [34].

Photovoltaic properties also strongly depend on the morphology of thin films, as shown in Figure 1(b) [78], and the morphology could be controlled by additives such as poly(methyl methacrylate) [59, 79, 80], phthalocyanines [81, 82] or polysilanes [83–86]. A large interfacial area between the perovskite layers and TiO_2 electron transport layers can increase carrier separation, which increases the short circuit current density. Smooth and homogeneous surfaces, interfaces, and grain boundary structures of perovskite layers are expected to result in improved open circuit voltages and fill factors [87, 88]. Improvement of hole transport layers [89, 90] and electron transport layers [91–94] are also important for the cell. In addition, enlargement of the cell area is especially mandatory to enable the use of such perovskite devices as actual commercial solar cell panels [95, 96].

An energy level diagram of $TiO_2/CH_3NH_3PbI_3$(Cl) photovoltaic cells is shown in Figure 1(c) [78]. The electronic charge generation is caused by light irradiation from the FTO substrate side. The TiO_2 layer receives the electrons from the $CH_3NH_3PbI_3$(Cl) crystal, and the electrons are transported to the FTO. The holes are transported to an Au electrode through spiro-OMeTAD. The $CH_3NH_3PbI_3$ crystals have perovskite structures with a cubic system (Pm3m) as shown in Figure 1(a), and provide structural transitions between a tetragonal to a cubic system at ~ 330 K [97–102]. The unit cell volume of the cubic system is bigger compared with that of the tetragonal system, which would be due to both thermal expansion of the unit cell and atomic disordering of iodine in the cubic structure [78, 99]. The Rietveld refinement [45, 103] and high-resolution electron microscopy structure analysis [104, 105] were also useful methods to study the perovskite structures, and the structural stabilities could be estimated from the tolerance factors [106–108] and first-principles calculation [109–112]. In the present article, the effects of adding NH_4Cl, PbI_2 or $PbCl_2$ to perovskite $CH_3NH_3PbI_3$(Cl) photovoltaic devices were described. NH_4Cl has surfactant properties, so is expected to promote a homogeneous morphology [113–115]. The doped Cl is expected to increase the carrier diffusion length in the perovskite crystals [74, 75]. The Cl-doped perovskite crystals using PbCl$_2$ are denoted as CH$_3$NH$_3$PbI$_3$(Cl) in the present study. The effects of a hot air blow method were also described. The effects of NH$_4$Cl, PbI$_2$ or PbCl$_2$ addition on the formation of perovskite compounds for photovoltaic cells were investigated by light-induced current density–voltage (*J–V*) measurements, incident photon-to current conversion efficiency (IPCE), X-ray diffraction (XRD), optical microscopy (OM), scanning electron microscopy (SEM), and energy dispersive X-ray spectroscopy (EDS).

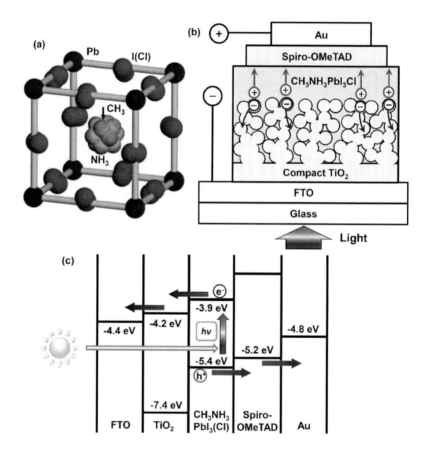

Figure 1. (a) Structure model, device structure, and energy level diagram of CH$_3$NH$_3$PbI$_3$(Cl) cells.

2. MATERIALS AND METHODS

All reagents and solvents were used as received. For the compact TiO$_2$ layer, 0.15 M and 0.30 M TiO$_2$ precursor solutions were prepared from titanium diisopropoxide bis(acetylacetonate) (Sigma-Aldrich, 0.055 mL and 0.11 mL, respectively) with 1- butanol (1 mL). For the mesoporous TiO$_2$ layer, TiO$_2$ paste was prepared using TiO$_2$ powder (Nippon Aerosil, P-25) with poly(ethylene glycol) (Nacalai Tesque, PEG #20000) in ultrapure water. The resulting dispersion was mixed with acetylacetone (Wako Pure Chemical Industries, 10 mL) and Triton X-100 (Sigma-Aldrich, 5 mL), stirred for 30 min, and then left to stand for 12 h to allow bubbles in the mixture to disperse.

For the perovskite layer, solutions containing CH$_3$NH$_3$I (Showa Chemical Co., Ltd.) and PbI$_2$ or PbCl$_2$ were prepared, and details are described in each section.

For the hole transport layer, two solutions of 2,2',7,7'-tetrakis [N,N-di(p- methoxyphenyl)amino]-9,9'-spirobifluorene (spiro-OMeTAD, Sigma-Aldrich, 36.1 mg) in chlorobenzene (Wako Pure Chemical Industries, 0.5 mL) and lithium bis(tri-fluoromethylsulfonyl)imide (Li-TFSI, Tokyo Chemical Industry, 260 mg) in acetonitrile (Nacalai Tesque, 0.5 mL) were prepared and stirred for 12 h. The former spiro-OMeTAD solution in chlorobenzene containing 4-tert- butylpyridine (Aldrich, 14.4 mL) was mixed with the latter Li-TFSI solution (8.8 mL), and the resulting solution was stirred for 30 min at 70°C.

Figure 2. shows a schematic illustration of the process used to fabricate the TiO$_2$/CH$_3$NH$_3$PbI$_3$(Cl) photovoltaic cells. Details of the basic fabrication process have been described previously [78, 116, 117], with the exception of details of the air blow procedure [118].

F-doped tin oxide (FTO) substrates were ultrasonically cleaned with acetone and methanol, and then dried under nitrogen gas. The 0.15 M TiO$_2$ precursor solution was spin-coated on the FTO substrate at 3000 rpm for 30 s, and the coated substrate was then heated to 125°C for 5 min in air to form a TiO$_x$ layer. The 0.30 M TiO$_2$ precursor solution was spin-coated onto the TiO$_x$ layer at 3000 rpm for 30 s, and the coated substrate was then heated at

125°C for 5 min. This process of coating with 0.30 M solution was performed a second time, and the resulting FTO substrate was then annealed at 500°C for 30 min to form a compact TiO_2 layer. For the mesoporous TiO_2 layer, the TiO_2 paste was spin-coated onto the substrate at 5000 rpm for 30 s. The substrate was then annealed at 120°C for 5 min, and then at 500°C for 30 min, to form a mesoporous TiO_2 layer.

Then, a solution containing $CH_3NH_3PbI_3(Cl)$ was introduced into the TiO_2 mesopores of the above coated substrate by the spin-coating at 2000 rpm for 60 s. The resulting substrate was annealed at 100 ~ 140°C for 10 min to form the perovskite layer [115]. Details of the perovskite formation are described in each section.

A hole transport layer was then prepared by spin-coating onto the perovskite layer. All procedures for preparing thin films were performed in ambient air. Finally, a gold (Au) thin film was evaporated onto the hole transport layer, as the top metal electrode. The layered structure of the solar cell was denoted $FTO/TiO_2/CH_3NH_3PbI_3(Cl)/spiro-OMeTAD/Au$, as shown in Figure 2.

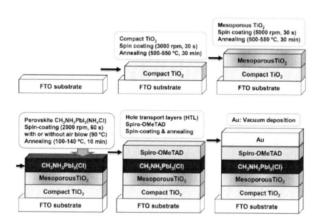

Figure 2. Schematic illustration of the process used to fabricate the $CH_3NH_3PbI_3(Cl)$ photovoltaic cells.

The J–V characteristics of the photovoltaic cells were measured under illumination at 100 mW cm^{-2}, using an AM 1.5 solar simulator (San-ei Electric, XES-301S). J–V measurements were performed using a source measure unit (Keysight, B2901A Precision SMU). The scan rate and sampling time were 0.08 V s^{-1} and 1 ms, respectively. Four cells were tested for each cell composition. The solar cells were illuminated through the sides of the FTO substrates, and the illuminated area was 0.090 cm^2.

IPCE of the cells were also measured (Enli Technology, QE-R). The microstructures of the cells were investigated using an X-ray diffractometer (Bruker, D2 PHASER), a transmission optical microscope (Nikon, Eclipse E600), and a scanning electron microscope (Jeol, JSM-6010PLUS/LA) equipped with EDS.

3. NH$_4$Cl Addition to CH$_3$NH$_3$PbI$_3$ Perovskite

For the preparation of the perovskite compounds, a solution of CH$_3$NH$_3$I (Showa Chemical Co., Ltd., 98.8 mg), PbI$_2$ (Sigma-Aldrich, 289.3 mg), and NH$_4$Cl (Wako Pure Chemicals Industries, Ltd.,) was prepared with a desired mole ratio in mixture of *N,N*-dimethylformamide (DMF, Nacalai Tesque, 225 µL) and γ-butyrolactone (Nacalai Tesque, 275 µL) at 60°C. Addition of the DMF to γ-butyrolactone is expected to improve the photovoltaic properties. The preparation compositions of the TiO$_2$/CH$_3$NH$_3$PbI$_3$(Cl) cells with a NH$_4$Cl additive are listed in Table 1 [115].

Table 1. Preparation composition of TiO$_2$/CH$_3$NH$_3$PbI$_3$(Cl) cells with NH$_4$Cl additive

Preparation composition	Amount of NH$_4$Cl additive (mg)
CH$_3$NH$_3$PbI$_3$	0
CH$_3$NH$_3$PbI$_3$Cl$_{0.15}$	5
CH$_3$NH$_3$PbI$_3$Cl$_{0.45}$	15
CH$_3$NH$_3$PbI$_3$Cl$_{0.75}$	25
CH$_3$NH$_3$PbI$_3$Cl$_{1.05}$	35
CH$_3$NH$_3$PbI$_3$Cl$_{1.35}$	45

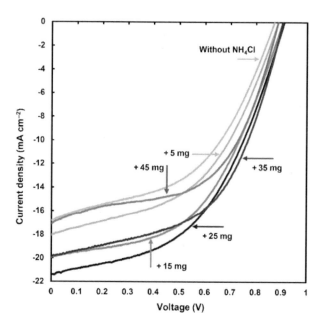

Figure 3. (a) *J–V* characteristic of $CH_3NH_3PbI_3(Cl)$ photovoltaic cells added with NH_4Cl.

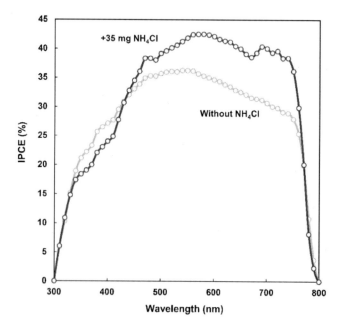

Figure 4. IPCE spectra of $CH_3NH_3PbI_3(Cl)$ cells.

Table 2. Measured photovoltaic parameters (top and average) of TiO$_2$/CH$_3$NH$_3$PbI$_3$Cl$_x$ cells

NH$_4$Cl (mg)	Jsc (mA cm^{-2})	Voc (V)	FF	η (%)	η_{ave} (%)
0	16.8	0.870	0.482	7.05	6.66
5	18.0	0.886	0.488	7.78	6.54
15	20.0	0.903	0.504	9.10	8.91
25	21.4	0.909	0.492	9.58	9.22
30	19.6	0.894	0.540	9.46	9.16
35	19.9	0.903	0.536	9.63	9.10
40	19.4	0.863	0.516	8.64	8.40
45	17.1	0.885	0.539	8.16	8.08

The J-V characteristics of the TiO$_2$/CH$_3$NH$_3$PbI$_3$(Cl)/spiro-OMeTAD photovoltaic cells under illumination are shown in Figure 3, which indicates an effect NH$_4$Cl addition to the CH$_3$NH$_3$PbI$_3$. The measured photovoltaic parameters of the TiO$_2$/CH$_3$NH$_3$PbI$_3$(Cl) cells are summarized in Table 2 [115]. The CH$_3$NH$_3$PbI$_3$ cell provided the highest power conversion efficiency (η) of 7.05%, and the average efficiency (η_{ave}) of three electrodes on the cells is 6.66%, as listed in Table 2. A short-circuit current density (J_{SC}) increased up to 21.4 mA cm^{-2} by an addition of 25 mg NH$_4$Cl, which would indicate an increase of carrier diffusion length. The highest efficiency was obtained for a cell added with 35 mg NH$_4$Cl, which provided the η_{max} of 9.63%, a fill factor (FF) of 0.536, a J_{SC} of 19.9 mA cm^{-2}, and an open-circuit voltage (V$_{oc}$) of 0.903 V. The highest average-efficiency was obtained for a cell added with 25 mg NH$_4$Cl, which provided an η_{ave} of 9.22%.

IPCE spectra of the CH$_3$NH$_3$PbI$_3$ and CH$_3$NH$_3$PbI$_3$(Cl) cells added with 30 mg NH$_4$Cl are shown in Figure 3. The CH$_3$NH$_3$PbI$_3$(Cl) device shows photoconversion efficiencies between 300 nm and 800 nm, which corresponds to an energy gap of 1.55 eV for the CH$_3$NH$_3$PbI$_3$. The IPCE was improved in the range of 450 – 750 nm by the addition of NH$_4$Cl to the CH$_3$NH$_3$PbI$_3$. In the present work, the energy gaps of the CH$_3$NH$_3$PbI$_3$(Cl) phase were almost constant even by the Cl-doping, which agreed well with the constant values of the open-circuit voltages.

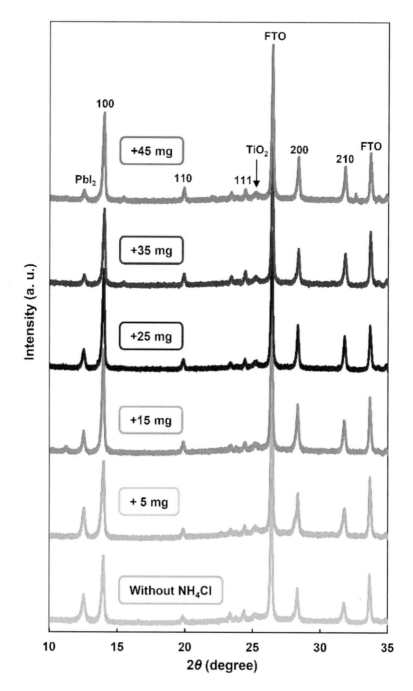

Figure 5. XRD patterns of $CH_3NH_3PbI_3(Cl)$ cells.

Effects of NH₄Cl or PbI₂ Additions to CH₃NH₃PbI₃ ... 309

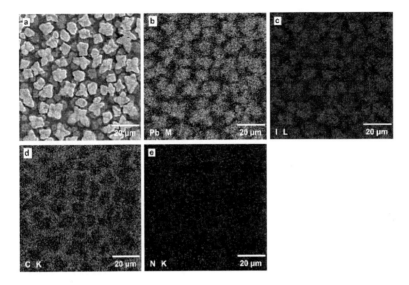

Figure 6. (a) SEM image of CH₃NH₃PbI₃. Elemental mapping images of (b) Pb M line, (c) I L line, (d) C K line, and (e) N K line.

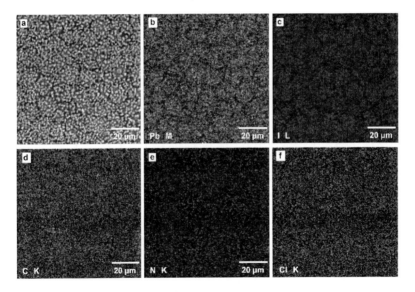

Figure 7. (a) SEM image of CH₃NH₃PbI₃(Cl) cell with NH₄Cl additive of 5 mg. Elemental mapping images of (b) Pb M line, (c) I L line, (d) C K line, (e) N K line, and (f) Cl L line.

XRD patterns of FTO/TiO$_2$/CH$_3$NH$_3$PbI$_3$(Cl) devices are shown in Figure 4. The diffraction peaks can be indexed by a cubic crystal system for the CH$_3$NH$_3$PbI$_3$(Cl) thin films. In addition to XRD peaks of the ordinary perovskite structure, broader diffraction peaks due to the PbI$_2$ compound appeared in the CH$_3$NH$_3$PbI$_3$(Cl) film, as shown in Figure 4. However, the NH$_4$Cl addition suppressed the formation of PbI$_2$, and peak intensities of 100 of the perovskite phase increased.

The photovoltaic property of the non-annealed cell had been measured in the previous work [78]. Although the PbI$_2$ compound was formed during annealing at 100°C, the perovskite structures changed from tetragonal into cubic phase, which resulted in the improvement of photovoltaic properties.

The CH$_3$NH$_3$PbI$_3$ crystals have a perovskite structure, and both I ions and CH$_3$NH$_3$ ions are disordered, which results in the disordered cubic structure [99, 100]. For the as-deposited CH$_3$NH$_3$PbI$_3$ thin film, only XRD peaks of CH$_3$NH$_3$PbI$_3$ were observed, and no XRD peak of PbI$_2$ was observed [78]. After annealing at 100°C for 15 min, the unit cell volume decreased and an XRD peak of PbI$_2$ appeared [78], which indicated a partial separation of PbI$_2$ from CH$_3$NH$_3$PbI$_3$. The XRD result of CH$_3$NH$_3$PbI$_3$ in Figure 4 also showed the existence of PbI$_2$ after annealing at 100°C for 15 min. This would indicate a partial separation of PbI$_2$ from CH$_3$NH$_3$PbI$_3$ after annealing, and the formation of PbI$_2$ was suppressed by the NH$_4$Cl addition, which would result in the increase of conversion efficiencies of the devices.

A SEM image of TiO$_2$/CH$_3$NH$_3$PbI$_3$ cell without NH$_4$Cl is shown in Figure 5(a). Perovskite crystals with sizes of 5 – 10 μm are observed at the surface of the mesoporous TiO$_2$, and the crystals have a star-like shape. Elemental mapping images of Pb, I, C, and N by SEM-EDS are shown in Figure 5(b–e), respectively. The elemental mapping images indicate the particles observed in Figure 5(a) correspond to the CH$_3$NH$_3$PbI$_3$ compound. The composition ratios of elements Pb, I, and C:N was calculated from the EDS spectrum using background correction by normalizing the spectrum peaks, as listed in Table 3. This result indicates that I might be deficient from the starting composition of CH$_3$NH$_3$PbI$_3$, and the deficient I might increase the hole concentration. In addition, carbon atoms are dispersed in the matrix.

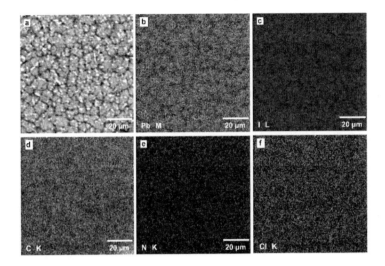

Figure 8. (a) SEM image of CH$_3$NH$_3$PbI$_3$(Cl) cell with NH$_4$Cl additive of 25 mg. Elemental mapping images of (b) Pb M line, (c) I L line, (d) C K line, (e) N K line, and (f) Cl L line.

Table 3. Measured compositions of TiO$_2$/CH$_3$NH$_3$PbI$_3$(Cl) cells

NH$_4$Cl (mg)	Pb (at.%)	I (at.%)	Cl (at.%)	C : N (at.%)
0	24.9	75.1	0.0	57.1 : 42.9
5	24.0	71.2	4.8	52.0 : 48.0
15	23.1	71.8	5.1	54.7 : 45.3
25	22.9	71.7	5.4	54.9 : 45.1
35	24.7	67.4	7.9	54.7 : 45.3
45	24.1	67.4	8.5	54.2 : 45.8

Figure 6(a) is a SEM image of TiO$_2$/CH$_3$NH$_3$PbI$_3$(Cl) cell with 5 mg NH$_4$Cl additive. The surface morphology was drastically changed by the addition of NH$_4$Cl to the CH$_3$NH$_3$PbI$_3$. The particle sizes are a few μm, and the crystals have a round shape. The composition ratios of metal elements and C:N were calculated from the EDS spectrum, as listed in Table 3, which indicates that I might be deficient from the starting composition of CH$_3$NH$_3$PbI$_3$Cl0.15. On the other hands, Cl element would be appropriately doped in the $CH_3NH_3PbI_3$, as listed in Table 3. In Figure 6(d), carbon atoms seems to be dispersed homogeneously.

Figure 7(a) is a SEM image of $CH_3NH_3PbI_3$(Cl) cell added with 25 mg NH_4Cl. By adding 25 mg NH_4Cl to the $CH_3NH_3PbI_3$, the surface morphology was extremely changed, and no special crystal shape is observed, which would be due to an effect of NH_4Cl addition. Elemental mapping images of Pb M line, I L line, C K line, N K line, and Cl K line are shown in Figure 7(b–f), respectively, and the images indicate the perovskite $CH_3NH_3PbI_3$ phase is dispersed homogeneously on the photovoltaic device.

These homogeneous surface structures would improve the photovoltaic properties, in addition to the doping effect of Cl at the I sites, as listed in Table 3. From the SEM-EDS results, site occupancies of I atom would also be less than 1, which might be due to the partial separation of PbI_2 from the $CH_3NH_3PbI_3$ phase. The composition of NH_3 also seem to be deficient compared with that of the CH_3, and it became almost constant by the NH_4Cl addition, which would indicate the NH_3 are doped at the vacant NH_3 positions. Three assumed mechanisms could be considered for the increase of the photoconversion efficiencies. The first mechanism is as follows: when a small amount of Cl was doped in the $CH_3NH_3PbI_3$ phase, the diffusion length of excitons would be lengthened by the doped Cl atoms [74, 75], which would result in the increase of the J_{SC} values. The second is as follows: the homogeneous surface and interfacial structures formed by adding NH_4Cl to the $CH_3NH_3PbI_3$, which improved the photovoltaic properties, especially the FF values. The third is as follows: the deficient NH_3 positions are filled with NH_3 from NH_4Cl, which would lead to the stable perovskite structure by suppression of PbI_2 separation. Further studies are mandatory for precise structure determination of the devices. The photovoltaic performance of the cell containing more than 35 mg was deteriorated, as listed in Table 2. This would be due to an increase of series resistance by the remained undoped Cl element in the perovskite phase and undetermined compounds, which were observed as small XRD peaks around 22 and 32.5° for NH_4Cl 45 mg added sample in Figure 5. As the NH_4Cl addition was effective in the present work, addition of KCl or $HC(NH_2)_2Cl$ could also be effective for the perovskite solar cells [119, 120].

4. PbCl₂ AND NH₄Cl ADDITION TO CH₃NH₃PbI₃

The widely used solution-based synthesis of the perovskite is by mixing an appropriate ratio of CH$_3$NH$_3$I and PbI$_2$. In addition, the CH$_3$NH$_3$PbI$_3$ perovskite can be formed with CH$_3$NH$_3$I and PbCl$_2$ as mixed precursors in a 3:1 molar ratio in DMF, followed by thermal annealing at around 100°C [121–123]. The reaction mechanism proposed for the CH$_3$NH$_3$PbI$_{3-x}$Cl$_x$ perovskite is as follows: 3CH$_3$NH$_3$I + PbCl$_2$ → CH$_3$NH$_3$PbI$_3$ + 2CH$_3$NH$_3$Cl (↑). During the reaction, 2CH$_3$NH$_3$Cl is generated as a byproduct. However, since there is still a small amount of Cl present in the final perovskite product, CH$_3$NH$_3$PbI$_{3-x}$Cl$_x$ can be expressed as CH$_3$NH$_3$PbI$_3$ or CH$_3$NH$_3$PbI$_3$(Cl). It was reported that the CH$_3$NH$_3$PbI$_{3-x}$Cl$_x$ contains both I and Cl.

For the perovskite layer, a solution containing CH$_3$NH$_3$I (Showa Chemical Co., Ltd., 190.7 mg) and PbCl$_2$ (Sigma-Aldrich, 111.2 mg) was prepared with a molar ratio of 3:1 in DMF (Nacalai Tesque, 0.5 mL) [124]. This solution was then stirred at 60°C for 24 h. Then, a solution containing CH$_3$NH$_3$PbI$_3$(Cl) was introduced into the TiO$_2$ mesopores of the above coated substrate by the spin-coating method [115–117]. For the spin-coating procedure, air blowing at a rate of 6 m s^{-1} and 3300 cm^3 s^{-1} was applied perpendicular to the substrate at a temperature of 90°C for 60 s. The resulting substrate was annealed at 140°C for 10 min to form the perovskite layer [125].

Figure 9 shows the *J–V* characteristics of the TiO$_2$/CH$_3$NH$_3$PbI$_3$(Cl)/ spiro- OMeTAD photovoltaic cells under illumination, in which the effects of NH$_4$Cl addition are evident. Forward and reverse scans are indicated by the dotted and solid lines, respectively. The measured photovoltaic parameters of the TiO$_2$/CH$_3$NH$_3$PbI$_3$(Cl) cells are summarized in Table 1, and reverse values are listed. Small hysteresis between the forward and reverse scans is observed for the *J–V* characteristics in Figure 9, and the degree of hysteresis decreases with NH$_4$Cl addition. A small amount of carriers may have been generated and transported into the TiO$_2$ layer during the *J–V* measurements under the light irradiation and subsequent current flow. In this case, the electrical resistance would decrease and the photocurrent would increase, resulting in the hysteresis. The perovskite would also exhibit hysteresis properties in its *J–V* characteristics [126].

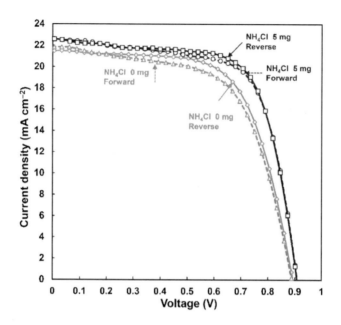

Figure 9. (a) *J–V* characteristic of $CH_3NH_3PbI_3(Cl)$ photovoltaic cells with and without NH_4Cl.

Table 4. Measured photovoltaic parameters of the $CH_3NH_3PbI_3(Cl)$ solar cells

NH_4Cl (mg)	J_{SC} (mA cm^{-2})	V_{OC} (V)	FF	η (%)	η_{ave} (%)	R_S (Ω cm^2)
0	21.5	0.893	0.647	12.41	12.37	6.56
1	22.2	0.912	0.651	13.19	12.89	6.18
3	22.5	0.917	0.651	13.40	12.13	6.06
5	22.6	0.908	0.685	14.03	13.74	4.70

The $CH_3NH_3PbI_3(Cl)$ cell without NH_4Cl provides a power conversion efficiency of 12.41% for the reverse scan [125]. The η_{ave} of four electrodes in identical cells is 12.37%, as shown in Table 4. The short-circuit current density, open-circuit voltage and fill factor are all higher for the NH_4Cl-added $CH_3NH_3PbI_3(Cl)$ cells, which results in the increases in their η values.

Adding 5 mg of NH₄Cl yields the highest efficiency CH₃NH₃PbI₃(Cl) cell, with a η of 14.03%, FF of 0.685, J_{SC} of 22.6 mA cm^{-2}, and V_{OC} of 0.908 V. The highest η_{ave} of 13.74% is obtained for this cell. The series resistance (R_S) also decreases with NH₄Cl addition (Table 4), leading to an increase in the J_{SC}. After standing for 2 weeks in ambient air, the decreases in efficiencies of cells containing NH₄Cl are small. An η of 13.19% is still obtained, as shown in Table 5.

Table 5. Measured photovoltaic parameters of the CH₃NH₃PbI₃(Cl) solar cells after 2 weeks

NH₄Cl (mg)	J_{SC} (mA cm^{-2})	V_{OC} (V)	FF	η (%)	R_S (Ω cm²)
0	20.5	0.870	0.647	11.51	4.61
1	21.6	0.892	0.640	12.34	4.51
3	22.1	0.893	0.668	13.19	4.63
5	21.5	0.893	0.643	12.33	4.29

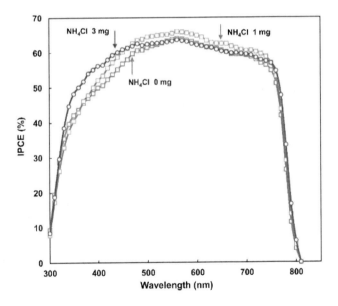

Figure 10. IPCE spectra of CH₃NH₃PbI₃(Cl) cells.

IPCE spectra of the $CH_3NH_3PbI_3(Cl)$ devices were shown in Figure 10. The $CH_3NH_3PbI_3(Cl)$ cells show photoconversion efficiencies between 320 and 810 nm, which corresponds to an energy gap of 1.53 eV for the $CH_3NH_3PbI_3$. The IPCE was improved in the range of 400 – 700 nm by adding NH_4Cl, which lead to increase of J_{SC} values. Although the J_{SC} values should agree with the integrated values of IPCE, the J_{SC} values in the present work are slightly lower than the integrated values of IPCE. In order to measure the IPCE of the perovskite solar cells, DC measurements mode is better. However, the lowest frequency of the lock-in-amplifier in the present work is 4 Hz (QE-R, Enli Technology), and the lower IPCE values tended to be measured compared with the actual IPCE values. Therefore, only the wavelength region can be evaluated for the IPCE data in the present work.

OM images of the $CH_3NH_3PbI_3(Cl)$ cells are shown in Figure 11. Microparticles with sizes of 5 – 10 μm are observed for the cell prepared without NH_4Cl, as shown in Figure 11(a). Adding NH_4Cl to the $CH_3NH_3PbI_3(Cl)$ decreases the particle size, as shown in Figure 11(b)–(d). In addition, networking structures with sizes of ~ 10 μm between microparticles are observed, especially in Figure 11(d). These networking microstructures could potentially improve photovoltaic properties.

A SEM image of the $CH_3NH_3PbI_3(Cl)$ cell without NH_4Cl is shown in Figure 12(a). Microparticles with sizes of 5 – 10 μm are observed on the surface of the mesoporous TiO_2, which correspond to those in Figure 11(a). The particles appear to have crystal facets. EDS elemental mapping images of the Pb M line, I L line, Cl K line, C K line, and N K line are shown in Figures 12(b)–12(f), respectively. These elemental mapping images indicate that the particles observed in Figure 12(a) correspond to the $CH_3NH_3PbI_3$ compound. The Pb and I compositions and C:N ratio were calculated from the raw EDS spectra (Figure 14) using background correction by normalizing the peaks, and are listed in Table 6. Although the EDS values contain some errors, these results seem to indicate that the composition of $CH_3NH_3PbI_3$ may be I deficient. In addition, EDS indicates that C is dispersed throughout the matrix.

Effects of NH₄Cl or PbI₂ Additions to CH₃NH₃PbI₃ ... 317

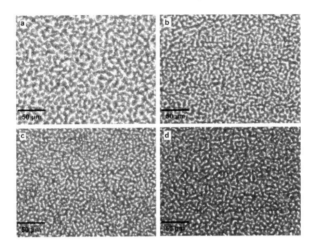

Figure 11. OM images of CH₃NH₃PbI₃(Cl) cells containing (a) 0 mg, (b) 1 mg, (c) 3 mg, and (d) 5 mg of NH₄Cl.

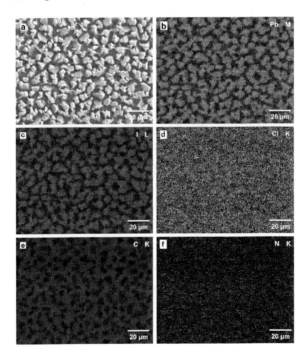

Figure 12. (a) SEM image of the CH₃NH₃PbI₃(Cl) cell without NH₄Cl, and corresponding elemental mapping images of the (b) Pb M line, (c) I L line, (d) Cl K line, (e) C K line, and (f) N K line.

Figure 13(a) shows a SEM image of the $CH_3NH_3PbI_3(Cl)$ cell containing 5 mg of NH_4Cl. The surface morphology changes upon adding NH_4Cl to the $CH_3NH_3PbI_3$. The average particle sizes is ~ 5 μm, and crystals are connected to each other with special facets. These networking surface structures could potentially improve the surface coverage and therefore the photovoltaic properties. Figures 13(b)–(f) show EDS elemental mapping images of Pb, I, Cl, C, and N, respectively. The elemental compositions and C:N ratios were calculated from the EDS spectra in Figure 14, and are also listed in Table 6. These data indicate that the $CH_3NH_3PbI_3$ starting composition may be I deficient. Cl would be doped into the $CH_3NH_3PbI_3$, as shown in Table 6. In Figure 13(e), C appears to be dispersed throughout the matrix. The EDS results indicate that the I site occupancy would be < 1, and that this I deficiency may increase the hole concentration. The networking surface structures are expected to affect the photovoltaic properties, in addition to the doping effect of Cl at I sites. The CH_3:NH_3 ratio in the $CH_3NH_3PbI_3(Cl)$ phase prepared by air blowing is almost 6:4. This is consistent with excess CH_3 compared with NH_3, and is caused by the air blowing procedure as shown in Table 6. Although the EDS values contain some errors, these results seem to indicate that the composition of $CH_3NH_3PbI_3$ may be I deficient. In addition, EDS indicates that C is dispersed throughout the matrix.

Figure 15(a)–(d) shows XRD patterns of the $FTO/TiO_2/CH_3NH_3PbI_3(Cl)$/spiro-OMeTAD/Au cells. The diffraction peaks can be indexed by a cubic crystal system for $CH_3NH_3PbI_3(Cl)$ perovskite thin films. The XRD pattern of a $CH_3NH_3PbI_3(Cl)$ cell prepared without air blowing or $PbCl_2$ is shown in Figure 15(e). The diffraction peaks of FTO and TiO_2 arise from the FTO substrate and TiO_2 mesoporous layer, respectively. For ordinary $CH_3NH_3PbI_3$ cells prepared at 100°C without air blowing, XRD peaks of PbI_2 have been reported [78] similarly to those in Figure 15(e). The intensities of the 100 and 200 peaks of the perovskite phase increase by more than 100 times with air blowing and NH_4Cl, as observed in Figure 15(b)–(d). The 100 and 200 reflections are sufficiently intense that the diffraction peaks of FTO and TiO_2 are not readily apparent.

Effects of NH₄Cl or PbI₂ Additions to CH₃NH₃PbI₃ ... 319

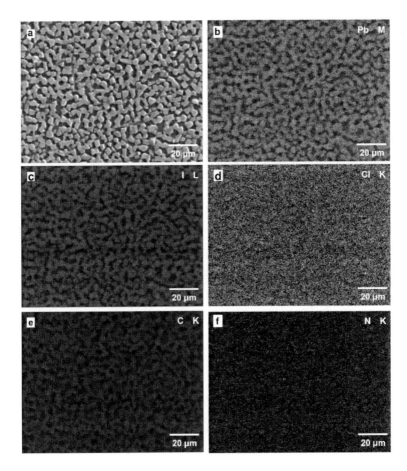

Figure 13. (a) SEM image of the CH₃NH₃PbI₃(Cl) cell containing 5 mg of NH₄Cl, and corresponding elemental mapping images of the (b) Pb M line, (c) I L line, (d) Cl K line, (e) C K line, and (f) N K line.

Table 6. Compositions of the solar cells as measured by EDS

NH₄Cl (mg)	Pb (at.%)	I (at.%)	Cl (at.%)	C : N (at.%)
0	24.3	71.7	4.0	57.1 : 42.9
1	24.0	71.4	4.7	52.0 : 48.0
3	24.4	71.1	4.6	54.7 : 45.3
5	24.1	70.2	5.7	54.9 : 45.1
5*	24.0	71.2	4.8	52.0 : 48.0

* No air blowing or PbCl₂.

The ratios of the 100 diffraction intensities (I_{100}) to the 210 diffraction intensities (I_{200}) of the perovskite crystals were calculated as I_{100}/I_{210}, from the XRD data in Figure 15 and Figure 16. The results are summarized in Table 7. If the $CH_3NH_3PbI_3$ cubic perovskite particles are randomly oriented, then the I_{100}/I_{210} value should be 1.81 [125]. For the cell prepared using air blowing and without NH_4Cl, the I_{100}/I_{210} is 61. This indicates that the (100) planes of the perovskite particles are preferentiality oriented parallel to the FTO substrate. Adding NH_4Cl to the cell results in the I_{100}/I_{210} increasing to 3600, which is 2000 times higher than the I_{100}/I_{210} of randomly oriented perovskite crystals. The cell prepared with 5 mg of NH_4Cl without air blowing or $PbCl_2$ has an I_{100}/I_{210} of 2.8. This indicates that most of the perovskite particles are randomly oriented.

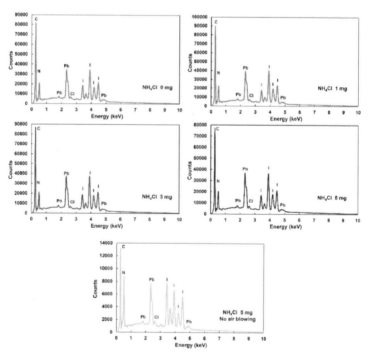

Figure 14. Raw EDX data of the present cells.

The ratios of the I_{100} intensities to the FTO substrate intensities (I_{FTO}) were also calculated for the perovskite crystals, and are also shown in Table 7. Using air blowing and NH_4Cl increases the I_{100}/I_{FTO}. These results indicate that preferential (100) crystal orientation occurs in films prepared using air blowing with NH₄Cl. By using a least squares method, the lattice constants were determined to be 6.274, 6.276, 6.276, 6.276, and 6.275 Å for the cells with NH₄Cl 0 mg, 1 mg, 3 mg, 5 mg, and 5 mg without air blow, respectively, and the lattice constants are almost constant for the cell.

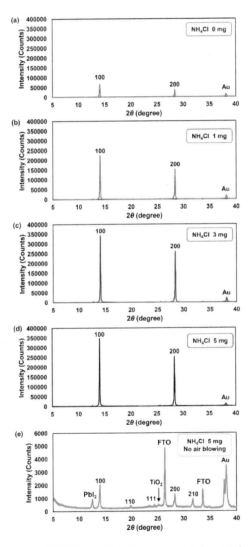

Figure 15. XRD patterns of CH₃NH₃PbI₃(Cl) cells containing (a) 0 mg, (b) 1 mg, (c) 3 mg, and (d) 5 mg of NH₄Cl, and (e) XRD pattern of the CH₃NH₃PbI₃(Cl) cell containing 5 mg of NH₄Cl without air blowing.

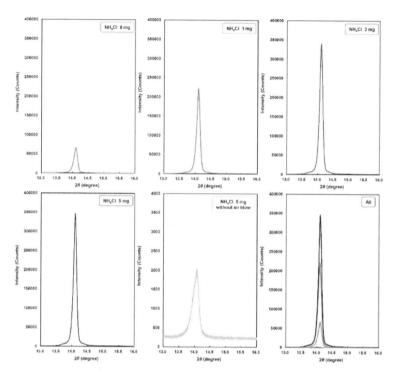

Figure 16. Enlarged XRD peaks of 100 for the present cells.

Two mechanisms are considered for the formation of the (100)-oriented $CH_3NH_3PbI_3(Cl)$ perovskite crystals. The first mechanism is crystal growth being promoted by air blowing during spin-coating of the perovskite layers. When the $CH_3NH_3PbI_3(Cl)$ solution crystalizes into perovskite particles, fast annealing with air flow accelerates the crystallization of $CH_3NH_3PbI_3(Cl)$. The (100) planes of the perovskite crystals have low surface tension, which results in the growth of (100)-oriented crystals. When highly-oriented crystals are formed, the amount of high angle grain boundaries decreases, which decreases the R_S and increases the V_{oc}. The second mechanism is the formation of a homogeneous morphology during annealing, as a result of the NH_4Cl surfactant. Networking structures are formed in cells containing NH_4Cl, as observed in Figure 11(d). This improves the surface coverage and carrier transport, which increases the FF and J_{SC}. The improved conversion efficiency can be described by these two mechanisms.

Cl doping at I sites also promotes the photovoltaic properties of the cells. Excess 2(CH$_3$NH$_3$Cl) could be vaporized from the starting composition of 3(CH$_3$NH$_3$I) + PbCl$_2$ [121–123]. A small amount of residual Cl is doped into the CH$_3$NH$_3$PbI$_3$ phase, as detected by EDS. The increases the exciton diffusion length [74, 75], which increases the J_{SC}. Further investigation is required regarding the specific nanostructures in the cell.

Table 7. Ratios of 100 diffraction intensities (I_{100}) to 210 diffraction intensities (I_{200}) and FTO substrate diffraction intensities (I_{FTO}), for the perovskite crystals

NH$_4$Cl (mg)	I_{100}/I_{210}	I_{100}/I_{FTO}
Calculation*	1.8	-
0	69	51
1	510	170
3	3400	280
5	3600	270
5**	2.8	0.40

* Calculated from randomly oriented cubic CH$_3$NH$_3$PbI$_3$ crystals.
** No air blowing or PbCl$_2$.

5. PbI$_2$ ADDITION TO CH$_3$NH$_3$PbI$_3$(Cl)

The purpose is to investigate the effect of PbI$_2$ addition to the CH$_3$NH$_3$PbI$_{3-x}$Cl$_x$ perovskite precursor solution. A small amount of PbI$_2$ would be effective for improvement of the perovskite solar cells [127–134], even for the present 3CH$_3$NH$_3$I + PbCl$_2$ system.

For the preparation of the perovskite compounds, mixed solutions of CH$_3$NH$_3$I (190.7 mg, Showa Chemical) and PbCl$_2$ (111.2 mg, Sigma-Aldrich) with 5 ~ 15% PbI$_2$ (Sigma-Aldrich) of PbCl$_2$ molar in DMF (Sigma-Aldrich, 500 mL) were prepared and stirred at 60°C. The solution of the perovskite phase was then introduced into the TiO$_2$ mesopores by spin-coating at 2000 rpm for 60 s three times followed by annealing at 140°C for 10 min [125] under 20°C and humidity of ~ 40%.

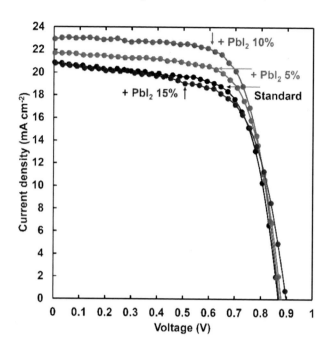

Figure 17. J–V characteristics of the present perovskite solar cells.

Table 8. Measured photovoltaic parameters (average and top) of the perovskite solar cells

Cell	J_{SC} (mA cm^{-2})	V_{OC} (V)	FF	η (%)	η_{ave} (%)
Standard	20.6	0.878	0.682	12.3	12.2
+ PbI$_2$ 5%	21.7	0.879	0.694	13.2	13.0
+ PbI$_2$ 10%	23.0	0.872	0.709	14.2	13.5
+ PbI$_2$ 15%	20.8	0.898	0.644	12.0	11.8

J–V characteristics under illumination recorded in the reverse scan and IPCE spectra of FTO/TiO$_2$/perovskite/spiro-OMeTAD/Au cells are shown in Figures 17 and 18, respectively. Measured photovoltaic parameters of the perovskite solar cells are summarized in Table 8. The perovskite solar cell without excess PbI$_2$ addition provided an average J_{SC} of 20.6 mA cm^{-2}, a V$_{oc}$ of 0.878 V, an FF of 0.682, and an η of 12.3%. The J_{SC} and FF were increased to 23.0 mA cm^{-2} and 0.709 by the PbI$_2$ addition, respectively. The η_{ave} of four electrodes was improved to 13.5%. The reproducibility of the devices was confirmed by fabricating more devices under the same environment. IPCE values were increased by the PbI₂ addition as shown in Figure 18, which lead to the increase of the current density of the cell.

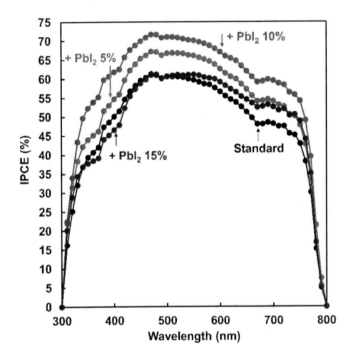

Figure 18. IPCE spectra of the present perovskite solar cells.

XRD patterns of the present perovskite solar cells are shown in Figure 19. Weak diffraction peaks of PbI₂ were observed for the perovskite solar cells, as observed in previously reported works [41, 134]. PbI₂ crystals were formed because Pb^{2+} and I^- exist excessively in the perovskite precursor solution. Intensities of a 100 peak of the perovskite crystals added with 10% PbI₂ was higher than that without excess PbI₂ addition. The measured XRD parameters of the perovskite solar cells are summarised in Table 9. The *d*-values and the crystallite sizes of the perovskite added with 10% PbI₂ slightly increased.

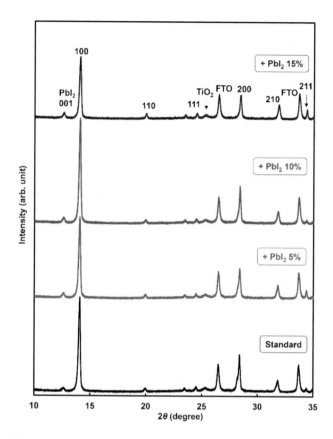

Figure 19. XRD patterns of the present perovskite solar cells.

Table 9. Measured XRD parameters of the perovskite solar cells

Cell	Index	2θ (deg)	d (Å)	FWHM (deg)	Crystallite size (Å)
Standard	Perovskite 100	14.07	6.288	0.194	519
	PbI$_2$ 001	12.65	6.993	0.221	428
+ PbI$_2$ 5%	Perovskite 100	14.07	6.289	0.188	546
	PbI$_2$ 001	12.65	6.993	0.201	490
+ PbI$_2$ 10%	Perovskite 100	14.06	6.292	0.187	555
	PbI$_2$ 001	12.63	7.005	0.232	399
+ PbI$_2$ 15%	Perovskite 100	14.08	6.286	0.204	482
	PbI$_2$ 001	12.65	6.994	0.205	475

Effects of NH₄Cl or PbI₂ Additions to CH₃NH₃PbI₃ ...

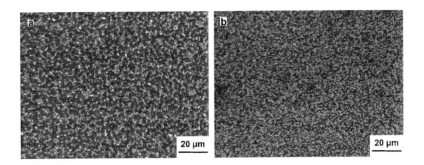

Figure 20. Optical microscopy images of TiO$_2$/CH$_3$NH$_3$PbI$_{3-x}$Cl$_x$-based solar cells for (a) standard and (b) the PbI$_2$ addition.

Optical micrographs of the present perovskite solar cells are shown in Figure 20(a) and (b), respectively. The perovskite crystals for the cell added with PbI$_2$ were smaller than that without the PbI$_2$ addition. However, the density of the perovskite crystals was increased by the PbI$_2$ addition. As a result, the surface coverage of the perovskite crystals was increased by the PbI$_2$ addition, and the grain boundary area for the cell with PbI$_2$ was less than that for the standard cell. SEM images of the present perovskite solar cells without and with excess PbI$_2$ are shown in Figures 21(a) and 22(a), respectively. Elemental mapping images of the Pb M line, I L line, C K line, N K line, and Cl K line obtained by SEM-EDS are shown in Figures 21(b)-21(f) and 22(b)-22(f), respectively. From the SEM image of Figure 21(a), grain sizes of the perovskite crystals for the cell without excess PbI$_2$ are ~ 5 μm, which is consistent with the results of optical microscopy shown in Figure 20(a). The perovskite crystals with different sizes were formed at the surface of the perovskite layer. Elemental mapping images of Figures 21(d) and 21(e) indicated the deviation in the distribution of the Pb and I at the perovskite surface, and there were lots of grain boundaries in the perovskite layer.

The grain sizes of the perovskite for the cell added with PbI$_2$ are ~ 1 μm, as observed in Figure 22(a). From elemental mapping images of Figure 22(d) and 22(e), the elements Pb and I were distributed homogeneously in the perovskite phase, and the CH$_3$NH$_3$PbI$_{3-x}$Cl$_x$ perovskite structure was formed. This resulted in the improvement of carrier generation because of decreasing the grain boundaries of the perovskite crystals. The measured composition of the perovskite solar cells are summarized in Table 10. The present SEM-EDX data did not show the absolute atomic ratios precisely, and the relative composition can be compared for the cells. It was indicated that more Pb was included in the perovskite layer by the PbI$_2$ addition.

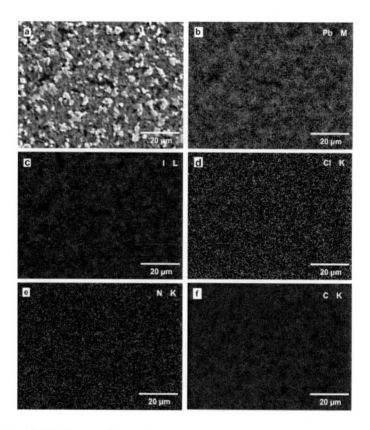

Figure 21. (a) SEM image of the cell without excess PbI$_2$, and corresponding elemental mapping images of the (b) Pb M line, (c) I L line, (d) Cl K line, (e) N K line, and (f) C K line.

Measured photovoltaic parameters of the perovskite solar cells after 11 and 14 days are summarized in Table 11. The solar cell added with excess PbI$_2$ were prepared at 500°C to form the TiO$_2$ layers, and the perovskite precursor solution was spin-coated twice. The result indicated that the stability of the perovskite solar cells was improved by the PbI$_2$ addition.

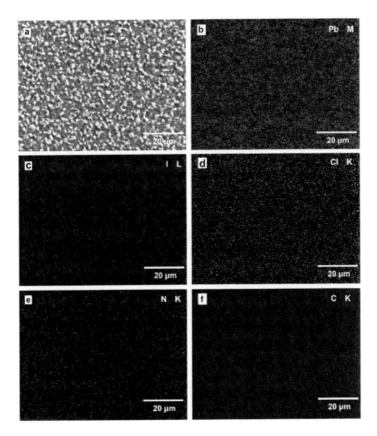

Figure 22. (a) SEM image of the cell with excess PbI$_2$, and corresponding elemental mapping images of the (b) Pb M line, (c) I L line, (d) Cl K line, (e) N K line, and (f) C K line.

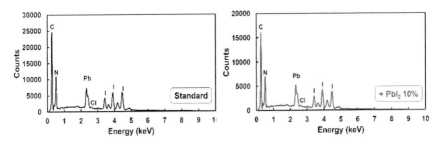

Figure 23. Raw EDS data of the present cells.

Table 10. Compositions of the solar cells as measured by EDS

Cell	Pb (at.%)	I (at.%)	Cl (at.%)	C : N (at.%)
Standard	23.6	71.3	5.0	51.7 : 48.3
+PbI₂ 10%	24.7	70.2	5.0	53.1 : 46.9

Table 11. Measured photovoltaic parameters of the perovskite solar cells after 14 days

Cell	J_{sc} (mA cm^{-2})	V_{oc} (V)	FF	η (%)
Standard	21.7	0.967	0.578	12.1
+ PbI₂ 10%	21.1	0.886	0.613	11.4
After 11 days				
Standard	20.1	0.910	0.436	7.98
+ PbI₂ 10%	19.3	0.918	0.684	12.1
After 14 days				
Standard	18.1	0.821	0.386	5.75
+ PbI₂ 10%	20.2	0.900	0.667	12.1

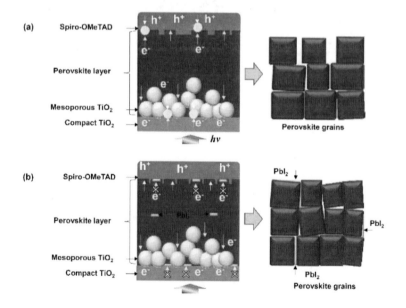

Figure 24. Assumed models of carrier transport of (a) standard cell and (b) the cell added with PbI₂.

Assumed models for carrier transport in the perovskite layers is illustrated in Figure 23. The electronic charge is generated by light irradiation from the FTO substrate side. Separated electrons are transported to the TiO$_2$ layers and separated holes are also transported to the spiro-OMeTAD layer. The recombination between a hole and an electron was suppressed by the PbI$_2$ formation. In addition, the PbI$_2$ is a p-type semiconductor, and the bandgap energies are 2.3 ~ 2.6 eV [134]. Therefore, PbI$_2$ could also work as a hole transport layer in the perovskite phase, and the IPCEs were improved by the PbI$_2$ addition.

CONCLUSION

The effects of NH$_4$Cl addition to perovskite CH$_3$NH$_3$PbI$_3$ precursor solutions on photovoltaic properties were investigated. TiO$_2$/CH$_3$NH$_3$PbI$_3$(Cl)-based photovoltaic devices were fabricated, and the microstructures of the devices were investigated by XRD and SEM-EDS. J–V characteristics and IPCE were improved by the NH$_4$Cl addition to perovskite CH$_3$NH$_3$PbI$_3$ precursor solutions. The structural analysis also indicated the formation of a homogeneous microstructure by NH$_4$Cl addition, which improved the FF values and photoconversion efficiencies. The IPCE spectrum of the CH$_3$NH$_3$PbI$_3$(Cl) cell was also improved by the NH$_4$Cl addition, and showed effective carrier- generation in the range of 300 and 800 nm.

The effects of NH$_4$Cl and PbCl$_2$ addition using an air blow method on CH$_3$NH$_3$PbI$_3$(Cl) perovskite solar cells were also investigated. The J–V characteristics were improved by introducing an appropriate amount of NH$_4$Cl and air blowing. The resulting cell conversion efficiency was 14%. Microstructure analysis indicated that highly (100)-orientated perovskite crystals with dense grains were formed when using NH$_4$Cl and air blowing. The I_{100}/I_{210} ratio of the perovskite crystals was 2000 times higher than that of randomly oriented grains. The (100)-oriented cells were stable during storage for two weeks in ambient air. Formation of the (100)-oriented perovskite crystals was due to the crystal growth accelerated by air blowing during spin-coating. The highly-oriented crystals with decreased grain boundaries could decrease the R_S and increase the V_{oc}. In addition, formation of a homogeneous morphology with networking structures was promoted by the NH$_4$Cl surfactant. Improvement of the surface coverage and carrier transport increased the FF and J_{SC}. As a result, these microstructures improved the conversion efficiency, and this NH$_4$Cl addition combined with an air blow method is an effective method for fabrication of highly crystalline-oriented homogeneous thin films.

By the PbI$_2$ addition, J_{SC} and FF values of the TiO$_2$/CH$_3$NH$_3$PbI$_{3-x}$Cl$_x$-based perovskite solar cells were also improved, and the conversion efficiency was increased to 14.2%. Models of the carrier transport in the perovskite layer were proposed from the results of the microstructural analysis, and they indicated improved carrier transport at the interface by the effect of PbI$_2$ addition. The stability of the cell was improved by forming PbI$_2$ crystals in the perovskite layer, and the reproducibility of the devices was confirmed under the same environment.

In summary, the effects of adding NH$_4$Cl, PbI$_2$ or PbCl$_2$ to perovskite CH$_3$NH$_3$PbI$_3$(Cl) photovoltaic devices were described in the present article. The microstructures of the perovskite compounds were controlled by these additives and hot air blowing, which was found to be effective for the improvement of the perovskite solar cells.

ACKNOWLEDGMENTS

We thank Yuzuru Miyazawa, Atsushi Suzuki and Hiroki Tanaka for providing assistance with experiments. This work was partly supported by Satellite Cluster Program of the Japan Science and Technology.

REFERENCES

[1] Kojima, A.; Teshima, K.; Shirai, Y.; Miyasaka, T. *J. Am. Chem. Soc.*, 2009, 131, 6050–6051.

[2] Im, J. H.; Lee, C. R.; Lee, J. W.; Park, S. W.; Park, N. G. *Nanoscale*, 2011, 3, 4088–4093.
[3] Lee, M. M.; Teuscher, J.; Miyasaka, T.; Murakami, T. N.; Snaith, H. J. *Science*, 2012, 338, 643–647.
[4] Kim, H. S; Lee, C. R.; Im, J. H.; Lee, K. B.; Moehl, T.; Marchioro, A.; Moon, S. J.; Yum, J. H.; Humphry-Baker, R.; Moser, J. E. *Sci. Rep.*, 2012, 2, 591-1–7.
[5] Kojima, A.; Ikegami, M.; Teshima, K.; Miyasaka, T. *Chem. Lett.*, 2012, 41, 397–399.
[6] Chung, I.; Lee, B.; He, J. Q.; Chang, R. P. H.; Kanatzidis, M. G. *Nature*, 2012, 485, 486–489.
[7] Im, J. H.; Chung, J.; Kim, S. J.; Park, N. G. *Nanoscale Res. Lett.*, 2012, 7, 353-1–7.
[8] Grinberg, I.; West, D. V.; Torres, M.; Gou, G.; Stein, D. M.; Wu, L.; Chen, G.; Gallo, E. M.; Akbashev, A.; Davies, P. K. *Nature*, 2013, 503, 509–512.
[9] Stranks, S. D.; Eperon, G. E.; Grancini, G.; Menelaou, C.; Alcocer, M. J. P.; Leijtens, T.; Herz, L. M.; Petrozza, A.; Snaith, H. J. *Science*, 2013, 342, 341–344.
[10] Burschka, J.; Pellet, N.; Moon, S. J.; Humphry-Baker, R.; Gao, P.; Nazeeruddin, MK; Grätzel, M. *Nature*, 2013, 499, 316–320.
[11] Liu, M.; Johnston, M. B.; Snaith, H. J. *Nature*, 2013, 501, 395–398.
[12] Liu, D.; Kelly, T. L. *Nat. Photonics*, 2014, 8, 133–138.
[13] Wang, J. T. W.; Ball, J. M.; Barea, E. M; Abate, A.; Alexander-Webber, J. A.; Huang, J.; Saliba, M.; Mora-Sero, I.; Bisquert, J.; Snaith, H. J.; Nicholas, R. J. *Nano Letters*, 2014, 14, 724–730.
[14] Wojciechowski, K.; Saliba, M.; Leijtens, T.; Abate, A.; Snaith, H. J. *Energy Environ. Sci.*, 2014, 7, 1142–1147.
[15] Zhou, H.; Chen, Q.; Li, G.; Luo, S.; Song, T. B.; Duan, H. S.; Hong, Z.; You, J.; Liu, Y.; Yang, Y. *Science*, 2014, 345, 542–546.
[16] Jeon, N. J.; Noh, J. H.; Yang, W. S.; Kim, Y. C.; Ryu, S.; Seo, J.; Seok, S. I. *Nature*, 2015, 517, 476–480.

[17] Nie, W.; Tsai, H.; Asadpour, R.; Blancon, J. C.; Neukirch, A. J.; Gupta, G.; Crochet, J. J.; Chhowalla, M.; Tretiak, S.; Alam, M. A.; Wang, H. L.; Mohite, A. D. *Science*, 2015, 347, 522–525.

[18] Yang, W. S.; Noh, J. H.; Jeon, N. J.; Kim, Y. C.; Ryu, S.; Seo, J.; Seok, S. I. *Science*, 2015, 348, 1234–1237.

[19] Saliba, M.; Orlandi, S.; Matsui, T.; Aghazada, S.; Cavazzini, M.; Correa- Baena, J. P.; Gao, P.; Scopelliti, R.; Mosconi, E.; Dahmen, K. H.; De Angelis, F.; Abate, A.; Hagfeldt, A.; Pozzi, G.; Graetzel, M.; Nazeeruddin, M. K. *Nat. Energy*, 2016, 1, 15017-1–7.

[20] Bi, D.; Tress, W.; Dar, M. I.; Gao, P.; Luo, J.; Renevier, C.; Schenk, K.; Abate, A.; Giordano, F.; Baena, J. P. C.; Decoppet, J. D.; Zakeeruddin, S. M.; Nazeeruddin, M. K.; Grätzel, M.; Hagfeldt, A. *Sci. Adv.*, 2016, 2, e1501170-1–7.

[21] Bi, D.; Yi, C; Luo, J.; Décoppet, J. D.; Zhang, F.; Zakeeruddin, S. M.; Li, X.; Hagfeldt, A.; Grätzel, M. *Nat. Energy*, 2016, 1, 16142-1–5.

[22] Saliba, M.; Matsui, T.; Seo, J. Y.; Domanski, K.; Correa-Baena, J. P.; Nazeeruddin, M. K.; Zakeeruddin, S. M.; Tress, W.; Abate, A.; Hagfeldtd, A.; Grätzel, M. *Energy Environ. Sci.*, 2016, 9, 1989–1997.

[23] Saliba, M.; Matsui, T.; Domanski, Konrad; Seo, J. Y.; Ummadisingu, Amita; Zakeeruddin, S. M.; Correa-Baena, J. P.; Tress, W. R.; Abate, A.; Hagfeldt, Anders; Grätzel, M. *Science*, 2016, 354, 206–209.

[24] He, M.; Li, B.; Cui, X.; Jiang, B.; He, Y.; Chen, Y.; O'Neil, D.; Szymanski, P.; EI-Sayed, M. A.; Huang, J.; Lin, Z. *Nat. Commun.*, 2017, 8, 16045-1– 10.

[25] Shin, S. S.; Yeom, E. J.; Yang, W. S.; Hur, S.; Kim, M. G.; Im, J.; Seo, J.; Noh, J. H.; Seok, S. I. *Science*, 2017, 356, 167–171.

[26] Tang, Z.; Bessho, T.; Awai, F.; Kinoshita, T.; Maitani, M. M.; Jono, R.; Murakami, T. N.; Wang, H.; Kubo, T.; Uchida, S.; Segawa, H. *Sci. Rep.*, 2017, 7, 12183-1–6.

[27] Wang, J. M.; Wang, Z. K.; Li, M.; Zhang, C. C.; Jiang, L. L.; Hu, K. H.; Ye, Q. Q.; Liao, L. S. *Adv. Energy Mater.*, 2017, 7, 1701688-1–8.

[28] Arora, N.; Dar, M. I.; Hinderhofer, A.; Pellet, N.; Schreiber, F.; Zakeeruddin, S. M.; Grätzel, M. *Science*, 2017, 358, 768–771.

[29] Yang, W. S.; Park, B. W.; Jung, E. H.; Jeon, N. J.; Kim, Y. C.; Lee, D. U.; Shin, S. S.; Seo, J.; Kim, E. K.; Noh, J. H.; Seok, S. I. *Science*, 2017, 356, 1376–1379.

[30] Cho, Y.; Soufiani, A. M.; Yun, J. S.; Kim, J.; Lee, D. S.; Seidel, J.; Deng, X.; Green, M. A.; Huang, S.; Ho-Baillie, A. W. Y. *Adv. Energy Mater.*, 2018, 1703392-1–10.

[31] Xu, W.; Zheng, L.; Zhang, X.; Cao, Y.; Meng, T.; Wu, D.; Liu, L.; Hu, W.; Gong, X. *Adv. Energy Mater.*, 2018, 1703178-1–11.

[32] Zhao, Y.; Tan, H.; Yuan, H.; Yang, Z.; Fan, J. Z.; Kim, J.; Voznyy, O.; Gong, X.; Quan, L. N.; Tan, C. S.; Hofkens, J.; Yu, D.; Zhao, Qing; Sargent, E. H. *Nat. Commun.*, 2018, 9, 1607-1–10.

[33] Oku, T. Solar Cells - New Approaches and Reviews; Kosyachenko L. A. Chapter 3. Crystal structures of $CH_3NH_3PbI_3$ and related perovskite compounds used for solar cells; InTech: Rijeca, Croatia, 2015, 77–102.

[34] Oku, T. Solar Cells and Energy Materials; De Gruyter; Berlin, Germany, 2016.

[35] Oku, T.; Zushi, M.; Suzuki, K.; Ohishi, Y.; Matsumoto, T.; Suzuki, A. Nanostructured Solar Cells, Intech, Ed. Das N. Chapter 11: Fabrication and characterization of element-doped perovskite solar cells; InTech: Rijeca, Croatia, 2017, 217–243.

[36] Hao, F.; Stoumpos, C. C.; Cao, D. H.; Chang, R. P. H.; Kanatzidis, M. G. *Nat. Photonics*, 2014, 8, 489–494.

[37] Liao, W.; Zhao, D.; Yu, Y.; Shrestha, N.; Ghimire, K.; Grice, C. R.; Wang, C.; Xiao, Y.; Cimaroli, A. J.; Ellingson, R. J.; Podraza, N. J.; Zhu, K.; Xiong, R. G.; Yan, Y. *J. Am. Chem. Soc.*, 2016, 138, 12360–12363.

[38] Zhu, H. L.; Xiao, J.; Mao, J.; Zhang, H.; Zhao, Yong; Choy, W. C. H. *Adv. Funct. Mater.*, 2017, 2017, 1605469-1–8.

[39] Tavakoli, M. M.; Zakeeruddin, S. M.; Grätzel, M.; Fan, Zhiyong. *Adv. Mater.*, 2018, 2018, 1705998-1–9.

[40] Zuo, F.; Williams, S. T.; Liang, P. W.; Chueh, C. C.; Liao, C. Y.; Jen, A. K. Y. *Adv. Mater.*, 2014, 26, 6454–6460.

[41] Oku, T.; Ohishi, Y.; Suzuki, A. *Chem. Lett.*, 2016, 45, 134–136.

[42] Oku, T.; Ohishi, Y.; Suzuki, A.; Miyazawa, Y. *Metals*, 2016, 6, 147-1–13.
[43] Zhang, J.; Shang, M. H.; Wang, P.; Huang, X.; Xu, J.; Hu, Z.; Zhu, Y.; Han, L. *ACS Energy Lett.*, 2016, 1, 535–541.
[44] Oku, T.; Ohishi, Y.; Suzuki, A. *AIP Conf. Proc.*, 2017, 1807, 020007-1–5.
[45] Ando, Y.; Oku, T.; Ohishi, Y. *Jpn. J. Appl. Phys.*, 2018, 57, 02CE02-1–5.
[46] Shirahata, Y.; Oku, T. *Phys. Stat. Solidi A*, 2017, 214, 1700268-1–6.
[47] Shirahata Y.; Oku, T. *AIP Conf. Proc.*, 2017, 1807, 020008-1–6.
[48] Shirahata, Y.; Oku, T. *Mater. Res. Express*, 2018, 5, 055504-1–11.
[49] Tanaka, H.; Ohishi, Y.; Oku, T. *Jpn. J. Appl. Phys.*, 2018, 57, 08RE10-1–5.
[50] Tanaka, H.; Oku, T.; Ueoka, N. *Synth. Metals*, 2018, 244, 128–133.
[51] Oku, T.; Ohishi, Y.; Tanaka, H. *AIP Conf. Proc.*, 2018, 1929, 020010-1–8.
[52] Hamatani, T.; Shirahata, Y.; Ohishi, Y.; Fukaya, M.; Oku, T. *Adv. Mater. Phys. Chem.*, 2017, 7, 1–10.
[53] Hamatani, T.; Shirahata, Y.; Ohishi, Y.; Fukaya, M.; Oku, T. *AIP Conf. Proc.*, 2017, 1807, 020012-1–9.
[54] Hamatani, T.; Oku, T. *AIP Conf. Proc.*, 2018, 1929, 020018-1–8.
[55] Krishnamoorthy, T.; Ding, H.; Yan, C.; Leong, W. L.; Baikie, T.; Zhang, Z.; Sherburne, M.; Li, S.; Asta, M.; Mathews, N.; Mhaisalkarac, S. G. *J. Mater. Chem. A*, 2015, 3, 23829–23832.
[56] Ohishi, Y.; Oku, T.; Suzuki, A. *AIP Conf. Proc.*, 2016, 1709, 020020-1–8.
[57] Tanaka, H.; Ohishi, Y.; Oku, T. *AIP Conf. Proc.*, 2018, 1929, 020007-1–7.
[58] Zhao, W.; Yang, D.; Yang, Z.; Liu, S. *Mater. Today Energy*, 2017, 5, 205–213.
[59] Zhang, X.; Yin, J.; Nie, Z.; Zhang, Q.; Sui, N.; Chen, B.; Zhang, Y.; Qu, K.; Zhao, J.; Zhou, Huawei. *RSC Advances*, 2017, 7, 37419–37425.

[60] Taguchi, M.; Suzuki, A.; Tanaka, H.; Oku, T. *AIP Conf. Proc.*, 2018, 1929, 020012-1–8.
[61] Zhang, H.; Wang, H.; Williams, S. T.; Xiong, D.; Zhang, W.; Chueh, C. C.; Chen, W.; Jen, A. K. Y. *Adv. Mater.*, 2017, 1606608-1–8.
[62] Ueoka, N.; Ohishi, Y.; Shirahata, Y.; Suzuki, A.; Oku, T. *AIP Conf. Proc.*, 2017, 1807, 020009-1–9.
[63] Ueoka, N.; Oku, T.; Suzuki, A. *Nanosci. Nanoeng.*, 2017, 5, 25–30.
[64] Ueoka, N.; Oku, T.; Ohishi, Y.; Tanaka, H.; Suzuki, A.; Sakamoto, H.; Yamada, M.; Minami, S.; Tsukada, S. *AIP Conf. Proc.*, 2018, 1929, 020026-1–8.
[65] Ueoka, N.; Oku, T.; Suzuki, A.; Sakamoto, H.; Yamada, M.; Minami, S.; Miyauchi. *Jpn. J. Appl. Phys.*, 2018, 57, 02CE03-1–7.
[66] Kato, M; Suzuki, A; Ohishi, Y; Tanaka, H; Oku, T. *AIP Conf. Proc.*, 2018, 1929, 020015-1–7.
[67] Turren-Cruz, S. H.; Saliba, M.; Mayer, M. T.; Juárez-Santiesteban, H.; Mathew, X.; Nienhaus, L.; Tress, W.; Erodici, M. P.; Sher, M. J.; Bawendi, M. G.; Grätzel, M.; Abate, A.; Hagfeldt, A.; Correa-Baena, J. P. *Energy Environ. Sci.*, 2018, 11, 78–86.
[68] Zhao, W.; Yao, Z.; Yu, F.; Yang, D.; Liu, S. *Adv. Sci.*, 2018, 5, 1700131- 1–7.
[69] Umemoto, Y.; Suzuki, A.; Oku, T. *AIP Conf. Proc.*, 2017, 1807, 020011- 1–10.
[70] Peng, W.; Miao, X.; Adinol, V.; Alarousu, E.; Tall, O. E.; Emwas, A. H.; Zhao, C.; Walters, G.; Liu, J.; Ouellette, O.; J. Pan, Murali, B.; Sargent, E. H.; Mohammed, O. F.; Bakr, O. M. *Angew. Chem., Int. Ed.*, 2016, 55, 10686–10690.
[71] Jodlowski, A. D.; Roldán-Carmona, C.; Grancini, G.; Salado, M.; Ralaiarisoa, M.; Ahmad, S.; Koch, N.; Camacho, L.; de Miguel, G. Nazeeruddin, M. K. *Nat. Energy*, 2017, 2, 972–979.
[72] Suzuki, A.; Okada, H.; Oku, T. *AIP Conf. Proc.*, 2016, 1709, 020022-1–8.
[73] Suzuki, A.; Okada, H.; Oku, T. *Energies*, 2016, 9, 376-1–11.
[74] Shi, D.; Adinolfi, V.; Comin, R.; Yuan, M.; Alarousu, E.; Buin, A.; Chen, Y.; Hoogland, S.; Rothenberger, A.; Katsiev, K.; Losovyj, Y.;

Zhang, X.; Dowben, PA; Mohammed, O. F.; Sargent, E. H.; Bakr, O. M. *Science*, 2015, 347, 519–522.

[75] Dong, Q.; Fang, Y.; Shao, Y.; Mulligan, P.; Qiu, J.; Cao, L.; Huang, J. *Science*, 2015, 347, 967–970.

[76] Oku, T.; Suzuki, K.; Suzuki, A. *J. Ceram. Soc. Jpn*, 2016, 124, 234–238.

[77] Suzuki, K.; Suzuki, A.; Zushi, M.; Oku, T. *AIP Conf. Proc.*, 2015, 1649, 96–101.

[78] Oku, T.; Zushi, M.; Imanishi, Y.; Suzuki, A.; Suzuki, K. *Appl. Phys. Express*, 2014, 7, 121601-1–4.

[79] Wang, F.; Shimazaki, A.; Yang, F.; Kanahashi, K.; Matsuki, K.; Miyauchi, Y.; Takenobu, T.; Wakamiya, A.; Murata, Y.; Matsuda, K. *J. Phys. Chem. C*, 2017, 121, 1562–1568.

[80] Li, G.; Zhang, T.; Xu, F.; Zhao, Y. *Mater. Today Energy*, 2017, 5, 293–298.

[81] Okada, Y.; Suzuki, A.; Yamasaki, Y.; Oku, T. *AIP Conf. Proc.*, 2017, 1807, 020015-1–7.

[82] Suzuki, A.; Ueda, H.; Okada, Y.; Ohishi, Y.; Yamasaki, Y.; Oku, T. *Chem. Mater. Eng.*, 2017, 5, 234–242.

[83] Shirahata, Y.; Yamomoto, Y.; Suzuki, A.; Oku, T.; Fukunishi, S.; Kohno, K. *Phys. Stat. Solidi A*, 2017, 214, 1600591-1–7.

[84] Shirahata, Y.; Oku, T.; Fukunishi, S.; Kohno, K. *Mater. Sci. Applications*, 2017, 8, 209–222.

[85] Oku, T.; Nomura, J.; Suzuki, A.; Tanaka, H.; Fukunishi, S.; Minami, S.; Tsukada, S. *Int. J. Photoenergy*, 2018, 2018, 8654963-1–7.

[86] Taguchi, M.; Suzuki, A.; Oku, T.; Fukunishi, S.; Minami, S.; Okita, M. *Coatings* 2018, 8, 461-1–10.

[87] Lee, J. W.; Bae, S. H.; Marco, N. D.; Hsieh, Y. T.; Dai, Z.; Yang, Y. *Mater. Today Energy*, 2018, 7, 149–160.

[88] Fan, J.; Ma, Y.; Zhang, C.; Liu, C.; Li, W.; Schropp, R. E. I.; Mai, Y. *Adv. Energy Mater.*, 2018, 1703421-1–8.

[89] Suzuki, A.; Kida, T.; Takagi, T.; Oku, T. *Jpn. J. Appl. Phys.*, 2016, 55, 02BF01-1–5.

[90] Zhou, W.; Wen, Z.; Gao, P. *Adv. Energy Mater.*, 2018, 8, 1702512-1–28.
[91] Oku, T; Iwata, T; Suzuki, A. *Chem. Lett.*, 2015, 44, 1033–1035.
[92] Kanayama, M; Oku, T; Suzuki, A; Yamada, M; Sakamoto, H; Minami, S; Kohno, K. *AIP Conf. Proc.*, 2016, 1709, 020019-1–7.
[93] Oku, T.; Ueoka, N.; Suzuki, K.; Suzuki, A.; Yamada, M.; Sakamoto, H.; Minami, S.; Fukunishi, S.; Kohno, K.; Miyauchi, S. *AIP Conf. Proc.*, 2017, 1807, 020014-1–7.
[94] Saito, J.; Suzuki, A.; Akiyama, T.; Oku, T. *AIP Conf. Proc.*, 2017, 1807, 020010-1–5.
[95] Oku, T.; Matsumoto, T.; Suzuki, A.; Suzuki, K. *Coatings*, 2015, 5, 646–655.
[96] Higuchi H.; Negami, T. *Jpn. J. Appl. Phys.*, 2018, 57, 08RE11-1–6.
[97] Weber, D. *Z. Naturforsch. B*, 1978, 33, 1443–1445.
[98] Poglitsch, A.; Weber, D. *J. Chem. Phys.*, 1987, 87, 6373–6378.
[99] Onoda-Yamamuro, N.; Matsuo, T.; Suga, H. *J. Phys. Chem. Solids*, 1990, 51, 1383–1395.
[100] Mashiyama, H.; Kurihara, Y.; Azetsu, T. *J. Korean Phys. Soc.*, 1998, 32, S156–S158.
[101] Baikie, T.; Fang, Y.; Kadro, J. M.; Schreyer, M.; Wei, F.; Mhaisalkar, S. G.; Grätzel, M.; Whitec, T. J. *J. Mater. Chem. A*, 2013, 1, 5628–5641.
[102] Kawamura, Y.; Mashiyama, H.; Hasebe, K. *J. Phys. Soc. Jpn.*, 2002, 71, 1694–1697.
[103] Ando, Y.; Ohishi, Y.; Suzuki, K., Suzuki, A.; Oku, T. *AIP Conf. Proc.*, 2018, 1929, 020003-1–8.
[104] Oku, T. *Nanotechnol. Rev.*, 2012, 1, 389–425.
[105] Oku, T. *Nanotechnol. Rev.*, 2014, 3, 413–444.
[106] Hoefler, S. F.; Trimmel, G.; Rath, T. *Monatsh Chem.*, 2017, 148, 795–826.
[107] Tanaka, H.; Oku, T.; Ueoka, N. *Jpn. J. Appl. Phys.*, 2018, 57, 08RE12-1–9.
[108] Travis, W.; Glover, E. N. K.; Bronstein, H.; Scanlon, D. O. ; Palgrave, RG. *Chem. Sci.*, 2016, 7, 4548–4556.

[109] Körbel, S.; Marques, M. A. L.; Botti, S. *J. Mater. Chem. C*, 2016, 4, 3157- 3167.
[110] Suzuki, A.; Oku, T. *Jpn. J. Appl. Phys.*, 2018, 57, 02CE04-1–7.
[111] Suzuki, A.; Oku, T. *Heliyon*, 2018, 4, e00755-1–22.
[112] Suzuki, A.; Oku, T. *Appl. Surf. Sci.*, 2019, 483, 912–921.
[113] Zuo, C.; Ding, L. *Nanoscale*, 2014, 6, 9935–9938.
[114] He, J.; Chen, T. *J. Mater. Chem. A*, 2015, 3, 18514–18520.
[115] Oku, T.; Ohishi, Y.; Suzuki, A.; Miyazawa, Y. *J. Ceram. Soc. Jpn.*, 2017, 125, 303–307.
[116] Oku, T.; Kakuta, N.; Kobayashi, K.; Suzuki, A.; Kikuchi, K. *Prog. Nat. Sci. Mater. Int.*, 2011, 21, 122–126.
[117] Zushi, M.; Suzuki, A.; Akiyama, T.; Oku T. *Chem. Lett.*, 2014, 43, 916–918.
[118] Tanaka, H.; Ohishi, Y.; Oku, T. *AIP Conf. Proc.*, 2018, 1929, 020005-1–7.
[119] Machiba, H.; Oku, T.; Suzuki, A. *AIP Conf. Proc.*, 2019, 2067, 020009-1–7.
[120] Suzuki, A.; Kato, M.; Ueoka, N.; Oku, T. *J. Electron. Mater.*, 2019, 48, 3900–3907.
[121] Yu, H.; Wang, F.; Xie, F.; Li, W.; Chen, J.; Zhao, N. *Adv. Funct. Mater.*, 2014, 24, 7102–7108.
[122] Dualeh, A.; Tétreault, N.; Moehl, T.; Gao, P.; Nazeeruddin, M. K.; Grätzel, M. *Adv. Funct. Mater.*, 2014, 24, 3250–3258.
[123] McLeod, J. A.; Wu, Z.; Sun, B.; Liu, L. *Nanoscale*, 2016, 8, 6361–6368.
[124] Oku, T.; Ohishi, Y. *J. Ceram. Soc. Jpn.*, 2018, 126, 56–60.
[125] Oku, T.; Ohishi, Y.; Ueoka, N. *RSC Advances*, 2018, 8, 10389–10395.
[126] Snaith, H. J.; Abate, A.; Ball, J. M.; Eperon, G. E.; Leijtens, T.; Noel, N. K.; Stranks, S. D.; Wang, J. T. W.; Wojciechowski, K.; Zhang, W. *J. Phys. Chem. Lett.*, 2014, 5, 1511–1515.
[127] Zhang, T.; Guo, N.; Qian, G. Li, X.; Zhao, Y. *Nano Energy*, 2016, 26, 50–56.

[128] Ueoka, N.; Oku, T.; Tanaka, H.; Suzuki, A.; Sakamoto, H.; Yamada, M.; Minami, S.; Miyauchi, S.; Tsukada, S. *Jpn. J. Appl. Phys.*, 2018, 57, 08RE05-1–7.

[129] Ueoka, N.; Oku, T.; Suzuki, A. *Chem. Lett.*, 2018, 47, 528–531.

[130] Ueoka, N.; Oku, T. *ACS Appl. Mater. Interfaces*, 2018, 10, 44443−44451.

[131] Roldán-Carmona, C.; Gratia, P.; Zimmermann, I.; Grancini, G.; Gao, P.; Graetzel, M.; Nazeeruddin, M. K. *Energy Environ. Sci.*, 2015, 8, 3550– 3556.

[132] Wang, L.; McCleese, C.; Kovalsky, A.; Zhao, Y.; Burda, C. *J. Am. Chem. Soc.*, 2014, 136, 12205–12208.

[133] Guo, Y.; Shoyama, K.; Sato, W.; Matsuo, Y.; Inoue, K.; Harano, K.; Liu, C.; Tanaka, H.; Nakamura, E. *J. Am. Chem. Soc.*, 2015, 137, 15907–15914.

[134] Chen, Q.; Zhou, H.; Song, T. B.; Luo, S.; Hong, Z.; Duan, H. S.; Dou, L.; Liu, Y.; Yang, Y. *Nano Lett.*, 2014, 14, 4158–4163.

ABOUT THE EDITOR

Dr. Murali Banavoth is an Assistant Professor at School of Chemistry, University of Hyderabad, Telangana, India, leading the Solar Cells and Photonics Research Laboratory. He has obtained his bachelor's degree from Osmania University and Masters from University of Hyderabad. He has joined for PhD at Materials Research Centre, Indian Institute of Science, Bangalore under the Supervision of Prof. S B Krupanidhi. During his PhD, Dr. Banavoth has worked on the third generation Cu (In, Al)Se$_2$-CIAS, solar cells. Later he moved to King Abdullah University of Science and Technology (KAUST), KAUST-Solar Centre, Kingdom of Saudi Arabia and carried out cutting edge research in the broader fields of nanoscience and nanotechnology.

Dr. Banavoth has been working on the interdisciplinary combinations of material synthesis, optimisations of device architectures by tuning the optoelectronic properties. His research interests include functional materials for future cost-effective integrated optoelectronic devices. Currently his research group works on the advanced functionality and fundamental understanding of processes that take place during photon absorption/emission at the active layer, surfaces/interfaces in the inorganic (CZTS/CIAS/CIGS), Organic (BHJ), Hybrid perovskite thin film/nanostructured photovoltaics (PV) using both the vacuum and soft chemistry approaches.

Dr. Banavoth is the recipient of Telangana state-Parthibha Puraskar Award, Junior and Senior Research Fellowships from Council of Scientific and Industrial Research-New Delhi. He is also the recipient of several international and national awards such as Young Scientist Award from Indian Science Congress (ISCA)-2017, Young Scientist Award from International Society for Energy Environment and Sustainability (ISEES)-2017, Young Scientist Award from BRICS-2017, Young Scientist Award from Scientific Planet Society (SPS)-2018, Young Scientist Award from Telangana Academy of Sciences (TAS)-2018, Startup grant from University Grants Commission (UGC), Early Career Research Award (ECRA) from Department of Science and Technology (DST)-2018, National Research award in Nanoscience and Nanotechnology-Young Career Award-2019 from the Nano Mission, DST, Government of India, NASI-Platinum Young Scientist Award-2019 from the National Academy of Sciences, India, etc.

INDEX

π

π-spacer, 185

A

absorber doping, 148, 153, 154
additive, 28, 105, 143, 227, 274, 280, 305, 309, 311
air blow, 215, 299, 300, 302, 303, 313, 318, 319, 320, 321, 322, 323, 331, 332
air blowing, 215, 299, 313, 318, 319, 320, 321, 322, 323, 331, 332
analogue, 187
Antimony, vi, 137, 269, 272, 280, 287
associated exciton radius, 191

B

band gap, viii, xii, 6, 14, 16, 31, 33, 44, 46, 47, 63, 88, 90, 91, 94, 95, 102, 107, 122, 126, 127, 128, 129, 131, 132, 133, 134, 135, 136, 137, 138, 141, 143, 144, 148, 170, 173, 177, 183, 186, 187, 188, 190, 192, 196, 197, 208, 245, 250, 270, 279, 293

bandgap energies, 183, 331
bar coating, 210
binding energy, 44, 82, 189, 190, 192, 231, 272, 276, 278, 281
bismuth, vi, 137, 168, 269, 272, 277, 278, 280, 282, 286, 287, 288, 290, 291, 292, 293, 294, 295, 296
bromide, 3, 95, 98, 99, 119, 122, 125, 127, 128, 129, 132, 158, 162, 164, 165, 194, 206, 213
brush printing, 210

C

carrier collection, 196, 292, 296
carrier diffusion length, 45, 191, 258, 271, 301, 307
carrier generation, 42, 159, 188, 192, 193, 327
carrier separation, 188, 301
carrier transport, 111, 279, 322, 330, 331, 332
CH_3NH_3I, 19, 68, 121, 165, 212, 215, 217, 218, 219, 220, 226, 228, 245, 246, 255, 303, 305, 313, 323
$CH_3NH_3PbI_3$, vi, 19, 24, 26, 27, 28, 32, 33, 35, 38, 40, 45, 48, 51, 52, 55, 57, 61, 65,

66, 67, 68, 69, 70, 72, 74, 111, 113, 120, 121, 122, 130, 141, 142, 143, 144, 145, 147, 156, 157, 165, 174, 196, 199, 203, 204, 206, 218, 219, 221, 226, 227, 231, 234, 235, 242, 245, 246, 255, 259, 261, 262, 267, 299, 300, 301, 302, 303, 304, 305, 306, 307, 308, 309, 310, 311, 312, 313, 314, 315, 316, 317, 318, 319, 320, 321, 322, 323, 327, 331, 332, 335

charge injection, 176, 185, 194, 195, 196
charge transfer, xii, 5, 14, 17, 24, 42, 176, 178, 183, 185, 192, 193, 194, 195
chemical hardness, 192, 197
chloride, 121, 122, 125, 126, 180, 194, 227, 245
conduction band, xii, 3, 24, 42, 43, 82, 132, 134, 135, 138, 185, 244, 250
conversion efficiency, vii, xi, xiii, 3, 5, 10, 14, 16, 44, 49, 58, 59, 63, 102, 121, 195, 286, 300, 302, 322, 331, 332
Coulomb and exchange interaction, 190
Coulomb attraction, 188
coupling, 64, 96, 114, 125, 158, 159, 193, 194, 252
crystallite, 325, 326
crystallization, 121, 122, 132, 136, 215, 218, 225, 232, 260, 261, 274, 277, 293, 322
cubic, 45, 72, 120, 129, 130, 133, 135, 160, 180, 181, 183, 187, 188, 242, 246, 285, 286, 295, 301, 310, 318, 320, 323
cubic structure, 133, 181, 188, 285, 301, 310
current collector, 195
current density, 41, 44, 299, 301, 302, 325
current efficiency, xii
current-voltage characteristics, 46, 118, 140, 142

D

dielectric constant, 188, 191, 287
diffraction intensities, 320, 323
diffusion, vii, xi, xii, xiii, 4, 42, 44, 46, 49, 52, 55, 72, 80, 81, 89, 90, 91, 93, 95, 103, 107, 113, 121, 122, 125, 138, 140, 141, 144, 145, 146, 147, 148, 150, 151, 152, 153, 154, 157, 174, 189, 196, 204, 205, 208, 231, 276, 284, 287, 295, 300, 301, 312, 323
diffusion length, vii, 4, 42, 44, 49, 52, 95, 121, 122, 138, 141, 144, 145, 146, 147, 148, 150, 151, 152, 153, 154, 157, 174, 208, 231, 276, 284, 287, 295, 300, 301, 312, 323
dimethylformamide (DMF), xi, 215, 217, 218, 227, 246, 251, 274, 281, 305, 313, 323
dip coating, 210, 221
dissociation rate, 189, 190
doctor blade coating, 209, 210, 211, 212, 228, 230
doping effect, 272, 312, 318
driving force, 2, 188, 193, 194, 252
d-value, 325

E

electrode, 3, 9, 13, 22, 55, 56, 58, 61, 69, 71, 155, 160, 209, 211, 212, 217, 219, 220, 221, 223, 226, 235, 238, 247, 248, 249, 251, 252, 253, 254, 257, 258, 262, 263, 264, 266, 267, 286, 301, 304
electron capture, 193
electron collection efficiency, 195
electron delivery, 195
electron mass, 191
electron transport layer, xii, 12, 212, 244, 301

electron-hole, 3, 24, 52, 98, 102, 121, 122, 140, 145, 150, 157, 174, 176, 185, 188, 189, 190, 192, 231, 295
elemental mapping image, 310, 316, 317, 318, 319, 327, 328, 329
energy gap, 88, 172, 176, 307, 316
energy losses, 139, 142, 145, 147
equivalent circuit, 146, 147
exciton, 7, 44, 125, 126, 158, 160, 165, 166, 176, 185, 188, 189, 190, 191, 192, 193, 196, 199, 205, 208, 231, 272, 276, 278, 281, 323
exciton-spin-orbit-photon interaction, 189, 205
exiton, 176

F

fill factor, xii, 41, 42, 44, 141, 147, 194, 217, 237, 276, 301, 307, 314
flexible perovskite solar cells (FPSCs), vi, 53, 234, 236, 241, 242, 243, 244, 245, 248, 249, 250, 252, 253, 254, 255, 256, 257, 258, 260, 262, 263, 264, 266, 268
flexible substrates and durability, 242
fluorine doped tin oxide (FTO), xii, 8, 32, 33, 34, 35, 40, 124, 128, 211, 215, 217, 220, 226, 258, 301, 303, 304, 305, 310, 318, 320, 323, 324, 331
formamidium (FA), 29, 123, 133, 141, 142, 280, 300
fractal interfaces, 78
functionality, 104, 175, 182, 183, 185, 194, 343

G

Geometrical fill factor (GFF), xii, 209, 222, 223, 224, 226, 227
geometrical structure, 180

grain boundary, 17, 95, 301, 322, 327, 328, 332

H

halide perovskite, vii, 50, 51, 53, 54, 58, 59, 64, 66, 68, 71, 72, 95, 109, 110, 111, 112, 114, 118, 121, 126, 127, 131, 134, 154, 155, 156, 157, 158, 159, 161, 164, 165, 166, 169, 170, 171, 172, 173, 175, 176, 178, 185, 186, 187, 192, 194, 197, 233, 234, 239, 245, 259, 263, 267, 268, 270, 287, 289, 290, 293, 294, 296
halide substitution, 121, 154
harvesting efficiency, xii, 193
highly-oriented crystal, 322, 331
Hoke effect, 78, 95, 98
hole concentration, 310, 318
hole transport layer, xii, 68, 170, 212, 232, 244, 283, 296, 301, 303, 304, 331
hybrid perovskite, 19, 44, 56, 59, 60, 67, 70, 78, 98, 102, 103, 108, 111, 118, 120, 121, 127, 132, 154, 156, 157, 169, 172, 174, 177, 178, 207, 235, 268, 290, 291, 343
hysteresis, 32, 46, 47, 48, 60, 71, 72, 73, 74, 95, 120, 156, 199, 221, 236, 252, 257, 313

I

I_{100}/I_{210}, 320, 323, 331
inactive area, 210, 224
incident photon conversion, xii, 193
incident photon conversion to current efficiency (IPCE), xii, 193, 195, 196, 197, 198, 302, 305, 306, 307, 315, 316, 324, 325, 331
inkjet printing, 210, 216, 219, 220, 224, 234, 235, 255, 259, 267

inorganic perovskites, viii, 2, 118, 127, 128, 242
interface, 7, 12, 43, 51, 56, 59, 73, 87, 88, 89, 90, 103, 128, 156, 165, 168, 189, 194, 201, 205, 212, 217, 235, 238, 239, 244, 252, 267, 274, 332
ionic radii, 131, 172, 178, 181, 182, 271, 272

J

J–V characteristics, 304, 313, 324, 331

K

kinetic coefficients, 119
kinetics, vi, 63, 91, 100, 101, 160, 175, 189, 192, 204, 261

L

Lanthanides, 136
laser patterning, 223, 232, 238
lattice constant, 95, 321
lead, vi, 21, 44, 50, 51, 52, 53, 54, 55, 57, 58, 59, 63, 67, 68, 71, 72, 74, 75, 78, 111, 113, 114, 118, 119, 122, 126, 127, 128, 129, 130, 131, 132, 133, 134, 135, 136, 137, 153, 154, 155, 156, 158, 159, 160, 161, 162, 164, 165, 166, 167, 168, 170, 171, 172,174, 177, 196, 199, 201, 202, 203, 206, 212, 218, 230, 231, 233, 239, 249, 258, 259, 261, 268, 269, 270, 271, 276, 278, 282, 286, 288, 289, 290, 291, 292, 293, 294, 295, 296, 300, 312, 316, 325
least squares method, 321
lifetime, xiv, 4, 6, 16, 42, 95, 122, 138, 155, 176, 177, 183, 191, 196, 199, 245, 268, 270, 281, 283, 294

light filtering, 196
low unoccupied molecular orbital (LUMO), xii, 26, 40, 43, 176, 183, 184, 185, 189, 190, 194

M

methylammonium (MA), 14, 44, 53, 67, 118, 119, 122, 123, 131, 133, 137, 141, 142, 156, 187, 192, 194, 212, 215, 227, 270, 276, 280, 282, 292, 293, 294, 300, 301
microstructure, 155, 299, 331
mixed cation, 119, 132, 220, 289
mixed halide, 95, 118, 120, 121, 126, 132, 169, 178, 234, 290
mixed organic-inorganic perovskites, 118
module, viii, 208, 209, 211, 215, 222, 223, 224, 225, 226, 227, 228, 229, 232, 233, 237, 238, 248, 271
module concept, viii, 208, 209, 223, 229
Molecular orbital (HOMO), xi, xii, 28, 40, 43, 176, 183, 184, 185, 186, 189, 190, 275

N

networking microstructure, 316
NH_4Cl, vi, 121, 299, 300, 301, 305, 306, 307, 309, 310, 311, 312, 313, 314, 315, 316, 317, 318, 319, 320, 321, 322, 323, 331, 332
numerical solution, 139, 141, 143

O

octahedral, 137, 178, 180, 182, 187, 188
open-circuit voltage, xiii, xiv, 32, 39, 46, 60, 95, 131, 154, 194, 217, 276, 292, 307, 314

Index   349

optical microscopy, 299, 302, 327
optoelectronic, vii, viii, 6, 26, 33, 69, 74, 119, 121, 123, 124, 131, 136, 154, 158, 161, 162, 165, 166, 171, 175, 177, 182, 183, 186, 187, 188, 199, 203, 204, 205, 241, 250, 266, 269, 270, 271, 274, 278, 279, 280, 283, 287, 288, 289, 291, 293, 295, 343
organic-inorganic perovskites, xii, 2, 61, 79, 80, 82, 85, 86, 87, 91, 107, 124, 242
oriented perovskite crystal, 320, 331
orthorhombic, 33, 129, 180, 181, 183
overall efficiency, xiii, 194

222, 232, 236, 241, 242, 252, 254, 260, 261, 262, 263, 264, 265, 266, 267, 270, 307, 314
power conversion efficiency, vii, 2, 48, 60, 63, 94, 118, 119, 191, 200, 207, 208, 241, 242, 263, 270, 307, 314
printing technology, 208, 219, 220, 256
PSC, viii, xiii, 8, 10, 12, 13, 26, 27, 28, 32, 34, 35, 49, 50, 118, 121, 123, 139, 140, 141, 142, 143, 144, 145, 147, 148, 152, 177, 179, 183, 185, 189, 191, 193, 195, 199, 208, 210, 211, 213, 215, 217, 220, 222, 223, 225, 226, 227, 228, 229, 242, 243, 247, 249, 255, 257

P

parasitic resistance(s), 143, 146, 147
$PbCl_2$, 19, 215, 217, 245, 301, 303, 313, 318, 319, 320, 323, 331, 332
PbI_2, vi, 14, 20, 56, 68, 121, 132, 157, 165, 168, 215, 218, 219, 220, 225, 227, 228, 232, 245, 246, 249, 256, 266, 271, 299, 300, 301, 303, 305, 310, 312, 313, 318, 323, 324, 325, 326, 327, 328, 329, 330, 331, 332
PCE, vii, viii, xiii, 14, 16, 19, 20, 21, 41, 44, 45, 49, 50, 60, 119, 121, 124, 128, 129, 130, 131, 132, 133, 134, 136, 139, 140, 141, 143, 145, 147, 149, 153, 154, 193, 195, 196, 208, 211, 213, 215, 217, 218, 219, 220, 221, 223, 226, 227, 228, 230, 242, 245, 246, 247, 248, 249, 250, 251, 252, 254, 255, 256, 270, 272, 274, 276, 277, 279, 280, 281, 282, 283, 285, 286, 288
photoanode, 22, 23, 24, 195, 249
photon interaction, 189
pinhole, 208, 209, 215, 223, 224, 229, 231
power, vii, xiii, 2, 5, 16, 39, 41, 42, 44, 47, 48, 53, 60, 63, 79, 94, 118, 119, 126, 128, 129, 140, 191, 194, 200, 207, 208,

Q

quantum reactivity indice, 176

R

radiation and degradation of solar cells, 78
recombination, 3, 4, 8, 9, 11, 17, 22, 24, 32, 41, 42, 46, 51, 67, 69, 74, 78, 83, 86, 87, 88, 90, 91, 93, 94, 95, 103, 119, 126, 130, 132, 138, 140, 141, 143, 145, 151, 158, 160, 167, 176, 185, 188, 195, 196, 199, 223, 231, 243, 245, 248, 253, 254, 270, 272, 273, 279, 283, 289, 294, 331
reproducibility, 4, 14, 73, 113, 120, 213, 220, 225, 277, 280, 285, 325, 332
Roll-to-roll, 17, 50, 208, 209, 210, 211, 213, 214, 216, 221, 222, 224, 225, 227, 229, 237, 241, 242, 247, 252, 255, 256, 258, 267
roll-to-roll processing, 224, 237, 241, 242, 255, 256, 258, 267
Rydberg analogous, 191

S

Saha-Langmuir equation, 189
scale-free network, 78, 104, 107
screen printing, 208, 210, 216, 220, 221, 226, 256
SEM-EDS, 310, 312, 327, 331
sensitiser, 176
series connection, 210, 222, 223, 224, 237, 248
series resistance, xiii, 26, 27, 41, 45, 122, 133, 253, 276, 312, 315
shear force, 222
sheet resistance, 209, 247
short-circuit current density, xii, 39, 41, 194, 217, 307, 314
Slot-die coating, 208, 209, 210, 213, 214, 215, 222, 223, 230, 233, 242, 256
solar cell architecture, 2, 249
spiro-OMeTAD, 12, 23, 24, 27, 28, 32, 33, 138, 217, 248, 275, 277, 286, 301, 303, 304, 307, 318, 324, 331
spray coating, 19, 210, 216, 217, 234, 242, 255
substitution, 118, 119, 120, 128, 130, 131, 132, 133, 134, 136, 137, 153, 154, 166, 171, 172, 173, 182, 183, 203, 282
surface coverage, 122, 196, 215, 217, 219, 228, 251, 318, 322, 327, 332
surface energy, 221, 228
surface tension, 36, 217, 322
symmetry, 137, 154, 180, 183, 242
synergetics, viii, 78, 80, 99, 100, 114
synthesis, viii, 2, 17, 18, 19, 28, 49, 79, 122, 125, 134, 159, 163, 164, 166, 175, 178, 196, 203, 249, 279, 287, 293, 294, 313, 343

T

tandems, 5, 15, 16, 70, 78, 102, 106, 107, 202
tetragonal, 45, 180, 181, 188, 245, 246, 247, 285, 301, 310
tin, xii, 40, 52, 63, 72, 131, 132, 133, 170, 171, 199, 202, 206, 250, 258, 259, 263, 271, 290, 300, 303
tolerance factor, 130, 170, 175, 176, 181, 182, 183, 285, 301
Transition dipole moment, 197
transition matrix element, 189
Transition Metals, 135
transport model, 119

V

vacuum permittivity, 188, 191
viscosity, 36, 196, 210, 214, 221, 225, 233

X

x-ray spectroscopy (EDS), 302, 305, 310, 311, 316, 318, 319, 323, 329, 330

Related Nova Publications

ETHANOL AS A GREEN ALTERNATIVE FUEL: INSIGHT AND PERSPECTIVES

EDITORS: Helen Treichel, Sérgio Luiz Alves Júnior, Gislaine Fongaro and Caroline Müller

SERIES: Renewable Energy: Research, Development and Policies

BOOK DESCRIPTION: The *"Ethanol as a Green Alternative Fuel"* book shows present and future scenarios about bioethanol and perspective in their chain, considering the economic and environmental impact mitigations approach.

HARDCOVER ISBN: 978-1-53615-719-2
RETAIL PRICE: $230

ADVANCES IN BIO-FUEL PRODUCTION

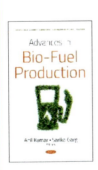

EDITORS: Anil Kumar and Sarika Garg

SERIES: Renewable Energy: Research, Development and Policies

BOOK DESCRIPTION: In this book, detailed production of biofuels from non-conventional bio-feedstocks and advanced biofuels production have been discussed.

HARDCOVER ISBN: 978-1-53614-671-4
RETAIL PRICE: $195

To see a complete list of Nova publications, please visit our website at www.novapublishers.com

Related Nova Publications

BIOGAS: PRODUCTION AND PROPERTIES

AUTHOR: James G. Speight

SERIES: Renewable Energy: Research, Development and Policies

BOOK DESCRIPTION: This book will introduce the reader to the fundamentals of biogas production, properties, and uses. The chapters focus on biogas as a renewable energy source, biogas production, the chemistry and engineering aspects of anaerobic digestion, landfill gas, biomass gasification, biogas upgrading technology, among other topics.

HARDCOVER ISBN: 978-1-53615-278-4
RETAIL PRICE: $230

EFFECTIVENESS OF INVESTMENT TO RENEWABLE ENERGY SOURCES IN SLOVAKIA

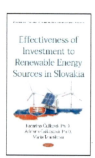

AUTHORS: Katarína Čulková, Ph.D., Adriana Csikósová, Ph.D. and Mária Janošková, Ph.D.

SERIES: Renewable Energy: Research, Development and Policies

BOOK DESCRIPTION: Renewable energy sources (RES) are considered as sources of the future. The European Union as a whole is dependent on the import of primary energy sources – around 50%.

EBOOK ISBN: 978-1-53614-689-9
RETAIL PRICE: $82

To see a complete list of Nova publications, please visit our website at www.novapublishers.com